THE STANDARD MODEL
OF QUANTUM PHYSICS
IN CLIFFORD ALGEBRA

THE STANDARD MODEL
OF QUANTUM PHYSICS
IN CLIFFORD ALGEBRA

Claude Daviau
Agrégé de Mathématiques, Docteur ès Sciences

Jacques Bertrand
Ingénieur, Ecole Polytechnique

World Scientific

NEW JERSEY · LONDON · SINGAPORE · BEIJING · SHANGHAI · HONG KONG · TAIPEI · CHENNAI · TOKYO

Published by

World Scientific Publishing Co. Pte. Ltd.
5 Toh Tuck Link, Singapore 596224
USA office: 27 Warren Street, Suite 401-402, Hackensack, NJ 07601
UK office: 57 Shelton Street, Covent Garden, London WC2H 9HE

British Library Cataloguing-in-Publication Data
A catalogue record for this book is available from the British Library.

ISBN 978-981-4719-86-5

Foreword

After Galileo, mathematics became the language of physics. After Maxwell, all laws of electricity, magnetism and light have been united in a synthesis around the electromagnetic field. This field is a function of space and time with value into a vector space containing the six parameters of the field. Maxwell laws are not invariant under the group of transformations that rules mechanics. From the invariance of the velocity of light, Einstein obtained his Relativity, which replaces the group of spacial rotations by a greater group, the Lorentz group, acting on space and time. The electromagnetic field becomes a tensor field. The next major change in physics was the discovery by Einstein and de Broglie of the quantum world. Einstein got the quantification of light, and de Broglie generalized the quantum wave to the movement of any material particle. Unhappily the first wave equation obtained by Schrödinger was not relativistic. Despite immense new results obtained in the quantum domain, quantum physics still suffers from this restricted starting point.

The true wave equation for the electron was obtained immediately after by Dirac. This wave equation is relativistic and moreover it gives the spin $1/2$ property of the electron. It is the starting point of the present work. Three points show the novelty of this approach: The mass term of the Dirac equation is replaced by a slightly different mass term, and this is enough to solve the problem of unphysical negative energy. The true frame allowing us to describe the Dirac wave is the Clifford algebra of the 3-dimensional space. This frame must replace the old frame of quantum physics, based on the complex field that is only the even part of the Clifford algebra of a plane, a 2-D software instead of the 3-D space algebra. The third and main novelty is the replacement of the Lorentz group by a greater group linked to the 3-D geometry.

Forty years after the beginning of the quantum odyssey, the Dirac theory was revisited by Hestenes and a few researchers. I was one of them. We used the Clifford algebra of the space-time. The Dirac wave is a function of space-time with value into the even part of the Clifford algebra. And this part is isomorphic to the Clifford algebra of space. This explains why the authors are able to put the Dirac theory in their 3-D frame. Since a combination of specific linear solutions is generally not a right solution in physical atoms, a non linear Dirac equation was obtained which led to more consistent results, such as a scale factor appearing in the form invariance. A much greater diversity of tensors is available in the Clifford frame.

The authors study next the consequences of the greater invariance group. It rules all electromagnetic laws. The space-time frame allows us to extend the wave to include both the electron and its neutrino. This allows us to include the gauge group $U(1) \times SU(2)$ of electro-weak interactions. Space-time algebra is enough to include leptons and anti-leptons. When all associations of right and left waves were explored, it appeared that the gauge invariance is compatible with a mass term that was not known in the Weinberg–Salam model.

The same method to go from the electron to the electron + neutrino pair is next used again to get a wave including the electron, its neutrino, and quarks u and d of the first generation with their three color states. The wave accounts for all particles of the first generation and also for their antiparticles. A wave equation with a mass term is obtained, both form invariant under the greater geometric group and gauge invariant under the $U(1) \times SU(2) \times SU(3)$ gauge group of the standard model. This invariance is not approximate but exact, and this is also a complete novelty in quantum physics. The method used to get the extended wave is equivalent to adding two dimensions to the space-time. The quantum wave is then a function of space and time with value into the full Clifford algebra $Cl_{1,5}$ of this extended space-time.

Since this approach is compatible with a mass term it is interesting to study the possibility of reconciling the quantum world with inertia and gravitation. A first attempt is made here.

One of the main consequences of the greater invariance is a double link between the wave and its Lagrangian formalism. Fermat was the first to understand that light moves in a way that minimizes the duration of its propagation. Next all laws of mechanics were understood as governed by such a principle of minimum. The de Broglie wave was obtained by uniting these two principles of minimum. Today all quantum mechanics is issued from such a mechanism, a Lagrangian density linked to the quantum wave. Calculus of variations allows us to get the wave equation from this Lagrangian density. The novelty is that when the wave equation is read in an invariant form, in Clifford algebra, the Lagrangian density is exactly the real scalar part of the wave equation. This is true for the electron in the space algebra, true for the pair electron+neutrino in the space-time algebra, and true again for the extended wave electron+neutrino+quarks in the Clifford algebra of the extended space-time.

This approach also explains why there are three generations and four kinds of neutrinos, the last one experiencing only gravitational interactions. It reinforces the standard model that today describes the quantum world. No insight beyond the standard model, no new particles with exotic properties are awaited. The only new possibilities are the magnetic monopoles. They are fully compatible with classical electromagnetism and quantum physics. Actually their existence is yet proved and this book reports the first experiments on these fermionic monopoles.

Roger Boudet,
Université de Provence,
Av. de Servian,
Bassan,
France.

Contents

Introduction

To see where the standard model that today rules quantum physics comes from we first return to its beginning. When the idea of a wave associated to the movement of a particle was born, Louis de Broglie was following the consequences of the construction by A. Einstein of his theory of relativity [29]. The first wave equation found by Schrödinger [56] was not relativistic, and could not be the true wave equation. At the same time the spin of the electron was discovered. This remains the main change from pre-quantum physics, since the spin $1/2$ has no classical equivalent. Pauli gave a wave equation for a non-relativistic equation with spin. This equation was the starting point of the attempt made by Dirac [34] to get a relativistic wave equation for the electron. The Dirac equation was a very great success. Until now, it is still considered as the wave equation for each particle with spin $1/2$, electrons and also positrons, muons and anti-muons, neutrinos and quarks.

This wave equation was intensively studied by Louis de Broglie and his students. He published a first book in 1934 [30] explaining how this equation gives in the case of the hydrogen atom the quantification of energy levels, all awaited quantum numbers, the true number of quantum states, the true energy levels and the Landé factors. The main novelty in physics coming with the Dirac theory is the fact that the wave does not have vector or tensor properties under a Lorentz rotation. The wave is a spinor and transforms very differently. It results from this transformation that the Dirac equation is form invariant under Lorentz rotations. This form invariance is the starting point of our study. It is the central thread of this book.

The Dirac equation was built from the Pauli equation. It is based on 4×4 complex matrices, which were constructed from the Pauli matrices.

Many years after this first construction, D. Hestenes [39] used the Clifford algebra of space-time to get a different form of the same wave equation. Tensors which are constructed from the Dirac spinors appear differently and the relations between these tensors are more easily obtained.

One of the parameters of the Dirac wave, the Yvon–Takabayasi angle [58], was completely different from all classical physics. G. Lochak understood that this angle allows a second gauge invariance and he found a wave equation for a magnetic monopole from this second gauge invariance [46]. He showed that a wave equation with a nonlinear mass term was possible for his magnetic monopole. When this mass term is null, the wave is made of two independent Weyl spinors.

This mass term is compatible with the electric gauge ruling the Dirac equation. So it can replace the linear term of the Dirac equation of the electron [7]. A nonlinear wave equation for the electron was awaited by de Broglie, because it was necessary to link the particle to the wave. But this does not explain how to choose the nonlinearity. And the nonlinearity is a formidable problem in quantum physics: quantum mechanics is a linear theory. It is by solving the linear wave equation that the quantification of energy levels and quantum numbers are obtained in the hydrogen atom. If you start from a nonlinear wave equation, you will not usually be able to even get quantification and quantum numbers.

Nevertheless the study of this nonlinear wave equation began in the case where the Dirac equation is its linear approximation. In this case the wave equation is homogeneous. It is obtained from a Lagrangian density which differs from the Lagrangian of the linear theory only by the mass term. Therefore many results are similar. For instance the dynamics of the electron are the same, and the electron follows the Lorentz force.

Two formalisms were available, the Dirac formalism with 4×4 complex matrices, and the real Clifford algebra of space-time. A matrix representation links these formalisms. Since the hydrogen case gave the main result, a first attempt was made to solve the nonlinear equation in this case. Heinz Krüger gave a precious tool [44] by finding a way to separate the spherical coordinates. Moreover the beginning of this resolution by separation of variables was the same in the case of the linear Dirac equation and in the case of the nonlinear homogeneous equation. But then there was a great difficulty: The Yvon–Takabayasi angle is null in the $x^3 = 0$ plane. This angle is a complicated function of an angular variable and of the radial variable. Moreover for any solution with a variable radial polynomial, circles exist where the Yvon–Takabayasi angle is not defined. In the vicinity of these circles this angle is not small and the solutions of the Dirac equation have no reason to be linear approximations of solutions of the nonlinear homogeneous equation. Finally it was possible to compute [8] another orthonormal set of solutions of the Dirac equation, which have everywhere a well defined and small Yvon–Takabayasi angle. These solutions are linear approximations of the solutions of the nonlinear equation. The existence of this set of orthonormalized solutions is a powerful argument for our nonlinear wave equation.

When you have two formalisms for the same theory the question necessarily comes: which is the best one? Comparing the advantages of these formalisms, the possibility of a third formalism which could be the true one arose. A third formalism is really available [10] to read the Dirac theory: it is the Clifford algebra Cl_3 of the 3-dimensional space used by W.E. Baylis [2]. This Clifford algebra is isomorphic, as real algebra, to the complex matrix algebra generated by Pauli matrices. Quantum physics knew this formalism very early, since these Pauli matrices were invented to get the first wave equation with spin $1/2$. Until now this formalism is also used to get the form invariance of the Dirac equation. Having then three formalisms for the same theory, the question was, once more: which is the true one?

The best choice was necessarily coming from the Lorentz invariance of the wave equation. Therefore a complete study, from the start, of this form invariance of the Dirac theory was made [12]. This problem was a classical one, treated by many books, but always with mathematical flaws. The reason is that two different Lie groups may have the same Lie algebra. The Lie algebra of a Lie group is the algebra generated by infinitesimal operators of the Lie group. Quantum mechanics uses only these infinitesimal operators and it is then very difficult to avoid ambiguities. But it is possible to avoid any infinitesimal operator. And working without them it is then easy to see the main novelty coming from quantum physics: the fundamental invariance group is greater than expected. This group is the 8-dimensional Lie group Cl_3^* of the invertible elements in Cl_3.

This is a major change in our understanding of physical laws. It is the direct generalization of the Einstein's point of view: rules applying to all physical laws are the same in any domain. All physical laws, in gravitation, electromagnetism, weak or strong interactions, have the same kind of invariance. Physical consequences are similar to those arising from the replacement of the Galileo group by the Lorentz group: fewer invariant quantities, grouping of other ones, new concepts. Among these new concepts we can cite the numeric dimension and the existence of two space-time manifolds: the relative one is 4-dimensional and the other one is 6-dimensional. Both manifolds have the same and unique dimension of time. The physical space and the physical time are both oriented and the invariance group conserves both orientations.

In several articles and in three previous books [15] [16] [23] several consequences of this larger invariance group were presented. This invariance group governs not only the Dirac theory, but also all electromagnetic laws, with or without magnetic monopoles, with or without photons. Moreover this form invariance is also the rule for electro-weak and strong interactions [17]. The form invariance will also be the key that allows us to get inertia.

Because it is impossible to read this book without knowledge of the Cl_3 algebra the first chapter presents Clifford algebras at an elementary level.

Chapter 2 reviews the Dirac equation, first with Dirac matrices, where we get a mathematically correct form of the relativistic invariance of the theory. Since the beginning of relativistic quantum mechanics this necessitates the use of the space algebra Cl_3. Next we explain the form of the Dirac equation in this simple frame and we review the relativistic form invariance of the Dirac wave. We explain with the tensors without derivative that the classical matrix formalism is deficient and must no longer be used. We review plane waves. We present the invariant form of the wave equation. Its scalar part is the Lagrangian density, this is another true novelty with many consequences. Finally we present the charge conjugation of quantum field theory in this frame.

Chapter 3 introduces our homogeneous nonlinear equation and explains why this equation is better than the Dirac equation which is its linear approximation. We review its two gauge invariances. We explain why plane waves have only positive energy. The form of the spinor wave and the form of its relativistic invariance introduce the dilation generated by the wave from an intrinsic space-time manifold onto the usual relative space-time manifold, the main geometric novelty of quantum physics. The link between the wave of the particle and the wave of the antiparticle coming from relativistic quantum mechanics gives a charge conjugation where only the differential term of the wave equation changes sign. This makes the CPT theorem trivial and it is also a powerful argument for the simplified mass term. We get the quantization of the energy in the case of the hydrogen atom and all results of the linear theory with this homogeneous nonlinear wave equation.

Chapter 4 presents the invariance of electromagnetic laws under Cl_3^*, the group of the invertible elements in Cl_3, for the Maxwell–de Broglie electromagnetism with massive photons. This allows the definition of a numeric dimension for each physical quantity, allowing us to discriminate easily covariance, invariance, contra-variance and half-variance. Next we study electromagnetism with magnetic monopoles. We explain how the photon of de Broglie may be obtained with real components by an antisymmetric product of spinors.

Chapter 5 presents several consequences of these novelties. The anisotropy of the intrinsic space-time explains why we see muons and tauons as well as electrons, their similarities and their differences. The intrinsic manifold has a torsion whose components were calculated for plane waves. The mass term is linked to this torsion. Next we present the building of the de Broglie's wave of a system of electrons as a wave in ordinary space-time, and not in a configuration space where space and time do not have the same status, with value onto the space algebra. We present as a counter-example a wave equation without Lagrangian formalism and we solve this wave equation in the hydrogen case. We present in our frame the three other photons of Lochak's theory.

Chapter 6 is devoted to our main progress since [16]. We start from the covariant derivative of the electro-weak gauge theory in the frame of the space-time algebra, first for the lepton case, electron+neutrino. We get both the form invariance under Cl_3^* and the gauge invariance under the gauge group of electro-weak interactions. We extend to the lepton wave the geometric transformation linked to the wave in the case of a lone electron. We get an identity allowing the existence of an inverse and of a wave equation with mass term for electron+neutrino. Similarly to the case of the lone electron, the real scalar part is simply $\mathcal{L} = 0$ where $\mathcal{L}$ is the Lagrangian density giving the whole wave equation by the variational calculus. The wave equation gives both the homogeneous non-linear equation studied in chapter 3 and the electro-weak gauge invariance previously studied. It is well-known that the standard model has great difficulty with the mass of the electron: the mass term of the Dirac equation links left wave to right wave. In the electro-weak interaction these left and right waves act differently. Therefore the standard model first cancels the mass of the electron then puts it back with a very complicated mechanism of spontaneously broken symmetry. We get here a wave equation with a mass term, both form invariant (and consequently relativistic invariant) and gauge invariant under the $U(1) \times SU(2)$ gauge group of electro-weak interactions. All these results could not have been obtained from the linear Dirac equation. A double link exists between the wave equation and the Lagrangian density; this generalizes to the electron + neutrino case the double link that we previously saw in the electron case. Only one other numeric equation is simple amongst the 14 ones: the law of conservation of the current. This one is the sum of the current of the electron and the current of the neutrino.

Chapter 7 uses the $Cl_{1,5}$ Clifford algebra to extend the gauge to strong interactions. Even though our aim is the same as in [17], we use here a different Clifford algebra, because we need the link between the wave of the particle and the wave of the antiparticle that is used in the standard model of electro-weak and strong interactions. We get exactly the $U(1) \times SU(2) \times SU(3)$ gauge group in this frame. In addition to the standard model we understand the reason of the insensitivity of leptons to strong interactions. We extend to the $Cl_{1,5}$ frame the form invariance of the gauge interactions. This induces the use of a complex 6-dimensional space-time into which the usual 4-dimensional space-time is well separated from two supplementary dimensions. Next the nullity of right waves of the neutrino and of quarks induces another remarkable identity. It implies that the full wave of the lepton and the three colored quarks has an invertible value. This identity allows a wave equation with a mass term, which is both form invariant and gauge invariant. The wave equation has a mass term generalizing the mass term of the electron+neutrino case, with a proper mass for the leptonic part and a proper mass for the quark part of the complete wave. The double link between the wave equation and the Lagrangian density explains why a Lagrangian formalism is always present in the standard model of the quantum theory. The wave equation is form invariant under the Cl_3^* group generalizing the relativistic invariance. It is also gauge invariant under precisely the $U(1) \times SU(2) \times SU(3)$ gauge group of the standard model.

The geometric transformation that the wave defines at each point of the space-time is generalized when we extend the wave to the electron+neutrino+quarks case. The transformation is still affine but the orthogonality is lost. The equality between $Cl_{1,5}$ and $Cl_{5,1}$ mixes the vector and the pseudo-vector parts of the Clifford algebra.

Chapter 8 is devoted to magnetic monopoles. We explain Russian experiments and our French experiments. We make their results more precise, particularly the wavelength. We apply our study of electro-weak interactions to the case of the magnetic monopole.

Chapter 9 presents how we can get the inertial phenomena from the wave equations previously obtained, making the form invariance of the wave equation local. Inertial frames are not only those in which there are no additional forces, they are also those in which the double link between the wave equation and Lagrangian density exists. Next we link the existence of the density of probability to the principle of equivalence of inertial and gravitational mass. The normalization of the wave inducing a symmetrical momentum-energy tensor, gravitation is then ruled by the Einstein–Ricci curvature tensor.

Chapter 10 presents our conclusions about the major change explained here in our way of seeing the standard model of our physical universe and its insertion into a theory of all interactions.

For the works at E.C.N., thanks to Didier Priem for his efficiency, his inventiveness and his kindness, and thanks to Guillaume Racineux who constantly supported us.

Chapter 1

Clifford algebras

This chapter presents what a Clifford algebra is, then we study the algebra of an Euclidean plane and the algebra of 3-dimensional physical space which is also the algebra of the Pauli matrices. We put the space-time and the relativistic invariance there. Then we present the space-time algebra and the Dirac matrices. We finally present the Clifford algebra of a 6-dimensional space-time, needed by the electro-weak and strong interactions.

It is quite usual in a physics book to put the mathematics into appendices even if they are necessary to understand the main part of the book. As it is impossible to understand the part containing physics without the Clifford algebras, we make here again a complete presentation[1] of this necessary tool. Why is this mathematical tool necessary? The physics of light and quantum physics use waves, therefore they use trigonometric functions, then also the exponential function, then necessarily products. The addition of vectors is not enough, a multiplication must be used. Then we must consider two operations, addition and multiplication of vectors. The mathematical tool of Clifford algebra is ready to be used in this case.

We shall only speak here about Clifford algebras on the real field. Algebras on the complex field also exist and we could expect complex algebras to be key point for quantum physics. The main algebra used here is also an algebra on the complex field, but it is its structure of real algebra[2] which is useful.

---

1. Readers already in the know may do a quick review. On the contrary a complete lecture is strongly advised for each reader who really wants to understand the physics contained in the following chapters.

2. A Clifford algebra on the real field has components of vectors which are real numbers and which cannot be multiplied by i. A Clifford algebra on the complex field has components of vectors which are complex numbers and which can be multiplied by i.

Our aim is not to say everything about any Clifford algebra but simply to give to our lecturer tools to understand the next chapters of this book.

Another very interesting introduction to the subject is the book by Doran and Lasenby [35]. We could have suppressed this chapter if their book was not too devoted to space-time algebra, whilst we use here three Clifford algebras.

1.1   What is a Clifford algebra?

1 – It is an algebra [5] [15]. There are two operations, denoted $A + B$ and AB, such that for any A, B, C:

$$A + (B + C) = (A + B) + C \; ; \quad A + B = B + A$$
$$A + 0 = A \; ; \quad A + (-A) = 0 \tag{1.1}$$
$$A(B + C) = AB + AC \; ; \quad (A + B)C = AC + BC \; ; \quad A(BC) = (AB)C.$$

2 – The algebra contains a set of vectors, denoted with arrows, in which a scalar product exists and the internal Clifford multiplication $\vec{u}\vec{v}$ is supposed to satisfy for any vector $\vec{u}$:

$$\vec{u}\vec{u} = \vec{u} \cdot \vec{u}. \tag{1.2}$$

where [3] $\vec{u} \cdot \vec{v}$ is the usual notation for the scalar product of two vectors. This implies, since $\vec{u} \cdot \vec{u}$ is a real number, that the algebra contains vectors but also real numbers.

3 – Real numbers commute with any member of the algebra: if a is a real number and if A is any element in the algebra:

$$aA = Aa, \tag{1.3}$$

$$1A = A. \tag{1.4}$$

Such an algebra exists for all finite-dimensional linear spaces which are the ones that we need in physics.

The smaller one is unique, to within an isomorphism.

Remark 1: Equation (1.1) and Eq. (1.4) imply that the algebra is itself a linear space, not to be confused with the first one. If the initial linear space is n-dimensional, we get a Clifford algebra which is 2^n-dimensional. We shall see for instance in Sec. 1.3 that the Clifford algebra of the 3-dimensional physical space is an 8-dimensional linear space on the real field. We do not need to distinguish between the left or right linear space

---

3. This equality seems strange, but gives nice properties. We need these properties in the following chapters.

here, since real numbers commute with each element of the algebra. We also do not need to consider multiplication by a real number as a third operation, because it is a particular case of the multiplication.

Remark 2: If $\vec{u}$ and $\vec{v}$ are two orthogonal vectors, $(\vec{u} \cdot \vec{v} = 0)$, the equality $(\vec{u}+\vec{v}) \cdot (\vec{u}+\vec{v}) = (\vec{u}+\vec{v})(\vec{u}+\vec{v})$ implies $\vec{u} \cdot \vec{u} + \vec{u} \cdot \vec{v} + \vec{v} \cdot \vec{u} + \vec{v} \cdot \vec{v} = \vec{u}\vec{u} + \vec{u}\vec{v} + \vec{v}\vec{u} + \vec{v}\vec{v}$, so we get :

$$0 = \vec{u}\vec{v} + \vec{v}\vec{u} \; ; \quad \vec{v}\vec{u} = -\vec{u}\vec{v}. \tag{1.5}$$

Main change to usual rules on numbers, multiplication is not commutative: we must be as careful as with matrix calculations.

Remark 3: Addition is defined in the whole algebra, which contains numbers and vectors. So we will get sums of numbers and vectors: $3 + 5\vec{i}$ is **authorized**. It is perhaps strange or disturbing, but it is not different from $3 + 5i$. And everyone using complex numbers finally gets used to it.

Even sub-algebra: It's the sub-algebra generated by the products of an even number of vectors: $\vec{u}\vec{v}$, $\vec{e}_1\vec{e}_2\vec{e}_3\vec{e}_4$ and so on. The even sub-algebra built on an n-dimensional vector space is 2^{n-1}-dimensional.

Reversion: The reversion $A \mapsto \widetilde{A}$ changes the order of products. Reversion does not change numbers a nor vectors: $\tilde{a} = a$, $\tilde{\vec{u}} = \vec{u}$, and we get, for any $\vec{u}$ and $\vec{v}$, A and B:

$$\widetilde{\vec{u}\vec{v}} = \vec{v}\vec{u} \; ; \quad \widetilde{AB} = \tilde{B}\tilde{A} \; ; \quad \widetilde{A+B} = \tilde{A} + \tilde{B}. \tag{1.6}$$

1.2 Clifford algebra Cl_2 of a Euclidean plane

Cl_2 contains the real numbers and the vectors of a Euclidean plane, which read $\vec{u} = x\vec{e}_1 + y\vec{e}_2$, where $\vec{e}_1$ and $\vec{e}_2$ form a direct orthonormal basis of the plane: $\vec{e}_1{}^2 = \vec{e}_2{}^2 = 1$, $\vec{e}_1 \cdot \vec{e}_2 = 0$. Usually we set: $\vec{e}_1\vec{e}_2 = i$. The general element of the algebra is :

$$A = a + x\vec{e}_1 + y\vec{e}_2 + ib, \tag{1.7}$$

where a, x, y and b are real numbers. This is enough because:

$$\begin{aligned}
\vec{e}_1 i &= \vec{e}_1(\vec{e}_1\vec{e}_2) = (\vec{e}_1\vec{e}_1)\vec{e}_2 = 1\vec{e}_2 = \vec{e}_2, \\
\vec{e}_2 i &= -\vec{e}_1 \; ; \quad i\vec{e}_2 = \vec{e}_1 \; ; \quad i\vec{e}_1 = -\vec{e}_2, \\
i^2 &= ii = i(\vec{e}_1\vec{e}_2) = (i\vec{e}_1)\vec{e}_2 = -\vec{e}_2\vec{e}_2 = -1.
\end{aligned} \tag{1.8}$$

Two remarks must be made:

1 – The even sub-algebra Cl_2^+ is the set formed by all $a + ib$, so it is the complex field. We may say that complex numbers are underlying as

soon as the dimension of the linear space is greater than one. This even sub-algebra is commutative.

2 – The reversion is here the usual conjugation: $\tilde{i} = \widetilde{\vec{e}_1\vec{e}_2} = \vec{e}_2\vec{e}_1 = -i$. We get then, for any $\vec{u}$ and any $\vec{v}$ in the plane: $\vec{u}\vec{v} = \vec{u}\cdot\vec{v} + i\det(\vec{u},\vec{v})$.

To establish that $(\vec{u}\cdot\vec{v})^2 + [\det(\vec{u},\vec{v})]^2 = \vec{u}^2\vec{v}^2$, it is possible to use $\vec{u}\vec{v}\vec{v}\vec{u}$ which can be calculated in two ways, and we can use $\vec{v}\vec{v}$ which is a real number and commutes with anything in the algebra.

1.3   Clifford algebra Cl_3 of the physical space

Cl_3 contains [2] the real numbers and the vectors of the physical space which read: $\vec{u} = x\vec{e}_1 + y\vec{e}_2 + z\vec{e}_3$, where x, y and z are real numbers, and $\vec{e}_1$, $\vec{e}_2$ and $\vec{e}_3$ form an orthonormal basis:

$$\vec{e}_1\cdot\vec{e}_2 = \vec{e}_2\cdot\vec{e}_3 = \vec{e}_3\cdot\vec{e}_1 = 0 \; ; \; \vec{e}_1{}^2 = \vec{e}_2{}^2 = \vec{e}_3{}^2 = 1. \tag{1.9}$$

We let:

$$i_1 = \vec{e}_2\vec{e}_3 \; ; \; i_2 = \vec{e}_3\vec{e}_1 \; ; \; i_3 = \vec{e}_1\vec{e}_2 \; ; \; i = \vec{e}_1\vec{e}_2\vec{e}_3. \tag{1.10}$$

Then we get:

$$i_1^2 = i_2^2 = i_3^2 = i^2 = -1, \tag{1.11}$$

$$i\vec{u} = \vec{u}i \; ; \; i\vec{e}_j = i_j \,, \; j = 1,\, 2,\, 3. \tag{1.12}$$

To derive Eq. (1.11) we can use the same method we used to get Eq. (1.8). To derive Eq. (1.12) we may firstly use the fact that i commutes with each $\vec{e}_j$.

The general element of Cl_3 reads: $A = a + \vec{u} + i\vec{v} + ib$. This gives $1 + 3 + 3 + 1 = 8 = 2^3$ dimensions for Cl_3.

Several remarks:

1 – The center of Cl_3 is the set of $a+ib$, the only elements which commute with every other one in the algebra. It is isomorphic to the complex field.

2 – The even sub-algebra Cl_3^+ is the set of $a + i\vec{v}$, isomorphic to the quaternion field. Therefore quaternions are implicitly present in calculations as soon as the dimension of the linear space is greater than or equal to three. This even sub-algebra is not commutative.

3 – $\tilde{A} = a + \vec{u} - i\vec{v} - ib$; The reversion is the conjugation, for complex numbers but also for the quaternions contained in Cl_3.

4 – $i\vec{v}$ is what is usually called "axial vector" or "pseudo-vector", whilst $\vec{u}$ is usually called vector. It is well known that it is specific to dimension 3.

5 – There are now four different terms with square -1, four ways to get complex numbers. Quantum theory is used to only one term with square -1. When complex numbers are used in quantum mechanics, it will then be necessary to ask the question of which i is used: $i = \sigma_1\sigma_2\sigma_3$ or $i_3 = \sigma_1\sigma_2$?

1.3.1 Cross-product, orientation

We use $\vec{u} \times \vec{v}$ to denote the cross-product of $\vec{u}$ and $\vec{v}$. Using coordinates in the basis $(\vec{e}_1, \vec{e}_2, \vec{e}_3)$, we can easily establish for any $\vec{u}$ and $\vec{v}$:

$$\vec{u}\vec{v} = \vec{u} \cdot \vec{v} + i\,\vec{u} \times \vec{v}, \tag{1.13}$$

$$(\vec{u} \cdot \vec{v})^2 + (\vec{u} \times \vec{v})^2 = \vec{u}^2\vec{v}^2. \tag{1.14}$$

We use $\det(\vec{u}, \vec{v}, \vec{w})$ to denote the determinant whose columns contain the components of vectors $\vec{u}$, $\vec{v}$, $\vec{w}$, in the basis $(\vec{e}_1, \vec{e}_2, \vec{e}_3)$. Again using coordinates, it is possible to establish, for any $\vec{u}$, $\vec{v}$, $\vec{w}$:

$$\vec{u} \cdot (\vec{v} \times \vec{w}) = \det(\vec{u}, \vec{v}, \vec{w}), \tag{1.15}$$

$$\vec{u} \times (\vec{v} \times \vec{w}) = (\vec{w} \cdot \vec{u})\vec{v} - (\vec{u} \cdot \vec{v})\vec{w}, \tag{1.16}$$

$$\vec{u}\vec{v}\vec{w} = i\det(\vec{u}, \vec{v}, \vec{w}) + (\vec{v} \cdot \vec{w})\vec{u} - (\vec{w} \cdot \vec{u})\vec{v} + (\vec{u} \cdot \vec{v})\vec{w}. \tag{1.17}$$

From Eq. (1.15) it follows that $\vec{u} \times \vec{v}$ is orthogonal to $\vec{u}$ and $\vec{v}$. Equation (1.14) allows us to calculate the length of $\vec{u} \times \vec{v}$, and Eq. (1.15) gives its orientation. We recall that a basis $(\vec{u}, \vec{v}, \vec{w})$ is said to be direct, or to have the same orientation as $(\vec{e}_1, \vec{e}_2, \vec{e}_3)$ if $\det(\vec{u}, \vec{v}, \vec{w}) > 0$, and to be inverse, or to have other orientation if $\det(\vec{u}, \vec{v}, \vec{w}) < 0$. Equation (1.17) allows us to establish that, if $B = (\vec{u}, \vec{v}, \vec{w})$ is any orthonormal basis, then $\vec{u}\vec{v}\vec{w} = i$ if and only if B is direct, and $\vec{u}\vec{v}\vec{w} = -i$ if and only if B is inverse. So i is strictly linked to the orientation of the physical space. To change i to $-i$ is equivalent to changing the space orientation (it is the same for a plan). The fact that i rules the orientation of the physical space will play an important role in the next chapters.

1.3.2 Pauli algebra

The Pauli algebra, introduced in physics as early as 1926 to account for the spin of the electron, is the algebra of 2×2 complex matrices. It is identical (isomorphic) to Cl_3, but only as algebras on the real[4] field.

---

4. The dimension of the Pauli algebra is 8 on the real field, but only 4 on the complex field.

Identifying complex numbers to scalar matrices and the e_j to the Pauli matrices σ_j is [5] enough. So, z being any complex number, we have

$$z = \begin{pmatrix} z & 0 \\ 0 & z \end{pmatrix}, \tag{1.18}$$

$$\vec{e}_1 = \sigma_1 = \begin{pmatrix} 0 & 1 \\ 1 & 0 \end{pmatrix} ; \ \vec{e}_2 = \sigma_2 = \begin{pmatrix} 0 & -i \\ i & 0 \end{pmatrix} ; \ \vec{e}_3 = \sigma_3 = \begin{pmatrix} 1 & 0 \\ 0 & -1 \end{pmatrix}. \tag{1.19}$$

This is fully compatible with all preceding calculations, because:

$$\sigma_1\sigma_2\sigma_3 = \begin{pmatrix} i & 0 \\ 0 & i \end{pmatrix} = i, \tag{1.20}$$

$$\sigma_1\sigma_2 = i\sigma_3 \ ; \ \sigma_2\sigma_3 = i\sigma_1 \ ; \ \sigma_3\sigma_1 = i\sigma_2. \tag{1.21}$$

And then the reverse is identical to the adjoint matrix:

$$\tilde{A} = A^\dagger = (A^*)^t. \tag{1.22}$$

Consequently we shall equally name Cl_3 "Pauli algebra" or "space algebra". This is not appreciated by a few users of Clifford algebras who have not understood the power of isomorphisms.

1.3.3   Three conjugations are used

$A = a + \vec{u} + i\vec{v} + ib$ is the sum of the even part $A_1 = a + i\vec{v}$ (quaternion) and the odd part $A_2 = \vec{u} + ib$. From this we define the conjugation (involutive automorphism) $A \mapsto \hat{A}$ as

$$\hat{A} = A_1 - A_2 = a - \vec{u} + i\vec{v} - ib. \tag{1.23}$$

This conjugation satisfies, for any element A and any B in Cl_3:

$$\widehat{A+B} = \hat{A} + \hat{B} \ ; \ \widehat{AB} = \hat{A}\hat{B}. \tag{1.24}$$

It is the main automorphism of this algebra, and each Clifford algebra has such an automorphism. From this conjugation and from the reversion we form the third conjugation:

$$\overline{A} = \hat{A}^\dagger = a - \vec{u} - i\vec{v} + ib \ : \ \overline{A+B} = \overline{A} + \overline{B} \ ; \ \overline{AB} = \overline{B}\,\overline{A}. \tag{1.25}$$

Composing, in any order, two of these three conjugations gives the third one. Only $A \mapsto \hat{A}$ preserves the order of products; $A \mapsto \overline{A}$ and $A \mapsto A^\dagger$

---

5. This identifying process may be considered a lack of rigor, but in fact it is frequent in mathematics. The same process allows us to include integer numbers into relative numbers, or real numbers into complex numbers. To go without this process implies very complicated notations. This identifying process considers the three σ_j as forming a direct basis of the physical space.

invert the order of products. Now a, b, c, d are any complex numbers and a^* is the complex conjugate of a. We can prove [6] that for any $A = \begin{pmatrix} a & b \\ c & d \end{pmatrix}$:

$$\widetilde{A} = A^\dagger = \begin{pmatrix} a^* & c^* \\ b^* & d^* \end{pmatrix} \;\; ; \;\; \widehat{A} = \begin{pmatrix} d^* & -c^* \\ -b^* & a^* \end{pmatrix} \;\; ; \;\; \overline{A} = \begin{pmatrix} d & -b \\ -c & a \end{pmatrix}, \quad (1.26)$$

$$A\overline{A} = \overline{A}A = \det(A) = ad - bc \;\; ; \;\; \widehat{A}A^\dagger = A^\dagger\widehat{A} = [\det(A)]^*,$$

$$i^\dagger = -i \;\; ; \;\; \widehat{i} = -i \;\; ; \;\; \overline{i} = i, \quad (1.27)$$

$$\sigma_j^\dagger = \sigma_j \;\; ; \;\; \widehat{\sigma}_j = -\sigma_j \;\; ; \;\; \overline{\sigma}_j = -\sigma_j.$$

1.3.4 *Gradient, divergence and curl*

In Cl_3 there exists one important differential operator, because all other operators may be made with it:

$$\vec{\partial} = \vec{e}_1\partial_1 + \vec{e}_2\partial_2 + \vec{e}_3\partial_3 = \begin{pmatrix} \partial_3 & \partial_1 - i\partial_2, \\ \partial_1 + i\partial_2 & -\partial_3 \end{pmatrix}, \quad (1.28)$$

with [7]

$$\vec{x} = x^1\vec{e}_1 + x^2\vec{e}_2 + x^3\vec{e}_3 \;\; ; \;\; \partial_j = \frac{\partial}{\partial x^j}. \quad (1.29)$$

The Laplacian is simply the square of $\vec{\partial}$:

$$\Delta = (\partial_1)^2 + (\partial_2)^2 + (\partial_3)^2 = \vec{\partial}\vec{\partial}. \quad (1.30)$$

Applied to a scalar a, $\vec{\partial}$ gives [8] the gradient; and applied to a vector $\vec{u}$ it gives both the divergence and the curl, also called rotational:

$$\vec{\partial}a = \text{grad}(a), \quad (1.31)$$

$$\vec{\partial}\vec{u} = \vec{\partial}\cdot\vec{u} + i\,\vec{\partial}\times\vec{u} \;\; ; \;\; \vec{\partial}\cdot\vec{u} = \text{div}(\vec{u}) \;\; ; \;\; \vec{\partial}\times\vec{u} = \text{rot}(\vec{u}). \quad (1.32)$$

---

6. The equality $A\overline{A} = \overline{A}A$ is general in Cl_3. The equality $A\overline{A} = \det(A)$ uses the identification between numbers and scalar matrices, or equivalently the inclusion of real numbers into the algebra.

7. This operator $\vec{\partial}$ is usually noted in quantum mechanics as a scalar product, for instance $\vec{\sigma}\cdot\vec{\nabla}$. From this useless scalar product many convoluted complications result. Simple notations fully simplify calculations.

8. We use here the notations grad, div and rot that are probably unusual for most of our lecturers, for two reasons: first these notations were in Louis de Broglie's books. Next we shall use the $\nabla = \sigma^\mu\partial_\mu$ symbol in the space-time algebra.

1.3.5 *Space-time in space algebra*

With

$$x^0 = ct \ ; \ \ \vec{x} = x^1 \vec{e}_1 + x^2 \vec{e}_2 + x^3 \vec{e}_3 \ ; \ \ \partial_\mu = \frac{\partial}{\partial x^\mu}, \tag{1.33}$$

we let [2] [53]

$$x = x^0 + \vec{x} = \begin{pmatrix} x^0 + x^3 & x^1 - ix^2 \\ x^1 + ix^2 & x^0 - x^3 \end{pmatrix}. \tag{1.34}$$

Then the space-time is made of the auto-adjoint part of the Pauli algebra

$$\hat{x} = \overline{x} = x^0 - \vec{x} \ ; \ \ x^\dagger = x, \tag{1.35}$$

$$\det(x) = x\hat{x} = x \cdot x = (x^0)^2 - \vec{x}^2 = (x^0)^2 - (x^1)^2 - (x^2)^2 - (x^3)^2. \tag{1.36}$$

The square of the pseudo-norm of any space-time vector [9] is then simply the determinant. Any element M of the Pauli algebra is the sum of a space-time vector v and of the product with i of another space-time vector w:

$$M = v + iw, \tag{1.37}$$

$$v = \frac{1}{2}(M + M^\dagger) \ ; \ \ v^\dagger = v, \tag{1.38}$$

$$iw = \frac{1}{2}(M - M^\dagger) \ ; \ \ w^\dagger = w. \tag{1.39}$$

Space-time vectors v and w are uniquely defined. We need two linked differential operators:

$$\nabla = \partial_0 - \vec{\partial} = \sigma^\mu \partial_\mu,$$

$$\hat{\nabla} = \partial_0 + \vec{\partial} \ ; \ \ \sigma^0 = \sigma_0 = 1 \ ; \ \ \sigma^j = -\sigma_j, \ j = 1, 2, 3. \tag{1.40}$$

They allow us to calculate the D'Alembertian:

$$\nabla \hat{\nabla} = \hat{\nabla} \nabla = (\partial_0)^2 - (\partial_1)^2 - (\partial_2)^2 - (\partial_3)^2 = \square \tag{1.41}$$

1.3.6 *Relativistic invariance*

If M is any nonzero element in Cl_3 and if R is the transformation from the space-time into itself, which to any x associates x' such that

$$x' = x'^0 + \vec{x}\,' = R(x) = MxM^\dagger, \tag{1.42}$$

we note, if $\det(M) \neq 0$:

$$\det(M) = re^{i\theta} \ , \ \ r = |\det(M)|. \tag{1.43}$$

---

9. We must notice that the pseudo-norm of the space-time metric comes not from a scalar product, a symmetric bilinear form, but from a determinant, an antisymmetric bilinear form. We are here very far from Riemannian spaces.

We get then:

$$(x'^0)^2 - (x'^1)^2 - (x'^2)^2 - (x'^3)^2 = \det(x') = \det(MxM^\dagger)$$
$$= re^{i\theta}\det(x)re^{-i\theta} = r^2[(x^0)^2 - (x^1)^2 - (x^2)^2 - (x^3)^2]. \quad (1.44)$$

Then R multiplies by r any space-time distance and is called "Lorentz dilation with ratio r". If we let, with the usual convention summing the up and down indices:

$$x'^\mu = R^\mu_\nu x^\nu, \quad (1.45)$$

we get for $M = \begin{pmatrix} a & b \\ c & d \end{pmatrix} \neq 0$:

$$2R^0_0 = |a|^2 + |b|^2 + |c|^2 + |d|^2 > 0. \quad (1.46)$$

x'^0 has then the same sign as x^0 at the origin: R conserves the arrow of time. Furthermore we get (the calculation is in [13] A.2.4):

$$\det(R^\mu_\nu) = r^4. \quad (1.47)$$

R conserves therefore the orientation of space-time and as it conserves the time orientation it conserves also the space orientation.

Let f be the application which associates R to M. Let M' be another matrix, with:

$$\det(M') = r'e^{i\theta'} \quad ; \quad R' = f(M') \quad ; \quad x'' = M'x'M'^\dagger. \quad (1.48)$$

We get

$$x'' = M'x'M'^\dagger = M'(MxM^\dagger)M'^\dagger = (M'M)x(M'M)^\dagger,$$
$$R' \circ R = f(M') \circ f(M) = f(M'M). \quad (1.49)$$

f is then a homomorphism. If we restrict M to $r \neq 0$, f is a homomorphism from the group $(Cl_3^*, \times)$ into the group $(D^*, \circ)$, where D^* is the set of dilations with nonzero ratio. These two groups are Lie groups. $(Cl_3^*, \times)$ is an 8-dimensional Lie group, because Cl_3 is 8-dimensional. On the other $(D^*, \circ)$ is only a 7-dimensional Lie group; one dimension is lost because the kernel of f is not reduced to the neutral element: let θ be any real number and let M be

$$M = e^{i\frac{\theta}{2}} = \begin{pmatrix} e^{i\frac{\theta}{2}} & 0 \\ 0 & e^{i\frac{\theta}{2}} \end{pmatrix} \quad ; \quad \det(M) = e^{i\theta}, \quad (1.50)$$

we then get:

$$x' = MxM^\dagger = e^{i\frac{\theta}{2}}xe^{-i\frac{\theta}{2}} = x. \quad (1.51)$$

$f(M)$ is therefore the identity and M belongs to the kernel of f, which is a group with only one parameter θ, and we get only 7 parameters in D^*: 6 angles defining a Lorentz rotation, plus the ratio of the dilation. For instance, if

$$M = e^{a+b\sigma_1} = e^a[\cosh(b) + \sinh(b)\sigma_1], \qquad (1.52)$$

then the R transformation defined in Eq. (1.42) satisfies

$$x' = MxM^\dagger = e^{a+b\sigma_1}(x^0 + x^1\sigma_1 + x^2\sigma_2 + x^3\sigma_3)e^{a+b\sigma_1}$$
$$= e^{2a}[e^{2b\sigma_1}(x^0 + x^1\sigma_1) + x^2\sigma_2 + x^3\sigma_3]. \qquad (1.53)$$

We then get

$$x'^0 + x'^1\sigma_1 = e^{2a}(\cosh(2b) + \sinh(2b)\sigma_1)(x^0 + x^1\sigma_1),$$
$$x'^0 = e^{2a}(\cosh(2b)x^0 + \sinh(2b)x^1), \qquad (1.54)$$
$$x'^1 = e^{2a}(\sinh(2b)x^0 + \cosh(2b)x^1), \qquad (1.55)$$
$$x'^2 = e^{2a}x^2, \qquad (1.56)$$
$$x'^3 = e^{2a}x^3. \qquad (1.57)$$

We can recognize R as the product of a Lorentz boost with velocity $v = c\tanh(2b)$ mixing the temporal component x^0 and the spacial component x^1 and of a homothety with ratio e^{2a}. Now if

$$M = e^{a+bi\sigma_1} = e^a[\cos(b) + \sin(b)i\sigma_1], \qquad (1.58)$$

then the R transformation defined in Eq. (1.42) satisfies

$$x' = MxM^\dagger = e^{a+bi\sigma_1}(x^0 + x^1\sigma_1 + x^2\sigma_2 + x^3\sigma_3)e^{a-bi\sigma_1}$$
$$= e^{2a}[x^0 + x^1\sigma_1 + e^{2bi\sigma_1}(x^2\sigma_2 + x^3\sigma_3)]. \qquad (1.59)$$

We then get

$$x'^2\sigma_2 + x'^3\sigma_3 = e^{2a}(\cos(2b) + \sin(2b)i\sigma_1)(x^2\sigma_2 + x^3\sigma_3),$$
$$x'^0 = e^{2a}x^0, \qquad (1.60)$$
$$x'^1 = e^{2a}x^1, \qquad (1.61)$$
$$x'^2 = e^{2a}(\cos(2b)x^2 + \sin(2b)x^3), \qquad (1.62)$$
$$x'^3 = e^{2a}(-\sin(2b)x^2 + \cos(2b)x^3). \qquad (1.63)$$

We can recognize R as the product of a rotation with axis Ox_1 and angle $2b$ and a homothety with ratio e^{2a}.

1.3.7 Restricted Lorentz group

If we impose now the condition $\det(M) = 1$, the set of M is called $SL(2, \mathbb{C})$, and Eq. (1.44) becomes:

$$(x'^0)^2 - (x'^1)^2 - (x'^2)^2 - (x'^3)^2 = (x^0)^2 - (x^1)^2 - (x^2)^2 - (x^3)^2. \quad (1.64)$$

R is then a Lorentz rotation and the set of the R is the Lorentz restricted group $\mathcal{L}_+^\uparrow$ (conserving space and time orientation). With Eq. (1.50) we get:

$$1 = e^{i\theta} \ ; \ \ \theta = k2\pi \ ; \ \ \frac{\theta}{2} = k\pi \ ; \ \ M = \pm 1. \quad (1.65)$$

Remark 1: Quantum mechanics never distinguishes M from R, it confuses $SL(2, \mathbb{C})$ and $\mathcal{L}_+^\uparrow$ and names "bi-valued representations of $\mathcal{L}_+^\uparrow$" the representations of $SL(2, \mathbb{C})$. This comes mainly from the use in quantum theory of infinitesimal rotations, so they work only in the vicinity of the origin of the group, then they work not in the group but in the Lie algebra of the group. And it happens that the Lie algebras of $SL(2, \mathbb{C})$ and of $\mathcal{L}_+^\uparrow$ are identical. $SL(2, \mathbb{C})$ is the covering group of $\mathcal{L}_+^\uparrow$. Globally $SL(2, \mathbb{C})$ and $\mathcal{L}_+^\uparrow$ are quite different, for instance any element reads e^A in $\mathcal{L}_+^\uparrow$ and this is false in $SL(2, \mathbb{C})$. It is therefore intolerable to have neglected for so long the fact that when an angle b is present in M, it is an angle $2b$ which is present in R.

Remark 2: We are forced to distinguish the group of the M from the group of the R, as soon as θ is not null, because these two groups do not have the same dimension and are not similar even in the vicinity of the origin.

Remark 3: $SL(2, \mathbb{C})$ contains as a subgroup the $SU(2)$ group of the unitary 2×2 complex matrices with determinant 1. The restriction of f to this subgroup is a homomorphism from $SU(2)$ into the $SO(3)$ group of rotations. The kernel of this homomorphism is also ± 1.

Remark 4: There are two non-equivalent homomorphisms from $(Cl_3^*, \times)$ into the group $(D^*, \circ)$. The second homomorphism $\widehat{f}$ is defined by

$$x' = \widehat{R}(x) = \widehat{M}x\overline{M} \ ; \ \ \widehat{R} = \widehat{f}(M). \quad (1.66)$$

1.4 Clifford algebra $Cl_{1,3}$ of the space-time

$Cl_{1,3}$ contains real numbers and space-time vectors $\mathbf{x}$,

$$\mathbf{x} = x^0\gamma_0 + x^1\gamma_1 + x^2\gamma_2 + x^3\gamma_3 = x^\mu\gamma_\mu. \quad (1.67)$$

The four γ_μ form an orthonormal basis [10] of space-time:

$$(\gamma_0)^2 = 1 \;\; ; \;\; (\gamma_1)^2 = (\gamma_2)^2 = (\gamma_3)^2 = -1 \;\; ; \;\; \gamma_\mu \cdot \gamma_\nu = 0 \; , \; \mu \neq \nu. \quad (1.68)$$

The general term in $Cl_{1,3}$ is a sum:

$$N = s + v + b + pv + ps, \quad (1.69)$$

where s is a real number, v is a space-time vector, b is a bivector, pv is a pseudo-vector and ps is a pseudo-scalar. There are $1 + 4 + 6 + 4 + 1 = 16 = 2^4$ dimensions on the real field because there are 6 independent bivectors $\gamma_{01} = \gamma_0\gamma_1$, γ_{02}, γ_{03}, γ_{12}, γ_{23}, γ_{31}, and $\gamma_{ji} = -\gamma_{ij}$, $j \neq i$ and 4 independent pseudo-vectors γ_{012}, γ_{023}, γ_{031}, γ_{123} and one pseudoscalar

$$ps = p\gamma_{0123} \;\; ; \;\; \gamma_{0123} = \gamma_0\gamma_1\gamma_2\gamma_3, \quad (1.70)$$

where p is a real number.

The even part of N is $s + b + ps$, the odd part is $v + pv$. The main automorphism is $N \mapsto \widehat{N} = s - v + b - pv + ps$.

The reverse of N is

$$\widetilde{N} = s + v - b - pv + ps. \quad (1.71)$$

Amongst the 16 generators of $Cl_{1,3}$, 10 have square -1 and 6 have square 1:

$$1^2 = \gamma_{01}{}^2 = \gamma_{02}{}^2 = \gamma_{03}{}^2 = \gamma_0{}^2 = \gamma_{123}{}^2 = 1,$$
$$\gamma_1{}^2 = \gamma_2{}^2 = \gamma_3{}^2 = \gamma_{12}{}^2 = \gamma_{23}{}^2 = \gamma_{31}{}^2$$
$$= \gamma_{012}{}^2 = \gamma_{023}{}^2 = \gamma_{031}{}^2 = \gamma_{0123}{}^2 = -1. \quad (1.72)$$

Remark 1: If we use the $+$ sign for the space, then we get 10 generators with square 1 and 6 with square -1. The two Clifford algebras are not identical. And yet there was until this work no known physical reason to prefer one to the other algebra.

Remark 2: The even sub-algebra $Cl_{1,3}{}^+$, formed by all the even elements $N = s + b + ps$, is 8-dimensional and is isomorphic to Cl_3. We shall see this in the next subsection by using the Dirac matrices.

The privileged differential operator, in $Cl_{1,3}$, is:

$$\partial = \gamma^\mu \partial_\mu \;\; ; \;\; \gamma^0 = \gamma_0 \;\; ; \;\; \gamma^j = -\gamma_j \; , \; j = 1,\, 2,\, 3. \quad (1.73)$$

It satisfies:

$$\partial\partial = \square = (\partial_0)^2 - (\partial_1)^2 - (\partial_2)^2 - (\partial_3)^2. \quad (1.74)$$

---

10. Users of Clifford algebras are nearly equally divided between users of a $+$ sign for the time (Hestenes [39] [41]), and users of a $-$ sign for the time (Deheuvels [33]). It seems that no physical property of space-time shows a preference for one to the other. We use here a $+$ sign for the time which was Hestenes's choice and it is the better choice to include 4-dimensional space-time into the 6-dimensional space-time of chapter 7.

1.4.1 Dirac matrices

Most physicists do not use directly the Clifford algebra of space-time but they use [11] a matrix algebra, generated by the Dirac matrices. These matrices are not uniquely defined. The best way [12] to link $Cl_{1,3}$ to Cl_3 is to let:

$$\gamma_0 = \gamma^0 = \begin{pmatrix} 0 & I \\ I & 0 \end{pmatrix} \; ; \; I = \begin{pmatrix} 1 & 0 \\ 0 & 1 \end{pmatrix} \; ; \; \gamma^j = -\gamma_j = \begin{pmatrix} 0 & -\sigma_j \\ \sigma_j & 0 \end{pmatrix}. \tag{1.75}$$

Then we get:

$$\partial = \gamma^\mu \partial_\mu = \begin{pmatrix} 0 & \nabla \\ \hat{\nabla} & 0 \end{pmatrix}. \tag{1.76}$$

It is easy to satisfy:

$$\gamma_{0j} = \begin{pmatrix} -\sigma_j & 0 \\ 0 & \sigma_j \end{pmatrix} ; \; \gamma_{23} = \begin{pmatrix} -i\sigma_1 & 0 \\ 0 & -i\sigma_1 \end{pmatrix} ; \; \gamma_{0123} = \begin{pmatrix} iI & 0 \\ 0 & -iI \end{pmatrix}. \tag{1.77}$$

Isomorphism between $Cl_{1,3}^+$ and Cl_3: Let N be any even element. With:

$$N = a + Bi + ps \; ; \; Bi = u_1\gamma_{10} + u_2\gamma_{20} + u_3\gamma_{30} + v_1\gamma_{32} + v_2\gamma_{13} + v_3\gamma_{21},$$
$$ps = b\gamma_{0123}, \tag{1.78}$$
$$M = a + \vec{u} + i\vec{v} + ib \; ; \; \vec{u} = u_1\sigma_1 + u_2\sigma_2 + u_3\sigma_3,$$
$$\vec{v} = v_1\sigma_1 + v_2\sigma_2 + v_3\sigma_3, \tag{1.79}$$

Bi is a bivector and ps a pseudo-scalar in space-time. We get, with the choice made in Eq. (1.75) for the Dirac matrices:

$$N = \begin{pmatrix} M & 0 \\ 0 & \widehat{M} \end{pmatrix} \; ; \; \tilde{N} = \begin{pmatrix} \overline{M} & 0 \\ 0 & M^\dagger \end{pmatrix}. \tag{1.80}$$

Since the conjugation $M \mapsto \widehat{M}$ is compatible with the addition and the multiplication, the algebra of the M is exactly isomorphic to the algebra of the N. As N contains both M and $\widehat{M}$, the Dirac matrices combine both inequivalent representations of Cl_3^*.

---

11. Generally the matrix algebra used is $M_4(\mathbb{C})$, an algebra on the complex field. This algebra is 16-dimensional on the complex field and therefore it is also an algebra on the real field. It is 32-dimensional on the real field. This is enough to prove that $M_4(\mathbb{C}) \neq Cl_{1,3}$.

12. This choice of the Dirac matrices is not the choice used in the Dirac theory to calculate the solutions in the hydrogen atom case, but the choice used when high velocities and restricted relativity are required. It is also the usual choice in the electro-weak theory. We shall see that it is also a convenient choice to solve the wave equation for the hydrogen case.

1.5   Clifford Algebra $Cl_{1,5}$

We shall use also the Clifford algebra of a larger space-time with a 5-dimensional space later in chapter 7. The general element $\underline{x}$ of this larger space-time reads

$$\underline{x} = x^a L_a = x^\mu L_\mu + x^4 L_4 + x^5 L_5, \tag{1.81}$$

$$x^a L_a = \sum_{n=0}^{n=5} x^n L_n \; ; \quad x^\mu L_\mu = \sum_{n=0}^{n=3} x^n L_n, \tag{1.82}$$

$$(\underline{x})^2 = (x^0)^2 - (x^1)^2 - (x^2)^2 - (x^3)^2 - (x^4)^2 - (x^5)^2. \tag{1.83}$$

We link the preceding $Cl_{1,3}$ space-time algebra to this greater algebra by using the following matrix representation, $\mu = 0, 1, 2, 3$:

$$L_\mu = \begin{pmatrix} 0 & \gamma_\mu \\ \gamma_\mu & 0 \end{pmatrix} \; ; \quad L_4 = \begin{pmatrix} 0 & -I_4 \\ I_4 & 0 \end{pmatrix} \; ; \quad L_5 = \begin{pmatrix} 0 & \mathbf{i} \\ \mathbf{i} & 0 \end{pmatrix}, \tag{1.84}$$

where I_4 is the identity matrix for 4×4 matrices and $\mathbf{i} = \gamma_{0123}$ (See Eq. (1.77)). We always use the matrix representation Eq. (1.75). We get

$$L_{\mu\nu} = L_\mu L_\nu = \begin{pmatrix} \gamma_{\mu\nu} & 0 \\ 0 & \gamma_{\mu\nu} \end{pmatrix}, \tag{1.85}$$

$$L_{\mu\nu\rho} = L_{\mu\nu} L_\rho = \begin{pmatrix} 0 & \gamma_{\mu\nu\rho} \\ \gamma_{\mu\nu\rho} & 0 \end{pmatrix}, \tag{1.86}$$

$$L_{0123} = L_{01} L_{23} = \begin{pmatrix} \gamma_{0123} & 0 \\ 0 & \gamma_{0123} \end{pmatrix} = \begin{pmatrix} \mathbf{i} & 0 \\ 0 & \mathbf{i} \end{pmatrix}, \tag{1.87}$$

$$L_{45} = L_4 L_5 = \begin{pmatrix} -\mathbf{i} & 0 \\ 0 & \mathbf{i} \end{pmatrix}, \tag{1.88}$$

$$L_{012345} = L_{0123} L_{45} = \begin{pmatrix} I_4 & 0 \\ 0 & -I_4 \end{pmatrix}. \tag{1.89}$$

The general term of this algebra reads

$$\Psi = \Psi_0 + \Psi_1 + \Psi_2 + \Psi_3 + \Psi_4 + \Psi_5 + \Psi_6 \; ; \quad \Psi_0 = s I_8, s \in \mathbb{R}, \tag{1.90}$$

$$\Psi_1 = \sum_{a=0}^{a=5} N^a L_a, \quad \Psi_2 = \sum_{0 \leqslant a < b \leqslant 5} N^{ab} L_{ab}, \quad \Psi_3 = \sum_{0 \leqslant a < b < c \leqslant 5} N^{abc} L_{abc},$$

$$\Psi_4 = \sum_{0 \leqslant a < b < c < d \leqslant 5} N^{abcd} L_{abcd}, \quad \Psi_5 = \sum_{0 \leqslant a < b < c < d < e \leqslant 5} N^{abcde} L_{abcde}, \tag{1.91}$$

$$\Psi_6 = p L_{012345}, \; p \in \mathbb{R}, \tag{1.92}$$

where N^{ind} are real numbers. We shall need in chapter 6:

$$P^+ = \frac{1}{2}(I_8 + L_{012345}) = \begin{pmatrix} I_4 & 0 \\ 0 & 0 \end{pmatrix},$$

$$P^- = \frac{1}{2}(I_8 - L_{012345}) = \begin{pmatrix} 0 & 0 \\ 0 & I_4 \end{pmatrix}. \tag{1.93}$$

Now we consider six elements of this algebra:

$$\Lambda_a = L_a L_{012345} ; \quad a = 0, 1, 2, 3, 4, 5. \tag{1.94}$$

We then get

$$\Lambda_\mu = \begin{pmatrix} 0 & -\gamma_\mu \\ \gamma_\mu & 0 \end{pmatrix}; \ \Lambda_4 = \begin{pmatrix} 0 & I_4 \\ I_4 & 0 \end{pmatrix}; \ \Lambda_5 = \begin{pmatrix} 0 & -\mathbf{i} \\ \mathbf{i} & 0 \end{pmatrix}, \tag{1.95}$$

$$L_{ab} = -\Lambda_{ab}, \ 0 \leqslant a \leqslant b \leqslant 5, \tag{1.96}$$

$$L_{abcd} = \Lambda_{abcd}, \ 0 \leqslant a < b < c < d \leqslant 5, \tag{1.97}$$

$$L_{012345} = -\Lambda_{012345} = \begin{pmatrix} I_4 & 0 \\ 0 & -I_4 \end{pmatrix}, \tag{1.98}$$

$$L_a = \Lambda_{012345} \Lambda_a. \tag{1.99}$$

These six Λ_a are the generators of the $L_{5,1}$ algebra since (1.96) implies

$$(\Lambda_0)^2 = -1; \ (\Lambda_a)^2 = 1, \ a = 1, 2, 3, 4, 5, \tag{1.100}$$

and the algebra generated by the Λ_a is a sub-algebra of the algebra generated by the L_a. Conversely Eq. (1.99) implies that we could start with the Λ_a and get the L_a from them. This explains how $Cl_{5,1} = Cl_{1,5}$. Vectors in $Cl_{1,5}$ are pseudo-vectors in $Cl_{5,1}$ and vice-versa and n-vectors in $Cl_{1,5}$ are (6-n)-vectors in $Cl_{5,1}$. Then the sums of vectors and pseudo-vectors that we shall need in chapter 7 are independent of the choice of the signature.

Chapter 2

Dirac equation

We present here in two different frames the Dirac wave equation for the electron. It is usually studied with the Dirac matrices. The relativistic invariance nevertheless requires us to use the space algebra, into which we rewrite all the Dirac theory. We get new tensors. We study the link between the invariant form of the Dirac equation and the Lagrangian density. We review the charge conjugation.

2.1 With the Dirac matrices

An important part of the standard model of quantum physics is the Dirac wave equation. This comes from the fact that electrons, neutrinos and quarks, are quantum objects with spin 1/2. The standard model explains the spin 1/2 as a relativistic consequence of the Dirac wave equation. This equation is intensively studied in this chapter and slightly modified in the following chapter. The modified equation shall be generalized in chapter 6 as a wave equation for a pair electron+neutrino. Next this equation shall also be generalized as a wave equation for all particles and antiparticles of the first generation.

The starting point of Dirac's work was the Pauli wave equation for the electron, which used a wave with two complex components mixed by Pauli matrices. The Schrödinger and Pauli wave equations include a first order time derivative, and second order derivatives for the space coordinates. This is inappropriate for a relativistic wave equation. So Dirac sought a wave equation with only first order derivatives, giving at the second order the equation for material waves. This necessitated the use of matrices as Pauli had done. Dirac [34] understood that more components were necessary. His wave equation proves that four components are enough. With the matrices in Eq. (1.75) this equation [1] reads

$$0 = [\gamma^\mu(\partial_\mu + iqA_\mu) + im]\psi \ ; \quad q = \frac{e}{\hbar c} \ ; \quad m = \frac{m_0 c}{\hbar}, \tag{2.1}$$

with the usual convention summing up and down indices. The A_μ are components of the space-time vector which is the exterior electromagnetic potential, e is the charge of the electron and m_0 is its proper mass. Even with a well defined signature for the space-time, the matrices of the theory are not uniquely defined. The choice we made in Eq. (1.75) allows us to use the Weyl spinors ξ and η which play a fundamental role, firstly for the relativistic invariance of the theory and secondly for the Lochak's theory for a magnetic monopole [48] or the electro-weak interactions in chapter 6. With them the wave ψ is the column-matrix:

$$\psi = \begin{pmatrix} \xi \\ \eta \end{pmatrix} \ ; \quad \xi = \begin{pmatrix} \xi_1 \\ \xi_2 \end{pmatrix} \ ; \quad \eta = \begin{pmatrix} \eta_1 \\ \eta_2 \end{pmatrix}. \tag{2.2}$$

The Dirac equation like the Schrödinger equation is a linear wave equation: it contains only partial derivatives and products by matrices, so linear combinations of solutions are also solutions of the wave equation.

The starting point of G. Lochak's monopole theory [48] is the existence in the case of a null proper mass of a double gauge invariance. This is possible because γ_{0123} anti-commutes with each of the four γ^μ matrices. He established that the Dirac equation could be invariant under this double gauge invariance if the mass term was replaced by a non linear mass term. And his gauge may be local if an adequate potential term is added:

$$[\gamma^\mu(\partial_\mu - \frac{ig}{\hbar c}B_\mu\gamma_5) + \frac{1}{2}\frac{m(\rho^2)c}{\hbar}(\Omega_1 - i\Omega_2\gamma_5)]\psi = 0, \tag{2.3}$$

where g is the charge of the monopole and the B_μ are the pseudo-potentials of Cabibbo and Ferrari. We will use later this mass term in a particular case.

---

1. First works about the Dirac equation [30] [34] use an imaginary temporal variable which allows us to use a $++++$ signature for space-time and avoids distinguishing covariant and contravariant indices. This brings also difficulties: the tensor components are either real or pure imaginary. It also hides the fact that matrices of the relativistic theory cannot be all hermitian. The algebra on the complex field generated by the Dirac matrices is the $M_4(\mathbb{C})$ algebra which is 16-dimensional on the complex field and 32-dimensional on the real field. Therefore this algebra cannot be isomorphic to the Clifford algebra of space-time, 16-dimensional on the real field, even if we have used in 1.4.1 a sub-algebra of $M_4(\mathbb{C})$ to represent $Cl_{1,3}$ and to link $Cl_{1,3}$ to Cl_3.

2.1.1 *Relativistic invariance*

With the notations and results of paragraph 1.3.6 the transformation R defined by Eq. (1.42) is a Lorentz dilation. With the N matrix and its reverse $\widetilde{N}$ in Eq. (1.80) we get with the R in Eq. (1.42) and Eq. (1.45), for any M and $\nu = 0, 1, 2, 3$ (a proof is in [13] A.2.2):

$$R^{\nu}_{\mu}\gamma^{\mu} = \widetilde{N}\gamma^{\nu}N. \tag{2.4}$$

We get also

$$\partial'_{\nu} = \frac{\partial}{\partial x'^{\nu}} \ ; \ \ \partial_{\mu} = R^{\nu}_{\mu}\partial'_{\nu} \ ; \ \ A_{\mu} = R^{\nu}_{\mu}A'_{\nu}, \tag{2.5}$$

and so we get:

$$\begin{aligned} 0 &= [\gamma^{\mu}(\partial_{\mu} + iqA_{\mu}) + im]\psi \\ &= [\gamma^{\mu}R^{\nu}_{\mu}(\partial'_{\nu} + iqA'_{\nu}) + im]\psi \\ &= [\widetilde{N}\gamma^{\nu}N(\partial'_{\nu} + iqA'_{\nu}) + im]\psi. \end{aligned}$$

If we restrict M to $SL(2, \mathbb{C})$, we get $M\overline{M} = \det(M) = 1$, so $\overline{M} = M^{-1}$ and $\widetilde{N} = N^{-1}$ which allows us to write

$$[\widetilde{N}\gamma^{\nu}N(\partial'_{\nu} + iqA'_{\nu}) + im]\psi = N^{-1}[\gamma^{\nu}(\partial'_{\nu} + iqA'_{\nu}) + im]N\psi. \tag{2.6}$$

So the Dirac theory supposes:

$$\psi' = N\psi, \tag{2.7}$$

and it gets

$$0 = [\gamma^{\mu}(\partial_{\mu} + iqA_{\mu}) + im]\psi = N^{-1}[\gamma^{\mu}(\partial'_{\mu} + iqA'_{\mu}) + im]\psi'. \tag{2.8}$$

This is why the Dirac equation is said to be form invariant under the Lorentz group. We must remark:

1 – Only transformations of the restricted Lorentz group $\mathcal{L}^{\uparrow}_{+}$ are obtained.

2 – The same γ^{μ} matrices appear in the two systems of coordinates, the x^{μ} system and the x'^{μ} system. Dirac matrices are independent of the used system; they do not depend on the moving observer seeing the wave. This is very important for the extension of the theory to general relativity and it is quite different from Hestenes' study [41] where the γ_{μ} form a variable basis of space-time and change from one observer to another.

3 – ξ and η change differently:

$$\psi' = \begin{pmatrix} \xi' \\ \eta' \end{pmatrix} = \begin{pmatrix} M & 0 \\ 0 & \widehat{M} \end{pmatrix}\begin{pmatrix} \xi \\ \eta \end{pmatrix} \ ; \ \ \xi' = M\xi \ ; \ \ \eta' = \widehat{M}\eta. \tag{2.9}$$

Left and right Weyl spinors are linked to each one of the two non-equivalent representations of $SL(2,\mathbb{C})$.

4 – Only one factor M or $\widehat{M}$ appears in these last relations, whilst two M factors are present in $x' = MxM^\dagger$. In case of a rotation the wave turns therefore only by θ when we rotate by 2θ.

5 – It is somewhat incorrect to say that the Dirac equation is relativistic invariant, while the equation is in fact form invariant under another group, $SL(2,\mathbb{C})$, which is not isomorphic to the Lorentz group.

Nevertheless for any relativistic object with a Dirac wave it is not possible to avoid the $SL(2,\mathbb{C})$ group, therefore it is impossible to avoid the algebra Cl_3 which contains this group. Now we shall explain how we are actually able to write the entire Dirac theory in the Cl_3 frame.

2.2   The wave with the space algebra

Since the Clifford algebra of the physical space is the Pauli algebra, we start again from Eq. (2.1) using the Weyl spinors ξ and η. With:

$$\vec{A} = A^1\sigma_1 + A^2\sigma_2 + A^3\sigma_3 \; ; \quad A = A^0 + \vec{A}, \tag{2.10}$$

and with Eq. (1.75) the Dirac equation reads

$$\begin{pmatrix} 0 & \nabla + iqA \\ \widehat{\nabla} + iq\widehat{A} & 0 \end{pmatrix} \begin{pmatrix} \xi \\ \eta \end{pmatrix} + im \begin{pmatrix} \xi \\ \eta \end{pmatrix} = 0. \tag{2.11}$$

This gives the following system, equivalent to the Dirac equation:

$$(\nabla + iqA)\eta + im\xi = 0, \tag{2.12}$$

$$(\widehat{\nabla} + iq\widehat{A})\xi + im\eta = 0. \tag{2.13}$$

We take the complex conjugate of Eq. (2.13), then we multiply on the left by [2] $-i\sigma_2$:

$$(-i\sigma_2)(\widehat{\nabla}^* - iq\widehat{A}^*)\xi^* - im(-i\sigma_2)\eta^* = 0. \tag{2.14}$$

But we have:

$$(-i\sigma_2)(\widehat{\nabla}^* - iq\widehat{A}^*) = (\nabla - iqA)(-i\sigma_2). \tag{2.15}$$

2. Whichever formalism is used to read the Dirac wave equation we can see that the third direction is privileged, and we shall explain this further. The 12 or 21 planes are also privileged, but indices 1 and 2 play the same role. When a i is added, it is the case with the electric interaction and the electric gauge invariance, then indices 1 and 2 do not play the same role because σ_1 is real while σ_2 is pure imaginary. Therefore the use of σ_2 here is necessary.

So Eq. (2.13) is equivalent to:

$$\nabla(-i\sigma_2\xi^*) + iqA(i\sigma_2\xi^*) + im(i\sigma_2\eta^*) = 0. \tag{2.16}$$

The system composed of Eqs. (2.12) (2.13) is consequently equivalent to one matrix [3] equation:

$$\nabla(\eta \quad -i\sigma_2\xi^*) + iqA(\eta \quad i\sigma_2\xi^*) + im(\xi \quad i\sigma_2\eta^*) = 0. \tag{2.17}$$

Now we let

$$\phi = \sqrt{2}(\xi \quad -i\sigma_2\eta^*) = \sqrt{2}\begin{pmatrix} \xi_1 & -\eta_2^* \\ \xi_2 & \eta_1^* \end{pmatrix}. \tag{2.18}$$

This gives

$$\widehat{\phi} = \sqrt{2}(\eta \quad -i\sigma_2\xi^*) = \sqrt{2}\begin{pmatrix} \eta_1 & -\xi_2^* \\ \eta_2 & \xi_1^* \end{pmatrix}, \tag{2.19}$$

and also

$$\phi\sigma_3 = \sqrt{2}(\xi \quad i\sigma_2\eta^*) \; ; \quad \widehat{\phi}\sigma_3 = \sqrt{2}(\eta \quad i\sigma_2\xi^*). \tag{2.20}$$

So Eq. (2.17) which is equivalent to the Dirac equation Eq. (2.1) reads

$$\nabla\widehat{\phi} + iqA\widehat{\phi}\sigma_3 + im\phi\sigma_3 = 0, \tag{2.21}$$

which we shall write with

$$\sigma_{12} = \sigma_1\sigma_2 = i\sigma_3 \; ; \quad \sigma_{21} = \sigma_2\sigma_1 = -i\sigma_3, \tag{2.22}$$

$$0 = \nabla\widehat{\phi} + qA\widehat{\phi}\sigma_{12} + m\phi\sigma_{12},$$
$$0 = \nabla\widehat{\phi}\sigma_{21} + qA\widehat{\phi} + m\phi. \tag{2.23}$$

Even if this equation seems very different from the well known form Eq. (2.1), it is necessary to insist on the fact that this wave equation [4] is strictly the Dirac equation.

---

3. This is possible because when we compute the product of two matrices we multiply each column of the right matrix by the left matrix. Terms between brackets in Eq. (2.17) are the column-matrices that we got separately in Eq. (2.12) and Eq. (2.16).

4. The indistinct i in quantum theory which is the generator of the gauge invariance is changed here into multiplication on the right by $\sigma_{12} = i_3$. This is interesting because i_3 is not the only element with square -1. In space algebra there are four independent terms with square -1. These terms generate a Lie algebra which is exactly the Lie algebra of the $SU(2) \times U(1)$ Lie group. Hestenes [40] was the first to use this Lie algebra and to compare with the Lie group of electro-weak theory. We shall see in chapter 6 that the gauge group of the electro-weak theory does not have these generators. Then the three dimensions of this $SU(2)$ group shall be available in chapter 5 to explain the existence of the three generations of fundamental fermions.

2.2.1   *Relativistic invariance*

Under a dilation R defined by any M matrix satisfying Eq. (1.42) and Eq. (1.43), we got in Eq. (2.9) $\xi' = M\xi$, $\eta' = \widehat{M}\eta$, and these relations are satisfied not only in the particular case $r = 1$ and $\theta = 0$. Furthermore, we get

$$-i\sigma_2\eta'^* = -i\sigma_2\widehat{M}^*\eta^* = \begin{pmatrix} 0 & -1 \\ 1 & 0 \end{pmatrix}\begin{pmatrix} d & -c \\ -b & a \end{pmatrix}\eta^* = \begin{pmatrix} b & -a \\ d & -c \end{pmatrix}\eta^*$$

$$= \begin{pmatrix} a & b \\ c & d \end{pmatrix}\begin{pmatrix} 0 & -1 \\ 1 & 0 \end{pmatrix}\eta^* = M(-i\sigma_2\eta^*). \tag{2.24}$$

So with

$$\phi' = \sqrt{2}(\xi' \quad -i\sigma_2\eta'^*) = \sqrt{2}\begin{pmatrix} \xi_1' & -\eta_2'^* \\ \xi_2' & \eta_1'^* \end{pmatrix}, \tag{2.25}$$

the formulae in Eq. (2.9) are equivalent to

$$\phi' = M\phi. \tag{2.26}$$

This signifies that the link between the Weyl spinors ξ, η and our ϕ is not only relativistic invariant, it is also invariant under the greater group Cl_3^* of the invertible elements in Cl_3. In addition, with

$$\nabla' = \sigma^\mu \partial_\mu' ; \quad \partial_\mu' = \frac{\partial}{\partial x'^\mu} ; \quad \sigma_0 = \sigma^0 = 1 ; \quad \sigma^j = -\sigma_j, \ j = 1, \ 2, \ 3, \tag{2.27}$$

we get (see [13] A.2.1), for any M:

$$\nabla = \overline{M}\nabla'\widehat{M}, \tag{2.28}$$

and the electric gauge invariance imposes then

$$qA = \overline{M}q'A'\widehat{M}, \tag{2.29}$$

which gives

$$\begin{aligned} 0 &= \nabla\widehat{\phi}\sigma_{21} + qA\widehat{\phi} + m\phi \\ &= \overline{M}\nabla'\widehat{M}\widehat{\phi}\sigma_{21} + q'\overline{M}A'\widehat{M}\widehat{\phi} + m\phi \\ &= \overline{M}(\nabla'\widehat{\phi}'\sigma_{21} + q'A'\widehat{\phi}') + m\phi. \end{aligned} \tag{2.30}$$

Form invariance under Cl_3^* of the Dirac equation signifies that we have

$$0 = \nabla'\widehat{\phi}'\sigma_{21} + q'A'\widehat{\phi}' + m'\phi' ; \quad \nabla'\widehat{\phi}'\sigma_{21} + q'A'\widehat{\phi}' = -m'\phi',$$

$$0 = \overline{M}(-m'\phi') + m\phi = -m'\overline{M}M\phi + m\phi$$

$$= (-m're^{i\theta} + m)\phi. \tag{2.31}$$

We get then the invariance of the wave equation under the Cl_3^* group if and only if

$$m = m're^{i\theta}. \tag{2.32}$$

Evidently in the case where we restrict to $r = 1$ and $\theta = 0$ we get $m' = m$.

2.2.2 *More tensors*

Tensorial densities of the Dirac theory appear very different when the Pauli algebra is used. We shall see this here only for tensors without derivatives. The 16 tensorial densities previously known are always presented as the only possible ones, as a result of the 16 dimensions of the algebra [5] generated by the Dirac matrices. But that is completely wrong! (Detailed calculations are in [15] appendix A). Invariants Ω_1 and Ω_2 satisfy

$$\det(\phi) = \Omega_1 + i\Omega_2 = \rho e^{i\beta}, \tag{2.33}$$

$$\Omega_1 = \overline{\psi}\psi \; ; \quad \overline{\psi} = \psi^\dagger \gamma_0 \; ; \quad \Omega_2 = -i\overline{\psi}\gamma_5\psi \; ; \quad \gamma_5 = -i\gamma_0\gamma_1\gamma_2\gamma_3. \tag{2.34}$$

So ρ is the modulus and the Yvon–Takabayasi angle β is [6] the argument of the determinant of ϕ. And ϕ is invertible if and only if $\rho \neq 0$. The calculation of components, using ξ and η, gives:

$$J = J^\mu \sigma_\mu = \phi\sigma_0\phi^\dagger \; ; \quad J^\mu = \overline{\psi}\gamma^\mu\psi, \tag{2.35}$$

$$K = K^\mu \sigma_\mu = \phi\sigma_3\phi^\dagger \; ; \quad K^\mu = \overline{\psi}\gamma^\mu\gamma_5\psi. \tag{2.36}$$

But now we may see immediately that these two space-time vectors which were known to be orthogonal and have opposite squares are part of a list (D_0, D_1, D_2, D_3) containing four space-time vectors:

$$D_0 = J \; ; \quad D_1 = \phi\sigma_1\phi^\dagger \; ; \quad D_2 = \phi\sigma_2\phi^\dagger \; ; \quad D_3 = K. \tag{2.37}$$

The components of D_1 and D_2 are not combinations of the 16 quantities known by the complex formalism. For a Lorentz dilation R defined by a M matrix, the four D_μ vectors transform in the same way:

$$D'_\mu = \phi'\sigma_\mu\phi'^\dagger = (M\phi)\sigma_\mu(M\phi)^\dagger = M\phi\sigma_\mu\phi^\dagger M^\dagger = MD_\mu M^\dagger. \tag{2.38}$$

The D_μ behave then as the space-time vectors x. We shall say that they are **contravariant**. They are also vectors with the same length. Furthermore, they are orthogonal and form a mobile basis of space-time:

$$2D_\mu \cdot D_\nu = D_\mu \widehat{D}_\nu + D_\nu \widehat{D}_\mu$$
$$= \phi\sigma_\mu\phi^\dagger \widehat{\phi}\widehat{\sigma}_\nu\overline{\phi} + \phi\sigma_\nu\phi^\dagger \widehat{\phi}\widehat{\sigma}_\mu\overline{\phi}$$
$$= \phi\sigma_\mu\rho e^{-i\beta}\widehat{\sigma}_\nu\overline{\phi} + \phi\sigma_\nu\rho e^{-i\beta}\widehat{\sigma}_\mu\overline{\phi}$$
$$= \rho e^{-i\beta}\phi(\sigma_\mu\widehat{\sigma}_\nu + \sigma_\nu\widehat{\sigma}_\mu)\overline{\phi} = \rho e^{-i\beta}\phi 2\delta_{\mu\nu}\overline{\phi}$$
$$= 2\delta_{\mu\nu}\rho e^{-i\beta}\phi\overline{\phi} = 2\delta_{\mu\nu}\rho e^{-i\beta}\rho e^{i\beta},$$

$$D_\mu \cdot D_\nu = \delta_{\mu\nu}\rho^2. \tag{2.39}$$

---

5. This incorrect idea is one of many consequences of the confusion between real algebras and complex algebras. The tensorial densities of the Dirac wave are real quantities, not complex quantities.

6. This explains the $\sqrt{2}$ factor that we put in Eq. (2.18) and Eq. (2.19).

Of course, as we use the space-time of the restricted relativity, with the choice of a $+$ sign for the time, we get

$$\delta_{00} = 1 \ ; \quad \delta_{11} = \delta_{22} = \delta_{33} = -1 \ ; \quad \delta_{\mu\nu} = 0 \ , \ \mu \neq \nu. \tag{2.40}$$

Among the ten relations in Eq. (2.39), only three were known and computed with difficulty by the formalism of Dirac matrices:

$$J^2 = \rho^2 \ ; \quad K^2 = -\rho^2 \ ; \quad J \cdot K = 0 \tag{2.41}$$

For the $S^{\mu\nu}$ tensor, we let:

$$S_3 = S^{23}\sigma_1 + S^{31}\sigma_2 + S^{12}\sigma_3 + S^{10}i\sigma_1 + S^{20}i\sigma_2 + S^{30}i\sigma_3, \tag{2.42}$$
$$S^{\mu\nu} = i\overline{\psi}\gamma^\mu\gamma^\nu\psi. \tag{2.43}$$

And we get:

$$S_3 = \phi\sigma_3\overline{\phi}. \tag{2.44}$$

We see immediately that S_3 is one of four analog terms

$$S_\mu = \phi\sigma_\mu\overline{\phi}. \tag{2.45}$$

We have already encountered S_0, because:

$$S_0 = \phi\sigma_0\overline{\phi} = \phi\overline{\phi} = \rho e^{i\beta} = \det(\phi). \tag{2.46}$$

With the 4 D_μ that each have 4 components, S_0 which has 2 components and the 3 S_j that each have 6 components, we get 36 components of tensors without derivatives, instead of only 16 from [7] the complex formalism.

Under a dilation R defined by a M matrix, the S_μ are transformed into

$$S'_\mu = \phi'\sigma_\mu\overline{\phi}' = M\phi\sigma_\mu\overline{M\phi} = M\phi\sigma_\mu\overline{\phi}\ \overline{M} = MS_\mu\overline{M}. \tag{2.47}$$

We get as a particular case:

$$\rho'e^{i\beta'} = S'_0 = MS_0\overline{M} = M\rho e^{i\beta}\overline{M} = \rho e^{i\beta}M\overline{M} = \rho e^{i\beta}re^{i\theta},$$
$$\rho' = r\rho \ ; \quad \beta' = \beta + \theta. \tag{2.48}$$

The formulae in Eq. (2.47) are completely different from formulae giving the transformation of two-ranked anti-symmetric tensors: $S'^{\rho\sigma} = R^\rho_\mu R^\sigma_\nu S^{\mu\nu}$. As R^ν_μ is quadratic in M and multiplies each space-time length by r, the presence of two R factors signifies a multiplication by r^2 while Eq. (2.47)

---

7. Using the linear space of the linear applications from Cl_3 into Cl_3, which is 64-dimensional, we can establish [11] that the 64 terms of a particular basis can be split into $28 = 8 \times 7/2$ terms forming a basis of the Lie algebra of the O(8) Lie group, and $36 = 9 \times 8/2$ terms which gives the 36 components of tensors without derivatives. The number 36 is not random. The 16 tensorial components previously known are the invariant ones under the electric gauge.

is quadratic in M and multiplies the length only by r. We can consider the two formalisms as equivalent only if we restrict the invariance to the $\mathcal{L}_+^\uparrow$ and $SL(2,\mathbb{C})$ groups, where $r^2 = r = 1$. To consider the invariance under the greater and consequently more restrictive, Cl_3^* group implies abandoning the formalism of Dirac matrices! The Pauli algebra is not only simpler than the algebra of Dirac matrices, it is the only formalism allowing us to formulate the greater group of invariance. Calculations with only Dirac matrices and without Clifford algebra are also as dangerous as calculations of relativistic physics with absolute space and time, without Lorentz transformations.

2.2.3 *Plane waves*

We study the simpler case, where the interaction with exterior fields is negligible. We can then take $A = 0$. The Dirac equation, in the Pauli algebra, is reduced to

$$\nabla\widehat{\phi}\sigma_{21} + m\phi = 0. \tag{2.49}$$

We consider a plane wave with a phasis φ such as:

$$\phi = \phi_0 e^{-\varphi\sigma_{12}} \; ; \quad \varphi = mv_\mu x^\mu. \tag{2.50}$$

We shall use the reduced speed space-time vector:

$$v = \sigma^\mu v_\mu, \tag{2.51}$$

ϕ_0 is a fixed term, which gives

$$\nabla\widehat{\phi}\sigma_{21} = \sigma^\mu\partial_\mu(\widehat{\phi}_0 e^{-\varphi\sigma_{12}})\sigma_{21} = -mv\widehat{\phi}. \tag{2.52}$$

Consequently the wave equation Eq. (2.49) is equivalent to

$$\phi = v\widehat{\phi}. \tag{2.53}$$

Conjugating, this is equivalent to

$$\widehat{\widehat{\phi}} = \widehat{v}\phi. \tag{2.54}$$

Combining now the two preceding equalities, we get

$$\phi = v(\widehat{v}\phi) = (v\widehat{v})\phi = (v \cdot v)\phi. \tag{2.55}$$

So we must have

$$1 = v \cdot v = v_0^2 - \vec{v}^{\,2}, \tag{2.56}$$

$$v_0^2 = 1 + \vec{v}^{\,2} \; ; \quad v_0 = \pm\sqrt{1 + \vec{v}^{\,2}}. \tag{2.57}$$

with *a priori* two possibilities for the sign. The minus sign implies a negative energy for the particle; this was at the beginning a disappointment for Dirac. It is impossible to suppress these troublesome negative energies. For instance they are necessary if we want to write any wave as a sum of plane waves using the Fourier transform. After the discovery of the positron, a particle with the same mass and a charge opposite to the charge of the electron, these plane waves with negative energy were associated to the positron, even though positrons seem to have a proper energy equal to, not opposite to the energy of the electron.

2.3   The Dirac equation in space-time algebra

Equation (2.21) and its conjugation give:
$$\nabla \widehat{\phi} = qA\widehat{\phi}\sigma_{21} + m\phi\sigma_{21} \; ; \;\; \widehat{\nabla}\phi = q\widehat{A}\phi\sigma_{21} + m\widehat{\phi}\sigma_{21}. \qquad (2.58)$$
We let now
$$\Psi = \begin{pmatrix} \phi & 0 \\ 0 & \widehat{\phi} \end{pmatrix} \; ; \;\; \boldsymbol{A} = \begin{pmatrix} 0 & A \\ \widehat{A} & 0 \end{pmatrix}, \qquad (2.59)$$
and we get
$$\boldsymbol{\partial}\Psi = \begin{pmatrix} 0 & \nabla\widehat{\phi} \\ \widehat{\nabla}\phi & 0 \end{pmatrix} = \begin{pmatrix} 0 & qA\widehat{\phi}\sigma_{21} + m\phi\sigma_{21} \\ q\widehat{A}\phi\sigma_{21} + m\widehat{\phi}\sigma_{21} & 0 \end{pmatrix}$$
$$= q\begin{pmatrix} 0 & A \\ \widehat{A} & 0 \end{pmatrix}\begin{pmatrix} \phi & 0 \\ 0 & \widehat{\phi} \end{pmatrix}\begin{pmatrix} \sigma_{21} & 0 \\ 0 & \sigma_{21} \end{pmatrix} + m\begin{pmatrix} 0 & \phi \\ \widehat{\phi} & 0 \end{pmatrix}\begin{pmatrix} \sigma_{21} & 0 \\ 0 & \sigma_{21} \end{pmatrix}. \qquad (2.60)$$
Equation (1.75) gives:
$$\gamma_{12} = \gamma_1\gamma_2 = \begin{pmatrix} 0 & \sigma_1 \\ -\sigma_1 & 0 \end{pmatrix}\begin{pmatrix} 0 & \sigma_2 \\ -\sigma_2 & 0 \end{pmatrix} = \begin{pmatrix} -\sigma_1\sigma_2 & 0 \\ 0 & -\sigma_1\sigma_2 \end{pmatrix} = \begin{pmatrix} \sigma_{21} & 0 \\ 0 & \sigma_{21} \end{pmatrix},$$
$$\Psi\gamma_0 = \begin{pmatrix} \phi & 0 \\ 0 & \widehat{\phi} \end{pmatrix}\begin{pmatrix} 0 & I \\ I & 0 \end{pmatrix} = \begin{pmatrix} 0 & \phi \\ \widehat{\phi} & 0 \end{pmatrix}, \qquad (2.61)$$
then Eq. (2.60) reads
$$\boldsymbol{\partial}\Psi = q\boldsymbol{A}\Psi\gamma_{12} + m\Psi\gamma_{012}, \qquad (2.62)$$
which is the Hestenes's form of the Dirac equation [41]. But the interpretation of Hestenes considers the four γ_μ as a basis of space-time, while the Dirac theory considers them as fixed. Since the relativistic form invariance of the Dirac wave comes from the implicit use of the Cl_3 algebra, we get with Eq. (1.76), Eq. (2.26) and Eq. (2.59)
$$\Psi' = \begin{pmatrix} \phi' & 0 \\ 0 & \widehat{\phi}' \end{pmatrix} = \begin{pmatrix} M\phi & 0 \\ 0 & \widehat{M\phi} \end{pmatrix} = \begin{pmatrix} M & 0 \\ 0 & \widehat{M} \end{pmatrix}\begin{pmatrix} \phi & 0 \\ 0 & \widehat{\phi} \end{pmatrix} = N\Psi. \qquad (2.63)$$

2.4   Invariant Dirac equation

The form invariance of the Dirac theory uses $\nabla = \overline{M}\nabla'\widehat{M}$. Since $\phi' = M\phi$ implies $\overline{\phi}' = \overline{\phi}\ \overline{M}$ the factor $\overline{M}$ on the left side of Eq. (2.30) induces considering a multiplication by $\overline{\phi}$ on the left side of the wave equation. When and where $\rho \neq 0$, ϕ is [8] invertible and if we multiply by $\overline{\phi}$ the Dirac equation is equivalent to

$$\overline{\phi}(\nabla\widehat{\phi})\sigma_{21} + \overline{\phi}qA\widehat{\phi} + m\overline{\phi}\phi = 0. \tag{2.64}$$

Under the Lorentz dilation R defined by an element M in Cl_3 by Eq. (1.42) we get Eq. (2.26), Eq. (2.28) and Eq. (2.29) which imply

$$\overline{\phi}(\nabla\widehat{\phi})\sigma_{21} = \overline{\phi}(\overline{M}\nabla'\widehat{M\phi})\sigma_{21} = \overline{\phi}'(\nabla'\widehat{\phi}')\sigma_{21}, \tag{2.65}$$

$$\overline{\phi}qA\widehat{\phi} = \overline{\phi}\ \overline{M}q'A'\widehat{M\phi} = \overline{\phi}'q'A'\widehat{\phi}'. \tag{2.66}$$

The two first terms of Eq. (2.64) are then form invariant and the mass term is also form invariant if we have

$$m\overline{\phi}\phi = m'\overline{\phi}'\phi' = m'\overline{\phi}\ \overline{M}M\phi = re^{i\theta}m'\overline{\phi}\phi, \tag{2.67}$$

which is equivalent to Eq. (2.32). This mass term reads

$$m\overline{\phi}\phi = m\Omega_1 + im\Omega_2. \tag{2.68}$$

it is then the sum of a scalar and a pseudo-scalar term. The second term of the invariant Dirac equation Eq. (2.64) has another peculiarity: it is a space-time vector because it is self-adjoint:

$$(\overline{\phi}qA\widehat{\phi})^{\dagger} = \widehat{\phi}^{\dagger}qA^{\dagger}\overline{\phi}^{\dagger} = \overline{\phi}qA\widehat{\phi}. \tag{2.69}$$

We can then let

$$\overline{\phi}qA\widehat{\phi} = V^0 + \vec{V} = V^{\mu}\sigma_{\mu}, \tag{2.70}$$

where V is a space-time vector. Only the first term of Eq. (2.64) is general, but we can also find its peculiarities with

$$\overline{\phi}(\nabla\widehat{\phi}) = \frac{1}{2}[\overline{\phi}(\nabla\widehat{\phi}) + (\overline{\phi}\nabla)\widehat{\phi}] + \frac{1}{2}[\overline{\phi}(\nabla\widehat{\phi}) - (\overline{\phi}\nabla)\widehat{\phi}], \tag{2.71}$$

$$\frac{1}{2}[\overline{\phi}(\nabla\widehat{\phi}) + (\overline{\phi}\nabla)\widehat{\phi}] = \frac{1}{2}\partial_{\mu}(\overline{\phi}\sigma^{\mu}\widehat{\phi}) = v = v^{\mu}\sigma_{\mu}, \tag{2.72}$$

$$\frac{1}{2}[\overline{\phi}(\nabla\widehat{\phi}) - (\overline{\phi}\nabla)\widehat{\phi}] = iw = iw^{\mu}\sigma_{\mu}, \tag{2.73}$$

---

8. This condition is not severe: we shall see in Appendix C that the inverse exists everywhere for each solution of the Dirac equation for the H atom; that is the most complicated calculation and the best success of the Dirac equation.

where v and w are two space-time vectors, because $v^\dagger = v$ and $(iw)^\dagger = -iw$. This gives

$$\overline{\phi}(\nabla\widehat{\phi})\sigma_{21} = (v + iw)\sigma_{21}$$
$$= (v^0 + v^1\sigma_1 + v^2\sigma_2 + v^3\sigma_3 + iw^0 + w^1 i\sigma_1 + w^2 i\sigma_2 + w^3 i\sigma_3)(-i\sigma_3)$$
$$= w^3 + v^2\sigma_1 - v^1\sigma_2 + w^0\sigma_3 + i(-v^3 + w^2\sigma_1 - w^1\sigma_2 - v^0\sigma_3).$$
$$(2.74)$$

Therefore the Dirac equation is equivalent to the system

$$0 = w^3 + V^0 + m\Omega_1, \qquad (2.75)$$
$$0 = v^2 + V^1, \qquad (2.76)$$
$$0 = -v^1 + V^2, \qquad (2.77)$$
$$0 = w^0 + V^3, \qquad (2.78)$$
$$0 = -v^3 + m\Omega_2, \qquad (2.79)$$
$$0 = w^2, \qquad (2.80)$$
$$0 = -w^1, \qquad (2.81)$$
$$0 = -v^0. \qquad (2.82)$$

First the gauge invariance concerns only four of the eight equations, those containing the V^μ. This is a consequence of the fact that classical electromagnetism is based on the absence of magnetic monopoles, as we will explain further. Less evident and of great importance in the Dirac theory, the first equation is exactly $\mathcal{L} = 0$ because (a detailed calculation is in appendix A.1):

$$\mathcal{L} = \frac{1}{2}[(\overline{\psi}\gamma^\mu(-i\partial_\mu + qA_\mu)\psi) + (\overline{\psi}\gamma^\mu(-i\partial_\mu + qA_\mu)\psi)^\dagger] + m\overline{\psi}\psi$$
$$= w^3 + V^0 + m\Omega_1. \qquad (2.83)$$

It is well known that by varying the $\mathcal{L}$ Lagrangian density we get the Dirac wave equation. Moreover the fact that the Dirac equation is homogeneous implies that $\mathcal{L} = 0$ when the Dirac equation is satisfied. Here we get the reciprocal situation, the equation $\mathcal{L} = 0$ is one of the wave equations and the Lagrangian formalism is a consequence [9] of the wave equation. The

---

9. Each law of movement, in classical mechanics and in electromagnetism, may be obtained from a Lagrangian mechanism. We know nowadays that this comes from the Lagrangian form and from the universality of quantum mechanics. But where does the Lagrangian form of quantum mechanics come from? Here we see this as totally determined since the Lagrangian density is the scalar part of the wave equation and since the Lagrangian formalism implies the wave equation.

four equations containing the symmetric part v of $\bar{\phi}(\widehat{\nabla\phi})$ are respectively, for indices 0, 3, 2, 1 and the D_μ of Eq. (2.37) (see A.2.6):

$$0 = \nabla \cdot D_0, \tag{2.84}$$

$$0 = \nabla \cdot D_3 + 2m\Omega_2, \tag{2.85}$$

$$0 = \nabla \cdot D_1 - 2qA \cdot D_2, \tag{2.86}$$

$$0 = \nabla \cdot D_2 + 2qA \cdot D_1. \tag{2.87}$$

Equation (2.84) is the equation of conservation of probability. It is now exactly one of the eight equations equivalent to the Dirac wave equation. Equation (2.85) is known as the relation of Uhlenbeck and Laporte. Equation (2.86) and Eq. (2.87) indicate that the D_1 and D_2 space-time vectors are not gauge invariant; the electric gauge transformation induces a rotation in the $D_1 - D_2$ plane. In spite of its peculiar aspect the invariant equation appears as the true wave equation since it is form invariant and since it has so many interesting aspects.

2.5 Charge conjugation

Many years after the discovery of electrons the positrons were also discovered. The only difference between these particles is the sign of the charge, negative for the electron, positive for the positron. From the Dirac equation of a particle Eq. (2.1), quantum theory gets the wave equation of the antiparticle as follows. The wave of the electron is denoted as ψ_e and the wave of the positron is denoted as ψ_p. We take the complex conjugate of Eq. (2.1):

$$0 = [\gamma^{\mu*}(\partial_\mu - iqA_\mu) - im]\psi_e^*. \tag{2.88}$$

Since Eq. (1.75) gives $\gamma_2\gamma^{\mu*} = -\gamma^\mu\gamma_2$, $\mu = 0, 1, 2, 3$, multiplying Eq. (2.88) by $i\gamma_2$ on the left we get

$$0 = -[\gamma^\mu(\partial_\mu - iqA_\mu) + im]i\gamma_2\psi_e^*. \tag{2.89}$$

Therefore, up to an arbitrary phase, quantum theory supposes

$$\psi_p = i\gamma_2\psi_e^*, \tag{2.90}$$

$$0 = [\gamma^\mu(\partial_\mu - iqA_\mu) + im]\psi_p. \tag{2.91}$$

Using Eq. (2.2) and indices e for the electron and p for the positron Eq. (2.90) reads

$$\begin{pmatrix} \xi_{1p} \\ \xi_{2p} \\ \eta_{1p} \\ \eta_{2p} \end{pmatrix} = \begin{pmatrix} 0 & 0 & 0 & 1 \\ 0 & 0 & -1 & 0 \\ 0 & -1 & 0 & 0 \\ 1 & 0 & 0 & 0 \end{pmatrix} \begin{pmatrix} \xi_{1e}^* \\ \xi_{2e}^* \\ \eta_{1e}^* \\ \eta_{2e}^* \end{pmatrix}, \tag{2.92}$$

$$\xi_{1p} = \eta_{2e}^*, \quad \xi_{2p} = -\eta_{1e}^*; \quad \eta_{1p} = -\xi_{2e}^*; \quad \eta_{2p} = \xi_{1e}^*. \tag{2.93}$$

Now with Eq. (2.38) and indices e for the electron and p for the positron we get

$$\widehat{\phi}_e = \sqrt{2} \begin{pmatrix} \eta_{1e} & -\xi_{2e}^* \\ \eta_{2e} & \xi_{1e}^* \end{pmatrix}; \quad \widehat{\phi}_p = \sqrt{2} \begin{pmatrix} \eta_{1p} & -\xi_{2p}^* \\ \eta_{2p} & \xi_{1p}^* \end{pmatrix}. \tag{2.94}$$

Then Eq. (2.90), which is equivalent to Eq. (2.93), is also equivalent to

$$\widehat{\phi}_p = \widehat{\phi}_e \sigma_1, \tag{2.95}$$

$$\phi_p = -\phi_e \sigma_1. \tag{2.96}$$

Charge conjugation is then involutive.

Chapter 3

The homogeneous nonlinear wave equation

We discuss the question of negative energies. We prove that with our wave equation all usual plane waves have a positive energy. We study the relativistic invariance, which introduces a greater invariance group, and a second space-time manifold. We discuss the charge conjugation. We explain how this nonlinear wave equation gets the quantification and the true results in the case of the H atom.

When Dirac was deriving his wave equation he was hoping to get a wave equation without the negative energies which came from the relativistic Klein–Gordon equation. But his wave equation also had solutions with negative energies since two signs are possible in Eq. (2.57). When positrons were discovered six years later, solutions with negative energy were associated to positrons and considered as a splendid success of the new wave equation. But the creation of an electron-positron pair necessitates an amount of energy: $+ 2 \times 511$ keV. The link between the positron and negative energy implies an interpretation in another theoretical frame and is now understood only with the second quantification and a complicated reasoning.

But it is possible to solve the problem of the non-physical negative energy with a simple modification to the Dirac equation: We replace the $\overline{\phi}\phi$ term in the invariant Dirac equation Eq. (2.64) by the modulus ρ of this term:

$$\overline{\phi}(\nabla\widehat{\phi})\sigma_{21} + \overline{\phi}qA\widehat{\phi} + m\rho = 0. \qquad (3.1)$$

Multiplying by the left by $\overline{\phi}^{-1}$ we get with $\rho = e^{-i\beta}\overline{\phi}\phi$ the equivalent equation

$$\nabla\widehat{\phi}\sigma_{21} + qA\widehat{\phi} + me^{-i\beta}\phi = 0. \qquad (3.2)$$

Equation (3.1) and Eq. (3.2) are the two main forms of the wave equation that we study in this chapter. We firstly obtained this wave equation from the wave equation for a magnetic monopole of G. Lochak Eq. (2.3), suppressing the potential term:

$$[\gamma^\mu \partial_\mu + \frac{1}{2}\frac{m(\rho^2)c}{\hbar}(\Omega_1 - i\Omega_2\gamma_5)]\psi = 0. \tag{3.3}$$

When we choose

$$\frac{1}{2}\frac{m(\rho^2)c}{\hbar} = \frac{im}{\rho},$$

this equation becomes:

$$[\gamma^\mu \partial_\mu + ime^{-i\beta\gamma_5}]\psi = 0. \tag{3.4}$$

Since the Yvon–Takabayasi β angle is electric gauge invariant, it is perfectly possible to add an electric potential term this gives [7]:

$$[\gamma^\mu(\partial_\mu + iqA_\mu) + ime^{-i\beta\gamma_5}]\psi = 0. \tag{3.5}$$

This wave equation is nonlinear, because β depends on the value of ψ. It is homogeneous, because if we multiply a solution ψ by a fixed real number k, β does not change, so $k\psi$ is also a solution of the equation. Our equation has many common properties with the Dirac equation. We must immediately say that, if β is null or negligible $ime^{-i\beta\gamma_5} \approx im$ then Eq. (3.5) has the Dirac equation as a linear approximation.

To write this equation in the Pauli algebra, we proceed as with the Dirac equation

$$\begin{pmatrix} 0 & \nabla + iqA \\ \widehat{\nabla} + iq\widehat{A} & 0 \end{pmatrix}\begin{pmatrix} \xi \\ \eta \end{pmatrix} + im\begin{pmatrix} e^{-i\beta}I & 0 \\ 0 & e^{i\beta}I \end{pmatrix}\begin{pmatrix} \xi \\ \eta \end{pmatrix} = 0. \tag{3.6}$$

This gives the following system, equivalent to Eq. (3.5):

$$(\nabla + iqA)\eta + ime^{-i\beta}\xi = 0, \tag{3.7}$$

$$(\widehat{\nabla} + iq\widehat{A})\xi + ime^{i\beta}\eta = 0. \tag{3.8}$$

Using the process explained in Sec. 2.2 the homogeneous nonlinear equation becomes [9]

$$\nabla\widehat{\phi} + qA\widehat{\phi}\sigma_{12} + me^{-i\beta}\phi\sigma_{12} = 0, \tag{3.9}$$

which is equivalent to Eq. (3.2) or to the invariant equation Eq. (3.1). The differential term $\overline{\phi}(\nabla\widehat{\phi})\sigma_{21}$ and the gauge term $\overline{\phi}qA\widehat{\phi}$ are those of the linear wave equation and the only change is in the mass term where $\overline{\phi}\phi = \Omega_1 + i\Omega_2$

is replaced by $\rho = \sqrt{\Omega_1^2 + \Omega_2^2}$. We therefore get instead of Eq. (2.75) to Eq. (2.82) and with notations of chapter 2 the system:

$$0 = w^3 + V^0 + m\rho, \tag{3.10}$$

$$0 = v^2 + V^1, \tag{3.11}$$

$$0 = -v^1 + V^2, \tag{3.12}$$

$$0 = w^0 + V^3, \tag{3.13}$$

$$0 = -v^3, \tag{3.14}$$

$$0 = w^2, \tag{3.15}$$

$$0 = -w^1, \tag{3.16}$$

$$0 = -v^0. \tag{3.17}$$

As with the Dirac equation, the scalar equation Eq. (3.10) gives the Lagrangian density:

$$0 = \mathcal{L} = \frac{1}{2}[(\overline{\psi}\gamma^\mu(-i\partial_\mu + qA_\mu)\psi) + (\overline{\psi}\gamma^\mu(-i\partial_\mu + qA_\mu)\psi)^\dagger] + m\rho$$

$$= w^3 + V^0 + m\rho. \tag{3.18}$$

Therefore the double link between wave equation and Lagrangian density is the same as with the linear wave equation. Similarly Eq. (3.17) is still the law of conservation of the probability density.

3.1 Gauge invariances

Since the differential term and the gauge term are the same and since the mass term is gauge invariant, the homogeneous nonlinear wave equation is also invariant under the electric gauge which reads in the Pauli algebra

$$\phi \mapsto \phi' = \phi e^{ia\sigma_3} \; ; \; A_\mu \mapsto A_\mu' = A_\mu - \frac{1}{q}\partial_\mu a. \tag{3.19}$$

The conservative current linked to the electric gauge invariance Eq. (3.19) by Noether's theorem is here also the probability current $J = D_0$, and Eq. (3.17) is exactly the conservation law Eq. (2.84).

But the homogeneous nonlinear equation allows a second, global, gauge invariance:

$$\phi \mapsto \phi' = e^{ia}\phi \; ; \; \overline{\phi} \mapsto \overline{\phi}' = e^{ia}\overline{\phi} \; ; \; \partial_\mu a = 0, \tag{3.20}$$

which gives

$$\rho e^{i\beta} = \phi\overline{\phi} \mapsto \rho' e^{i\beta'} = \phi'\overline{\phi}' = e^{2ia}\phi\overline{\phi} = \rho e^{i(\beta + 2a)},$$

$$\rho \mapsto \rho' = \rho \; ; \; \beta \mapsto \beta' = \beta + 2a. \tag{3.21}$$

We shall see in chapter 6 that this second gauge is part of the electro-weak gauge group. Since Eq. (3.20) is also the chiral [1] gauge coming from the magnetic monopole of G. Lochak [46] we get his result on the associated current: Noether's theorem implies the existence of another conservative current, $K = D_3$ (see [15] B.1.3), and this replaces the Uhlenbeck and Laporte relation Eq. (2.85) by the conservative law:

$$0 = \nabla \cdot D_3. \tag{3.22}$$

This is, along with the change in the Lagrangian density and the scalar equation, the only change: the 6 other equations are unchanged. Since the chiral gauge multiplies ϕ by e^{ia}, $\widehat{\phi}$ is multiplied by e^{-ia}, the Weyl spinor ξ is multiplied by e^{ia} and η is multiplied by e^{-ia} [48]. The generator i of the chiral gauge is exactly the i of the Pauli algebra which rules the orientation of the physical space (see Sec. 1.3.1). Since we have lost linearity the sum $\phi_1 + \phi_2$ of two solutions of Eq. (3.1) is not necessarily a solution of Eq. (3.1). But since the equation is homogeneous and invariant under the chiral gauge, if ϕ is a solution and z is any complex number then $z\phi$ is also a solution of Eq. (3.1). This property, true for the Schrödinger equation and the i of the electric gauge, is not true for the Dirac equation in Cl_3.

3.2 Plane waves

We repeat what has been done in Sec. 2.2.3 for the linear equation. Our equation is now reduced, for $A = 0$, to:

$$\nabla\widehat{\phi} + me^{-i\beta}\phi\sigma_{12} = 0. \tag{3.23}$$

If we consider a plane wave with a phase φ satisfying

$$\phi = \phi_0 e^{-\varphi\sigma_{12}} \ ; \ \ \varphi = mv_\mu x^\mu \ ; \ \ v = \sigma^\mu v_\mu, \tag{3.24}$$

where v is a fixed reduced speed and ϕ_0 is also a fixed term, we get:

$$\nabla\widehat{\phi} = \sigma^\mu \partial_\mu(\widehat{\phi_0} e^{-\varphi\sigma_{12}}) = -mv\widehat{\phi}\sigma_{12}. \tag{3.25}$$

Equation (3.23) is then equivalent to

$$\phi = e^{i\beta}v\widehat{\phi}, \tag{3.26}$$

or to

$$\widehat{\phi} = e^{-i\beta}\widehat{v}\phi, \tag{3.27}$$

---

1. The electric gauge multiplies ξ and η by the same factor e^{ia} while the chiral gauge multiplies ξ by e^{ia} and η by e^{-ia}. This gauge is a local one in Lochak's theory as well as in the electro-weak theory.

which implies

$$\phi = e^{i\beta} v(e^{-i\beta} \widehat{v}\phi) = v\widehat{v}\phi = (v \cdot v)\phi. \tag{3.28}$$

So, if ϕ_0 is invertible, we must take

$$1 = v \cdot v = v_0^2 - \vec{v}^2, \tag{3.29}$$

$$v_0^2 = 1 + \vec{v}^2 \; ; \quad v_0 = \pm\sqrt{1 + \vec{v}^2}, \tag{3.30}$$

which is the expected relation for the reduced speed of the particle. Furthermore, with the nonlinear equation, we have:

$$D_0 = \phi\phi^\dagger = (e^{i\beta} v\widehat{\phi})\phi^\dagger = e^{i\beta} v(\widehat{\phi}\phi^\dagger) = e^{i\beta} v\rho e^{-i\beta} = v\rho. \tag{3.31}$$

So we get

$$D_0^0 = \rho v^0, \tag{3.32}$$

and since D_0^0 and ρ are always positive, Eq. (3.30) is obtained only if

$$v_0 = \sqrt{1 + \vec{v}^2}. \tag{3.33}$$

This proves that the replacement of $\overline{\phi}\phi$ by ρ in the mass term of the invariant equation is enough to rid the Dirac theory of unphysical negative energies in the electron case.

3.3 Relativistic invariance

With a Lorentz dilation R with ratio $r = |\det(M)|$ satisfying

$$x' = R(x) = MxM^\dagger \; , \quad \det(M) = re^{i\theta} \; , \quad \phi' = M\phi$$
$$\nabla = \overline{M}\nabla'\widehat{M} \; ; \quad qA = \overline{M}q'A'\widehat{M}, \tag{3.34}$$

we have also

$$\rho' e^{i\beta'} = \det(\phi') = \phi'\overline{\phi'} = M\phi\overline{\phi}\;\overline{M} = M\rho e^{i\beta}\overline{M}$$
$$= M\overline{M}\rho e^{i\beta} = re^{i\theta}\rho e^{i\beta} = r\rho e^{i(\beta+\theta)}, \tag{3.35}$$

$$\rho' = r\rho, \tag{3.36}$$

$$\beta' = \beta + \theta. \tag{3.37}$$

And so we get:

$$0 = \overline{\phi}(\nabla\widehat{\phi})\sigma_{21} + \overline{\phi}qA\widehat{\phi} + m\rho$$
$$= \overline{\phi}\;\overline{M}\nabla'\widehat{M}\widehat{\phi}\sigma_{21} + \overline{\phi}\;\overline{M}q'A'\widehat{M}\widehat{\phi} + m\rho$$
$$= \overline{\phi'}(\nabla'\widehat{\phi'})\sigma_{21} + \overline{\phi'}q'A'\widehat{\phi'} + m\rho. \tag{3.38}$$

The homogeneous nonlinear equation is form invariant under Cl_3^*, the group of invertible elements in Cl_3, if and only if

$$m\rho = m'\rho',\tag{3.39}$$

$$m\rho = m'r\rho.\tag{3.40}$$

We get then the form invariance of the wave equation under $Cl_3^* = GL(2,\mathbb{C})$ if[2] and only if

$$m = m'r.\tag{3.41}$$

What is the significance of this equality for physics? If the true invariance group for the electromagnetism is not only the Lorentz group, and not even its covering group, but the greater group Cl_3^*, similar things to what happens when we go from Galilean physics to relativistic physics must also occur: there are fewer invariant quantities. The proper mass m_0 and ρ are both invariant under Lorentz rotations. Under Lorentz dilations induced by all M matrices, m and ρ are no longer separately invariant; it is the product $m\rho$ alone which is invariant

$$m\rho = m'r\rho = m'\rho'.\tag{3.42}$$

What does the invariance of $m\rho$ mean? It is the product of a reduced mass and a dilation ratio which is invariant. A reduced mass $m = m_0 c/\hbar$ is proportional to the inverse of a space-time length, which is a frequency. This is exactly what $E = h\nu$ says. Otherwise, the existence of Planck's constant is linked to the fact that m and ρ are not separately invariant, but only their product is. Or again, the existence of Planck's constant is linked to the invariance under the Cl_3^* group, greater than the invariance group of the restricted relativity. Somewhere we can say: the existence of Planck's constant was not fully understood from the physical point of view. To consider this greater invariance group will enable us to see things otherwise and to understand **why there is a Planck constant**.

The invariance of the $m\rho$ product has also another consequence. If we restrict the invariance to the subgroup of Lorentz rotations, m is invariant. Since the product $m\rho$ is a constant, this implies that ρ has a physically

---

2. The simplification that we see here, from Eq. (2.32), is a powerful argument for the homogeneous nonlinear equation. A factor $e^{i\theta}$ in the mass term is not annoying because $m\widehat{m} = |m|^2$. But it indicates a lack of symmetry, and it explains by itself why the greater group of invariance Cl_3^* was not previously seen. The form invariance of the electromagnetism, which we shall study in the next chapter, and the form invariance of the Dirac theory, are fully compatible only with the homogeneous nonlinear equation and this transformation of masses.

determined value. But if we multiply ψ or ϕ by a real constant k, ρ is multiplied by k^2. To say that ρ has a physically determined value is equivalent to saying that the wave is normalized, or that, in a way or another, there is a physical condition which fixes the amplitude of the wave. We shall use this in chapter 9.

With the wave equation Eq. (3.2) we get the invariance of the wave equation under Cl_3^* with the condition

$$m = m'r. \tag{3.43}$$

We had implicitly considered previously r and ρ on the same footing, this is natural since $\rho' = r\rho$. More generally: **There is no difference of structure between the M defining the dilation R and the ϕ wave, which are both complex 2×2 matrices, elements of the space algebra Cl_3. More precisely ϕ is a function from space-time with value into Cl_3. Consequently ϕ, as M, can define a Lorentz dilation D, with ratio ρ, by:**

$$D : y \mapsto x = \phi y \phi^\dagger. \tag{3.44}$$

And the components D_μ^ν of the four D_μ are just the 16 terms of the matrix of the dilation because

$$x = x^\mu \sigma_\mu = \phi y^\nu \sigma_\nu \phi^\dagger = y^\nu \phi \sigma_\nu \phi^\dagger = y^\nu D_\nu = y^\nu D_\nu^\mu \sigma_\mu; \ x^\mu = D_\nu^\mu y^\nu. \tag{3.45}$$

There is no difference between the product $M'M$ which gives the composition of dilations $R' \circ R$ and the product $M\phi$ which gives the transformation of the wave under a dilation, and this induces then a composition of dilations $R \circ D$:

$$x' = MxM^\dagger = M\phi y\phi^\dagger M^\dagger = (M\phi)y(M\phi)^\dagger = \phi' y\phi'^\dagger. \tag{3.46}$$

This signifies that the y introduced into Eq. (3.44) does not change, either seen by the observer of x or by the observer of x'. It is independent of the dilation defined by any M.

And since ϕ is function of x, the D dilation is also a function of x, and varies from point to point in space-time: y is not an element of the global space-time, only of the local space-time. So we must see y as the general element of the tangent space-time, at x, to a space-time manifold which depends only on the wave, not on the observer, which we will name **intrinsic manifold**. On the contrary the dilation depends on the observer: the observer of x sees D, the one of x' sees $D' = R \circ D$.

At each point of the space-time we have then not one but **two space-time manifolds**, and **two different affine connections**: the manifold of the x and x', for which each relativistic observer is associated to a Lorentzian tangent space-time, and another manifold, this of the y on which we will study a few properties. We shall generalize this result in chapter 6. The wave equations that we shall get for the electron+neutrino pair and for electron+neutrino+quarks are not a generalization of the linear wave equation but a generalization of the homogeneous nonlinear equation.

3.4   Charge conjugation

We start again from the usual link between the wave of the particle and the wave of the antiparticle, Eq. (2.90) in the frame of Dirac matrices and Eq. (2.95) in the space algebra. The homogeneous nonlinear equation Eq. (3.2) is

$$\nabla \widehat{\phi}_e \sigma_{21} + qA\widehat{\phi}_e + me^{-i\beta_e}\phi_e = 0. \tag{3.47}$$

We also have

$$\rho_e e^{i\beta_e} = \phi_e \overline{\phi}_e. \tag{3.48}$$

This gives

$$\rho_e e^{i\beta_e} = \phi_e \overline{\phi}_e = \phi_p(-\sigma_1)(\widehat{\phi}_p\sigma_1)^\dagger = -\phi_p\overline{\phi}_p = -\rho_p e^{i\beta_p}. \tag{3.49}$$

Therefore Eq. (3.47) reads

$$\nabla \widehat{\phi}_p \sigma_1 \sigma_{21} + qA\widehat{\phi}_p\sigma_1 + m(-e^{-i\beta_p})(-\phi_p\sigma_1) = 0. \tag{3.50}$$

Multiplying by σ_1 on the right this is equivalent to

$$-\nabla \widehat{\phi}_p \sigma_{21} + qA\widehat{\phi}_p + me^{-i\beta_p}\phi_p = 0. \tag{3.51}$$

Now multiplying by $\overline{\phi}_p$ on the left we get the invariant form of the wave equation for the positron:

$$-\overline{\phi}_p\nabla \widehat{\phi}_p\sigma_{21} + q\overline{\phi}_p A\widehat{\phi}_p + m\rho_p = 0. \tag{3.52}$$

This means that only the differential part of the wave equation is changed. Instead of the system Eq. (3.10) to Eq. (3.17) we get now

$$0 = -w^3 + V^0 + m\rho, \tag{3.53}$$

$$0 = -v^2 + V^1, \tag{3.54}$$

$$0 = v^1 + V^2, \tag{3.55}$$

$$0 = -w^0 + V^3, \tag{3.56}$$

$$0 = v^3, \tag{3.57}$$

$$0 = -w^2, \tag{3.58}$$

$$0 = w^1, \tag{3.59}$$

$$0 = v^0. \tag{3.60}$$

So the charge conjugation does not really change the sign of the charge nor the sign of the mass, only the sign [3] of the differential term of the wave equation. Only v^μ and w^μ change sign. Therefore the Lagrangian density which is the scalar part of the wave equation becomes

$$\mathcal{L} = -\frac{1}{2}[(\overline{\psi}\gamma^\mu(-i\partial_\mu - qA_\mu)\psi) + (\overline{\psi}\gamma^\mu(-i\partial_\mu - qA_\mu)\psi)^\dagger] + m\rho$$
$$= -w^3 + V^0 + m\rho. \tag{3.61}$$

The positive mass-energy of the positron is exactly opposite to the negative energy-coefficient of the stationary [4] state. The homogeneous nonlinear wave equation solves then the puzzle of the sign of the energy in a much more understandable way than second quantization: we have the negative coefficient E necessary to obtain the Fourier transformation, but the true density of energy is T_0^0 which remains positive.

Since Eq. (3.52) may be formally gotten from Eq. (3.1) by changing x^μ into $-x^\mu$ which is the PT transformation, the CPT theorem of quantum field theory is trivial.

Since the Dirac equation is the linear approximation of the homogeneous nonlinear wave equation, we get the Dirac equation of the positron from the nonlinear equation by changing the mass term. But we must account for the fact that $\beta_p \approx \pi$ and $\rho_p \approx -\Omega_{1p}$. The linear approximation of the homogeneous nonlinear wave equation is then:

$$0 = -\overline{\phi}_p \nabla\widehat{\phi}_p \sigma_{21} + q\overline{\phi}_p A\widehat{\phi}_p + m\rho = -\overline{\phi}_p \nabla\widehat{\phi}_p \sigma_{21} + q\overline{\phi}_p A\widehat{\phi}_p - m\Omega_{1p}, \tag{3.62}$$

$$0 = -\nabla\widehat{\phi}_p \sigma_{21} + qA\widehat{\phi}_p - m\phi_p. \tag{3.63}$$

We have, for the sign of E and T_0^0, the same results as for the nonlinear equation: E is negative but T_0^0 is positive. There is no longer a problem with the negative energy. Consequently we do not need the second quantization to solve the problem.

---

3. The electric gauge invariance is gotten, in place of Eq. (3.19), as $\phi_p \mapsto \phi_p' = \phi_p e^{ia\sigma_3}$ and $A_\mu \mapsto A_\mu' = A_\mu - \frac{1}{q}(-\partial_\mu a) = A_\mu - \frac{1}{-q}\partial_\mu a$. Therefore the positron appears as having a charge opposite to the charge of the electron. In fact it is not q but $\partial_\mu a$ which changes sign.

4. The study of plane waves in the case of the positron gives in place of Eq. (3.26): $\phi_p = -e^{i\beta_p}v\widehat{\phi}_p$, $\widehat{\phi}_p = -e^{-i\beta_p}\widehat{v}\widehat{\phi}_p$, $\phi_p = v\widehat{v}\phi_p$. We now get $D_0 = -v\rho_p$ and then $D_0^0 = -\rho_p v^0$. Therefore we get $v^0 = -\sqrt{1+\widehat{v}^2}$: v^0 and E are then negative.

3.5   The Hydrogen atom

Quantum mechanics got quantized energy levels by solving the Schrödinger equation in the case of the hydrogen atom, an electron "turning" around a proton. The quantification was a brilliant result, but the other results were not so good. The energy levels were not accurate, and the number of states for a principal quantum number **n** was n^2 when $2n^2$ states were awaited.

We put the detailed calculation in Appendix C, it is very beautiful but also very difficult. We resume here conclusions. Our study of the solutions for the homogeneous nonlinear equation proves that a family of solutions may exist, labeled by the same quantum numbers appearing in the linear Dirac theory, and that these solutions of the nonlinear equation are very close to the solutions of the linear equation because the Yvon–Takabayasi angle is everywhere defined and small. We recall here that the existence of the Yvon–Takabayasi angle is equivalent to the existence of the inverse. For these solutions the inverse exists everywhere.

Chapter 4

Invariance of electromagnetic laws

The group of Lorentz dilations induced by the Cl_3^* group is also the invariance group of electromagnetic laws. This is established for the electromagnetism of Maxwell–de Broglie, with photons, and for the electromagnetism with magnetic monopoles.

The laws of Maxwell's electromagnetism in the void are not invariant under the invariance group of mechanics. Putting at the center of his thought the invariance of the speed of light, Einstein replaced, for all physics, the invariance group of mechanics by a greater group, containing translations and rotations, but also the Lorentz transformations including space and time. When an invariance group is replaced by another, greater group, there are fewer invariants, for instance the mass and the impulse are no longer invariant; only the proper mass remains invariant. And there is a grouping of quantities, for instance the electric field and the magnetic field become parts of the same object, the electromagnetic tensor field. Energy and impulse become parts of the same impulse–energy vector.

The existence of particles with spin $1/2$ shows us that the group of Lorentz transformations is still too small, and we must use another greater group, $SL(2, \mathbb{C})$, itself a subgroup of the group $GL(2, \mathbb{C}) = Cl_3^*$ made of the invertible elements of the space algebra. Since Cl_3 is naturally linked to the geometry of the physical space and $SL(2, \mathbb{C})$ is not, this implies that Cl_3^* is the true invariance group, not only of the Dirac equation, but also of all physical laws and this is what we will look at now for electromagnetism.

4.1 Maxwell–de Broglie electromagnetism

Louis de Broglie worked out a quantum theory of light [31] [32] where the wave of the photon is built by fusion of two Dirac spinors. The electric field $\vec{E}$, the magnetic field $\vec{H}$, the electric potential V, and the potential vector $\vec{A}$ follow Maxwell's laws in the vacuum, supplemented by mass terms:

$$-\frac{1}{c}\frac{\partial \vec{H}}{\partial t} = \text{rot}(\vec{E}) \; ; \quad \text{div}(\vec{H}) = 0 \; ; \quad \vec{H} = \text{rot}(\vec{A}) \; ; \quad \frac{1}{c}\frac{\partial V}{\partial t} + \text{div}(\vec{A}) = 0$$

$$\frac{1}{c}\frac{\partial \vec{E}}{\partial t} = \text{rot}(\vec{H}) + k_0^2 \vec{A} \; ; \quad \text{div}(\vec{E}) = -k_0^2 V \; ; \quad \vec{E} = -\frac{1}{c}\frac{\partial \vec{A}}{\partial t} - \text{grad}(V)$$

$$(4.1)$$

The $k_0 = m_0 c/\hbar$ term contains the proper mass m_0 of the photon. That term is certainly very small, since there has been very little time dispersion for light emitted millions of years ago. But Louis de Broglie answered those who think the photon mass exactly null that no physical experiment can prove a quantity to be exactly, with an infinite accuracy, equal to another one. To write [1] these Maxwell–de Broglie equations into space algebra, we let:

$$x^0 = ct \; ; \quad A^0 = V \; ; \quad A = A^0 + \vec{A} \; ; \quad F = \vec{E} + i\vec{H}. \tag{4.2}$$

The seven equations Eq. (4.1) group together into only two equations:

$$F = \nabla \widehat{A}, \tag{4.3}$$

$$\widehat{\nabla} F = -k_0^2 \widehat{A}. \tag{4.4}$$

Because Eq. (4.3) reads:

$$\vec{E} + i\vec{H} = (\partial_0 - \vec{\partial})(A^0 - \vec{A}),$$

$$0 + \vec{E} + i\vec{H} + 0i = (\partial_0 A^0 + \vec{\partial} \cdot \vec{A}) + (-\partial_0 \vec{A} - \vec{\partial} A^0) + i\vec{\partial} \times \vec{A} + 0i. \tag{4.5}$$

This equation is equivalent to the system obtained by separating the scalar, vector, pseudo-vector and pseudo-scalar parts:

$$0 = \frac{1}{c}\frac{\partial V}{\partial t} + \text{div}(\vec{A}), \tag{4.6}$$

$$\vec{E} = -\frac{1}{c}\frac{\partial \vec{A}}{\partial t} - \text{grad}(V), \tag{4.7}$$

$$\vec{H} = \text{rot}(\vec{A}). \tag{4.8}$$

Similarly Eq. (4.4) gives:

$$(\partial_0 + \vec{\partial})(\vec{E} + i\vec{H}) = -k_0^2(A^0 - \vec{A}),$$

$$\partial_0 \vec{E} + i\partial_0 \vec{H} + \vec{\partial} \cdot \vec{E} + i\vec{\partial} \times \vec{E} + i(\vec{\partial} \cdot \vec{H} + i\vec{\partial} \times \vec{H}) = -k_0^2(A^0 - \vec{A}), \tag{4.9}$$

$$\vec{\partial} \cdot \vec{E} + \partial_0 \vec{E} - \vec{\partial} \times \vec{H} + i(\partial_0 \vec{H} + \vec{\partial} \times \vec{E}) + i\vec{\partial} \cdot \vec{H} = -k_0^2 A^0 + k_0^2 \vec{A}.$$

---

1. To read the electromagnetic field as $\vec{E} + i\vec{H}$ is archaic. We will remark that the i here is the generator of the chiral gauge, it is not the i of quantum mechanics, generator of the electric gauge.

Separating, as previously, the scalar, vector, pseudo-vector and pseudo-scalar parts, this is equivalent to:

$$\operatorname{div}(\vec{E}) = -k_0^2 V, \tag{4.10}$$

$$\frac{1}{c}\frac{\partial \vec{E}}{\partial t} - \operatorname{rot}(\vec{H}) = k_0^2 \vec{A}, \tag{4.11}$$

$$\frac{1}{c}\frac{\partial \vec{H}}{\partial t} + \operatorname{rot}(\vec{E}) = \vec{0}, \tag{4.12}$$

$$\operatorname{div}(\vec{H}) = 0. \tag{4.13}$$

These equations reduce to Maxwell equations in the vacuum, plus the Lorentz gauge condition, if the proper mass of the photon is null. We get then, in place of Eq. (4.4): $\widehat{\nabla}F = 0$.

4.1.1 *Invariance under Cl_3^* and numeric dimension*

With Maxwell–de Broglie electromagnetism the potential terms V and $\vec{A}$ are not simple tools for calculations, but are parts of physical quantities of the wave of the photon, as much as $\vec{E}$ and $\vec{H}$. How do these quantities vary under a rotation, vary under a Lorentz dilation with ratio not equal to 1?

Since Maxwell's laws of electromagnetism in the vacuum are invariant, not only under the group of Lorentz transformations, but also under the conformal group, which contains in addition inversions and dilations, we will suppose that, under a Lorentz dilation R with ratio r, generated by an M matrix satisfying Eq. (1.42), the electromagnetic field transforms as:

$$F' = MFM^{-1}. \tag{4.14}$$

If we write then M as $M = \sqrt{r}e^{i\frac{\theta}{2}}P$, where P is an element of $SL(2,\mathbb{C})$, we have $P^{-1} = \overline{P}$ and we get:

$$F' = \sqrt{r}e^{i\frac{\theta}{2}}PF\frac{1}{\sqrt{r}}e^{-i\frac{\theta}{2}}\overline{P} = PF\overline{P}. \tag{4.15}$$

which is the same transformation as if the dilation was induced only by P, that is to say it was a Lorentz transformation. So Eq. (4.14) is such that the electromagnetic field depends neither on r, nor on θ: the presence of the Cl_3^* group is as discreet as possible.

Equation (4.14) is form invariant if we have

$$\widehat{\nabla}'F' = -k_0'^2 \widehat{A}'. \tag{4.16}$$

We also have:
$$\nabla = \overline{M}\nabla'\widehat{M} \ ; \quad \widehat{\nabla} = M^{\dagger}\widehat{\nabla}'M \ ; \quad \widehat{\nabla}' = (M^{\dagger})^{-1}\widehat{\nabla}M^{-1}. \qquad (4.17)$$

We then get:
$$\begin{aligned}
-{k_0'}^2\widehat{A}' &= \widehat{\nabla}'F' = (M^{\dagger})^{-1}\widehat{\nabla}M^{-1}MFM^{-1} \\
&= (M^{\dagger})^{-1}\widehat{\nabla}FM^{-1} = (M^{\dagger})^{-1}(-k_0^2\widehat{A})M^{-1}. \qquad (4.18)
\end{aligned}$$

Moreover $k_0 = rk_0'$ since $m = rm'$ is required by the invariance of the homogeneous nonlinear [2] We then get:
$$\begin{aligned}
-{k_0'}^2\widehat{A}' &= (M^{\dagger})^{-1}(-r^2{k_0'}^2\widehat{A})M^{-1}, \\
\widehat{A}' &= (M^{\dagger})^{-1}re^{-i\theta}\widehat{A}re^{i\theta}M^{-1} \\
&= (M^{\dagger})^{-1}M^{\dagger}\widehat{M}\widehat{A}\widehat{M}MM^{-1} \\
&= \widehat{M}\widehat{A}\widehat{M}, \qquad (4.19) \\
A' &= MAM^{\dagger}. \qquad (4.20)
\end{aligned}$$

which means that, contrary to qA which transforms as ∇, A transforms as x and is "contravariant". Physically potential terms are linked to and move with sources. How can A be contravariant and qA covariant? This means:
$$qA = \overline{M}q'A'\widehat{M} = q'\overline{M}MAM^{\dagger}\widehat{M} = q're^{i\theta}Are^{-i\theta} = q'r^2A, \qquad (4.21)$$
that [3] is to say:
$$q = q'r^2. \qquad (4.22)$$

The electric charge, like the proper mass, is a relativistic invariant. The electric charge, like the mass, is not invariant under the complete group Cl_3^*, and varies when the ratio of the dilation is not equal to 1.

The transformation Eq. (4.14), and the contra-variance Eq. (4.20) of A which comes from, are compatible with the law Eq. (4.3) linking the field to the potentials, because this gives:
$$F' = \nabla'\widehat{A}', \qquad (4.23)$$
$$MFM^{-1} = M(\nabla\widehat{A})M^{-1} = M(\overline{M}\nabla'\widehat{M}\widehat{A})M^{-1}. \qquad (4.24)$$

---

2. This is the best indication that the true wave equation for the electron is not the Dirac linear equation, but the homogeneous nonlinear equation. Electromagnetism and wave equation of the electron are both Cl_3^* form invariant only with our wave equation.

3. We get used to lowering up indexes and raising down indexes of tensors. To do that we use the metric, and so we implicitly consider it as invariant. But if the space-time metric is invariant under the Lorentz group, it is not invariant under the greater group of dilations, so we no longer have the right to raise or lower indices of tensors. A covariant vector does not behave as a contravariant vector under a dilation. Therefore we are not allowed to treat ∇, covariant, as x, contravariant and to compute $T(\nabla)$ instead of $T(x)$, a common thing [45] in space-time algebra which we must also avoid.

But we have, with Eq. (4.20):

$$\widehat{A'} = \widehat{M}\widehat{A}\widehat{M} \; ; \; \widehat{M}\widehat{A} = \widehat{A'}\widehat{M}^{-1}, \tag{4.25}$$

and Eq. (4.24) gives

$$MFM^{-1} = M\overline{M}\nabla'\widehat{A'}\overline{M}^{-1}M^{-1}$$
$$= (M\overline{M})F'(M\overline{M})^{-1} = \det(M)F'(\det(M))^{-1} = F'. \tag{4.26}$$

A dilation is composed of a Lorentz rotation and a pure homothety with ratio $r > 0$. We know that c is invariant under Lorentz rotations. Since a speed is a ratio distance on time, and since these two terms are multiplied by the same ratio r of a pure homothety, the ratio distance on time is invariant. So we may suppose that the invariance of light speed is true not only under the Lorentz rotations, but also under all dilations induced by an element of Cl_3^*. The other essential invariant of the Dirac theory is the fine structure constant α, which is a pure number, and so cannot vary under a dilation, no more than it can under a Lorentz rotation. But we have:

$$q = \frac{e}{\hbar c} \; ; \; qe = \frac{e^2}{\hbar c} = \alpha = q'e' \; ; \; qe = q'r^2e = q'e'. \tag{4.27}$$

We get then

$$e' = r^2e. \tag{4.28}$$

We have now:

$$\alpha = \frac{e^2}{\hbar c} = \frac{e'^2}{\hbar' c} = \frac{r^4 e^2}{\hbar' c} \; ; \; e^2\hbar'c = \hbar c r^4 e^2, \tag{4.29}$$

which gives:

$$\hbar' = r^4\hbar. \tag{4.30}$$

We must then see $\hbar$ as a variable [4] term under a dilation with ratio $r \neq 1$.

$$\frac{m_0 c}{\hbar} = m = rm' = r\frac{m_0'c}{\hbar'} = r\frac{m_0'c}{r^4\hbar} = \frac{m_0'c}{r^3\hbar}, \tag{4.31}$$

gives:

$$m_0' = r^3 m_0. \tag{4.32}$$

This allows us to define:

---

4. To let $\hbar = 1$ is then a very bad habit that we must get rid of as soon as possible.

4.1.2   *Numeric dimension*

The **numeric dimension** of any physical quantity is the power of the ratio dilation r in the formula giving the transformation of this quantity under the dilation R defined by the element M in Cl_3. Equation (1.44) says that x has numeric dimension 1. Equation (2.7) implies that ϕ has numeric dimension $1/2$. Equation (2.28) implies that ∇ has numeric dimension -1. Equation (4.14) implies that the electromagnetic field (and this will be the same for all other gauge fields) has numeric dimension 0. This is also the case for any velocity and for the fine structure constant. Equation (4.28) says that an electric charge (and it is the same for a magnetic charge) has numeric dimension 2. From Eq. (4.32) a proper mass does not vary as an electric charge under a Lorentz dilation. An electric charge varies as a surface; a proper mass has numeric dimension 3 and varies as a volume. There is a geometrical difference between a mass and a charge. Equation (4.30) implies that the Planck factor has numeric dimension 4, that is the numeric dimension of a space-time volume. Next Eq. (3.44) implies that the numeric dimension of the generic element y of the intrinsic manifold is 0. This manifold is then a real manifold both in the physical and the mathematical point of view. With Eq. (2.33), Eq. (2.37) and Eq. (2.45) we can see that all tensorial densities without derivatives $\phi\sigma_\mu\phi^\dagger$ and $\phi\sigma_\mu\overline{\phi}$ have the same numeric dimension 1. The interpretation of these quantities made from the matrix Dirac theory as scalar, vector, bivector and so on, is then completely out of date.

Since a speed is multiplied by r^0, an acceleration is multiplied by r^{-1}: An acceleration has numeric dimension -1. Therefore a force is multiplied by r^2 and has a numeric dimension 2. This is coherent with the Lorentz force since the electromagnetic field is multiplied by r^0 and the charge by r^2. The probabilistic interpretation of the quantum wave says that the square of the wave, generalized as J^0, is a probability of density. This attributes to J_0 the numeric dimension -3, since a probability is a pure number with numeric dimension 0. We shall see in chapter 9 that the correct probability of density is $J^0/\hbar c$ which has the true numeric dimension -3 since J^0 has the numeric dimension 1 of x and $\hbar c$ has the numeric dimension 4. It is completely impossible to get the true physical laws with $\hbar = 1$ since $\hbar$ is not a constant under Cl_3^*, that is the invariance group of all physical laws.

4.2 Electromagnetism with magnetic monopoles

When Maxwell wrote his laws for magnetism, he supposed that magnetic fields come from magnetic charges, which we now call magnetic monopoles. Later this was forgotten, because for decades nobody was able to prove the existence of such monopoles. Finally teachers have presented the laws to their students as if magnetic monopoles could not exist. Nevertheless the laws of electromagnetism can easily be modified if magnetic monopoles exist. On top of the electric charge density ρ_e and the current density $\vec{j}$, a density of magnetic charge ρ_m and a density of magnetic current $\vec{k}$ exist. On top of the electric potential V and of the potential vector $\vec{A}$, a magnetic potential W and a magnetic potential vector $\vec{B}$ exist. The laws of electromagnetism with monopoles read:

$$\vec{E} = -\mathrm{grad}(V) - \frac{1}{c}\frac{\partial \vec{A}}{\partial t} + \mathrm{rot}(\vec{B}) \; ; \; \vec{H} = \mathrm{rot}(\vec{A}) + \mathrm{grad}(W) + \frac{1}{c}\frac{\partial \vec{B}}{\partial t}$$

$$0 = \partial_\mu A^\mu = \frac{1}{c}\frac{\partial V}{\partial t} + \mathrm{div}(\vec{A}) \; ; \; 0 = \partial_\mu B^\mu = \frac{1}{c}\frac{\partial W}{\partial t} + \mathrm{div}(\vec{B})$$

$$\mathrm{rot}(\vec{H}) - \frac{1}{c}\frac{\partial \vec{E}}{\partial t} = \frac{4\pi}{c}\vec{j} \; ; \; \mathrm{div}(\vec{E}) = 4\pi\rho_e$$

$$\mathrm{rot}(\vec{E}) + \frac{1}{c}\frac{\partial \vec{H}}{\partial t} = \frac{4\pi}{c}\vec{k} \; ; \; \mathrm{div}(\vec{H}) = -4\pi\rho_m. \tag{4.33}$$

We can see that these equations are equivalent to:

$$F = \nabla(\widehat{A + iB}), \tag{4.34}$$

$$\widehat{\nabla} F = \frac{4\pi}{c}(\widehat{j + ik}), \tag{4.35}$$

where we have let:

$$B = W + \vec{B} \; ; \; k = \rho_m + \vec{k}. \tag{4.36}$$

The calculation is identical to that made to establish Eq. (4.3) and Eq. (4.4). So it is very simple to go from electromagnetism without monopoles to electromagnetism with monopoles: it is enough to add a pseudo-vector, made of the magnetic potential and the magnetic potential vector to the space-time vector made of the electric potential and the potential vector, and to add a space-time pseudo-vector made of the density of magnetic charge and the density of magnetic current to the space-time vector made of the density of charge and density of current. The laws are exactly the same, and we cannot see why such potentials and current should be prohibited. Until now quantum physics has not been able to see the magnetic part of electromagnetism because no distinction was made between real and complex quantities: $A + iB$ was seen as only A with complex components and it was the same for $j + ik$.

The form invariance of the law Eq. (4.34) under the Cl_3^* group has evidently the same consequence for the two potentials, so B must be, as A, a contravariant vector:

$$B' = MBM^\dagger. \tag{4.37}$$

To look at what is implied by Eq. (4.35), we remark that we have:

$$\widehat{\nabla} = M^\dagger \widehat{\nabla}' M \; ; \quad F'M = MF, \tag{4.38}$$

so we have:

$$\frac{4\pi}{c} \widehat{(j + ik)} = \widehat{\nabla} F = M^\dagger \widehat{\nabla}' MF = M^\dagger \widehat{\nabla}' F'M = M^\dagger \frac{4\pi}{c} \widehat{(j' + ik')} M,$$

$$j + ik = \overline{M}(j' + ik') \widehat{M}, \tag{4.39}$$

$$j = \overline{M} j' \widehat{M} \; ; \quad k = \overline{M} k' \widehat{M}, \tag{4.40}$$

which means that the j and k vectors are covariant, transform as ∇. This is consistent with electrostatics, because a charge density is the quotient of a charge e on a volume dv, and because we have, under a dilation with ratio r:

$$\rho_e = \frac{e}{dv} \; ; \quad \rho'_e = \frac{e'}{dv'} = \frac{r^2 e}{r^3 dv} = \frac{\rho_e}{r} \; ; \quad \rho_e = r\rho'_e. \tag{4.41}$$

We may say consequently that the choice made for the transformation of the electromagnetic field under a dilation, even if it gives surprising results, with the variation of the charge, the proper mass and the Planck term, is consistent with all elementary laws of electricity and magnetism.

4.3   Back to space-time

Until now we mainly used the space algebra, because the relativistic invariance of the Dirac theory leads inevitably to this algebra and because the even sub-algebra of the space-time algebra is isomorphic to the space algebra. Nevertheless we shall see further that electro-weak interactions need the space-time algebra.

4.3.1 *From* Cl_3 *to* $Cl_{1,3}$

To go from the space-time algebra to the space algebra, all you need is to use only even [5] terms. To go from the space algebra to the space-time algebra the easiest way is to use the matrix representation Eq. (1.75). The wave, denoted ϕ in space algebra, is denoted Ψ in space-time algebra. We have gotten in Eq. (2.59) $\Psi = \begin{pmatrix} \phi & 0 \\ 0 & \widehat{\phi} \end{pmatrix}$. N of Eq. (1.80) is similarly an even element of the space-time algebra and also the electromagnetic field that we denote $\mathbf{F}$:

$$\mathbf{F} = \begin{pmatrix} F & 0 \\ 0 & \widehat{F} \end{pmatrix} = \begin{pmatrix} \vec{E} + i\vec{H} & 0 \\ 0 & -\vec{E} + i\vec{H} \end{pmatrix} = \begin{pmatrix} \vec{E} & 0 \\ 0 & -\vec{E} \end{pmatrix} + \begin{pmatrix} iI & 0 \\ 0 & -iI \end{pmatrix} \begin{pmatrix} \vec{H} & 0 \\ 0 & -\vec{H} \end{pmatrix}$$

$$= \mathbf{E} + \gamma_{0123}\mathbf{H}. \tag{4.42}$$

We can shorten the notations with:

$$\mathbf{i} = \gamma_{0123} \ ; \ \mathbf{F} = \mathbf{E} + \mathbf{iH}. \tag{4.43}$$

Odd elements of the space-time algebra are the product by γ_0 of an even element and read

$$\begin{pmatrix} P & 0 \\ 0 & \widehat{P} \end{pmatrix} \begin{pmatrix} 0 & I \\ I & 0 \end{pmatrix} = \begin{pmatrix} 0 & P \\ \widehat{P} & 0 \end{pmatrix}. \tag{4.44}$$

We have also used, in Eq. (1.76) and Eq. (2.59):

$$\partial = \gamma^\mu \partial_\mu = \begin{pmatrix} 0 & \nabla \\ \widehat{\nabla} & 0 \end{pmatrix} \ ; \ A = \begin{pmatrix} 0 & A \\ \widehat{A} & 0 \end{pmatrix}.$$

Similarly the magnetic potential reads:

$$\mathbf{B} = \begin{pmatrix} 0 & B \\ \widehat{B} & 0 \end{pmatrix}. \tag{4.45}$$

The reverse of an even element is:

$$\widetilde{N} = \begin{pmatrix} \overline{M} & 0 \\ 0 & M^\dagger \end{pmatrix} \ ; \ \widetilde{\Psi} = \begin{pmatrix} \overline{\phi} & 0 \\ 0 & \phi^\dagger \end{pmatrix}. \tag{4.46}$$

The reverse of an odd element is:

$$\widetilde{\mathbf{B}} = \begin{pmatrix} 0 & B^\dagger \\ \overline{B} & 0 \end{pmatrix}. \tag{4.47}$$

---

5. In a Clifford algebra on an n-dimensional linear space, the linear space of even elements and the linear space of odd elements are linear spaces with dimension 2^{n-1}. Since the product of two odd elements is even the linear space of odd elements is not a sub-algebra. Since the product of two even elements is even, the linear space of even elements is a sub-algebra. In the case of $Cl_{1,3}$ the even sub-algebra is isomorphic to Cl_3.

With tensorial densities without derivative of Sec. 2.2.2 we have:

$$\Psi\widetilde{\Psi} = \begin{pmatrix} \phi\overline{\phi} & 0 \\ 0 & \widehat{\phi}\phi^\dagger \end{pmatrix} = \begin{pmatrix} \rho e^{i\beta} & 0 \\ 0 & \rho e^{-i\beta} \end{pmatrix} = \begin{pmatrix} \Omega_1 + i\Omega_2 & 0 \\ 0 & \Omega_1 - i\Omega_2 \end{pmatrix}$$

$$= \Omega_1 + i\Omega_2 = \rho e^{i\beta}, \tag{4.48}$$

$$\mathbf{D}_\mu = \begin{pmatrix} 0 & \phi\sigma_\mu\phi^\dagger \\ \widehat{\phi\sigma_\mu\phi^\dagger} & 0 \end{pmatrix} = \begin{pmatrix} \phi & 0 \\ 0 & \widehat{\phi} \end{pmatrix}\begin{pmatrix} 0 & \sigma_\mu \\ \widehat{\sigma}_\mu & 0 \end{pmatrix}\begin{pmatrix} \overline{\phi} & 0 \\ 0 & \phi^\dagger \end{pmatrix} = \Psi\gamma_\mu\widetilde{\Psi}, \tag{4.49}$$

$$\mathbf{S}_k = \begin{pmatrix} \phi\sigma_k\overline{\phi} & 0 \\ 0 & -\widehat{\phi}\sigma_k\phi^\dagger \end{pmatrix} = \Psi\gamma_{k0}\widetilde{\Psi}. \tag{4.50}$$

We must also notice

$$\widetilde{\Psi}\Psi = \Psi\widetilde{\Psi} = \Omega_1 + i\Omega_2. \tag{4.51}$$

We saw in Eq. (2.62) how Hestenes reads the Dirac equation. Since we have:

$$\Psi\gamma_0 e^{\beta\mathbf{i}} = \begin{pmatrix} 0 & e^{-\beta i}\phi \\ e^{\beta i}\widehat{\phi} & 0 \end{pmatrix}, \tag{4.52}$$

the homogeneous nonlinear wave equation reads in space-time algebra:

$$\partial\Psi\gamma_{21} = m\Psi\gamma_0 e^{\beta\mathbf{i}} + q\mathbf{A}\Psi. \tag{4.53}$$

For a space-time dilation, with:

$$\mathbf{x} = \begin{pmatrix} 0 & x \\ \widehat{x} & 0 \end{pmatrix} \; ; \; \mathbf{x}' = \begin{pmatrix} 0 & x' \\ \widehat{x}' & 0 \end{pmatrix}, \tag{4.54}$$

and with Eq. (1.80) and Eq. (2.26), we get equalities

$$\mathbf{x}' = N\mathbf{x}\widetilde{N} \; ; \; \Psi' = N\Psi \; ; \; \partial = \widetilde{N}\partial'N. \tag{4.55}$$

4.3.2   *Electromagnetism*

The laws of Maxwell–de Broglie electromagnetism become:

$$\mathbf{F} = \partial\mathbf{A} \tag{4.56}$$

$$\partial\mathbf{F} = -k_0^2\mathbf{A} \tag{4.57}$$

because

$$\partial\mathbf{A} = \begin{pmatrix} 0 & \nabla \\ \widehat{\nabla} & 0 \end{pmatrix}\begin{pmatrix} 0 & A \\ \widehat{A} & 0 \end{pmatrix} = \begin{pmatrix} \nabla\widehat{A} & 0 \\ 0 & \widehat{\nabla}A \end{pmatrix} = \begin{pmatrix} F & 0 \\ 0 & \widehat{F} \end{pmatrix} = \mathbf{F}, \tag{4.58}$$

$$\partial\mathbf{F} = \begin{pmatrix} 0 & \nabla \\ \widehat{\nabla} & 0 \end{pmatrix}\begin{pmatrix} F & 0 \\ 0 & \widehat{F} \end{pmatrix} = \begin{pmatrix} 0 & \nabla\widehat{F} \\ \widehat{\nabla}F & 0 \end{pmatrix} = -k_0^2\mathbf{A}. \tag{4.59}$$

The laws of electromagnetism with magnetic monopoles become:

$$\mathbf{F} = \partial(\mathbf{A} + i\mathbf{B}) \tag{4.60}$$

$$\partial\mathbf{F} = \frac{4\pi}{c}(\mathbf{j} + i\mathbf{k}) \tag{4.61}$$

Since space-time algebra is well known and gives simpler results, what is the purpose of developing calculations in space algebra? We may think that space-time algebra is too simple, there is too much symmetry between space and time. We can use the services of space-time algebra as long as it is not necessary to distinguish space and time and as long as there is no zero space-time length. But it is also necessary to never forget that time is not space. Time flows only from past to future while we can go away and back in space. Under dilations generated by elements of the Cl_3^* group the orientation of time and space cannot change. There is no physical way to change the time orientation and there is no physical way to change the space orientation. P and T transformations of quantum fields are purely theoretical.

4.4   A real photon

The beginning of quantum physics was the invention by Einstein in 1905 of a theory of light with quanta of impulse–energy. After Newton's corpuscular theory, Huyghens' undulatory theory was imposed by Fresnel with his transversal waves. This undulatory theory allowed a synthesis including electromagnetism and optics. The next page of this story was the discovery of the wave associated to any moving particle by Louis de Broglie. When he had the Dirac equation, he returned to the initial problem of the wave of a corpuscular photon. A photon with a proper mass $m_0 < 10^{-52}$kg gives, for all observable radiations, a non-observable dispersion. The corpuscular nature of light explains Compton diffusion and it is compatible with the absorption and emission of light by electrons of atoms. It allows one to understand the radiation pressure and to calculate completely all kinds of Doppler effects. In the same time light has also the undulatory aspects of Fresnel's waves and we know since Einstein that the density of photons and the intensity of electromagnetic waves are proportional.

Louis de Broglie first tried to associate a Dirac wave to the photon, but it was impossible to associate an electromagnetic wave. From this first attempt he understood that the electromagnetic field of the photon must be associated to the change of state of the electron interacting with the photon. And the only processes of interaction between photons and matter are absorption and the photoelectric effect.

For his construction of the wave of a photon Louis de Broglie started from two Dirac spinors, one of a particle and one of an anti-particle, able to annihilate, giving then all impulse–energy to the exterior. He established also that electromagnetic quantities must be linear combinations of the wave components. In the frame of the initial formalism used by de Broglie his two spinors read

$$\psi = \begin{pmatrix} \psi_1 \\ \psi_2 \\ \psi_3 \\ \psi_4 \end{pmatrix} \; ; \; \varphi = \begin{pmatrix} \varphi_1 \\ \varphi_2 \\ \varphi_3 \\ \varphi_4 \end{pmatrix}. \tag{4.62}$$

They are solutions of the Dirac wave equation for a particle without charge, like a neutrino

$$\partial_0\psi = (\alpha_1\partial_1 + \alpha_2\partial_2 + \alpha_3\partial_3 + i\frac{m}{2}\alpha_4)\psi, \tag{4.63}$$

and of the wave equation for its antiparticle, similar to an antineutrino

$$\partial_0\varphi = (\alpha_1\partial_1 - \alpha_2\partial_2 + \alpha_3\partial_3 - i\frac{m}{2}\alpha_4)\varphi, \tag{4.64}$$

where

$$x^0 = ct \; ; \; \partial_\mu = \frac{\partial}{\partial x^\mu} \; ; \; m = \frac{m_0 c}{\hbar}, \tag{4.65}$$

$$\alpha_j\alpha_k + \alpha_k\alpha_j = 2\delta_{jk}. \tag{4.66}$$

It is well known that these matrix relations are not enough to define α_μ uniquely. We can choose different sets of α_μ matrices. We choose here a set working with the Weyl spinors and the relativistic invariance:

$$\alpha_j = \begin{pmatrix} -\sigma_j & 0 \\ 0 & \sigma_j \end{pmatrix}, \; j = 1,2,3 \; ; \; \alpha_4 = \begin{pmatrix} 0 & -I \\ -I & 0 \end{pmatrix} \; ; \; I = \begin{pmatrix} 1 & 0 \\ 0 & 1 \end{pmatrix}, \tag{4.67}$$

where σ_j are the Pauli matrices, and we let

$$\xi = \begin{pmatrix} \psi_1 \\ \psi_2 \end{pmatrix} = \begin{pmatrix} \xi_1 \\ \xi_2 \end{pmatrix} \; ; \; \eta = \begin{pmatrix} \psi_3 \\ \psi_4 \end{pmatrix} = \begin{pmatrix} \eta_1 \\ \eta_2 \end{pmatrix},$$

$$\zeta^* = \begin{pmatrix} \varphi_1 \\ \varphi_2 \end{pmatrix} = \begin{pmatrix} \zeta_1^* \\ \zeta_2^* \end{pmatrix} \; ; \; \lambda^* = \begin{pmatrix} \varphi_3 \\ \varphi_4 \end{pmatrix} = \begin{pmatrix} \lambda_1^* \\ \lambda_2^* \end{pmatrix}, \tag{4.68}$$

where a^* is the complex conjugate of a. With

$$\vec{\partial} = \sigma_1\partial_1 + \sigma_2\partial_2 + \sigma_3\partial_3,$$

$$\vec{\partial}^* = \sigma_1\partial_1 - \sigma_2\partial_2 + \sigma_3\partial_3, \tag{4.69}$$

the wave equation Eq. (4.63) is equivalent to the system

$$(\partial_0 + \vec{\partial})\xi + i\frac{m}{2}\eta = 0, \tag{4.70}$$

$$(\partial_0 - \vec{\partial})\eta + i\frac{m}{2}\xi = 0. \tag{4.71}$$

ξ and η are the Weyl spinors of the wave ψ and the wave equation of the anti-particle Eq. (4.64) is equivalent to the system

$$(\partial_0 + \vec{\partial}^*)\zeta^* - i\frac{m}{2}\lambda^* = 0,$$

$$(\partial_0 - \vec{\partial}^*)\lambda^* - i\frac{m}{2}\zeta^* = 0. \tag{4.72}$$

By complex conjugation we get

$$(\partial_0 + \vec{\partial})\zeta + i\frac{m}{2}\lambda = 0, \tag{4.73}$$

$$(\partial_0 - \vec{\partial})\lambda + i\frac{m}{2}\zeta = 0. \tag{4.74}$$

This system is identical to Eq. (4.70) and Eq. (4.71) if we replace ζ by ξ and λ by η. We let

$$\phi_1 = \sqrt{2}\begin{pmatrix} \xi_1 & -\eta_2^* \\ \xi_2 & \eta_1^* \end{pmatrix} \; ; \; \phi_2 = \sqrt{2}\begin{pmatrix} \zeta_1 & -\lambda_2^* \\ \zeta_2 & \lambda_1^* \end{pmatrix}. \tag{4.75}$$

which have their value in the Pauli algebra. Comparing the system Eq. (4.70)–Eq. (4.71) to the system Eq. (2.12)–Eq. (2.13) we see from Eq. (2.21) that this system is equivalent to the equation

$$\nabla\widehat{\phi_1} + \frac{m}{2}\phi_1\sigma_{12} = 0. \tag{4.76}$$

Similarly the system Eq. (4.73) – Eq. (4.74) is equivalent to

$$\nabla\widehat{\phi_2} + \frac{m}{2}\phi_2\sigma_{12} = 0. \tag{4.77}$$

The two spinors follow the same wave equation. This is consistent with the linear Dirac theory where charge conjugation changes the sign of the charge but does not change the sign of the mass.

De Broglie had no theory for the wave of a relativistic system of particles nor for the interaction between its two spinors. So he simply supposed that his two half-photons ψ and φ are linked, have the same energy and the same impulse [31]. They satisfy

$$\varphi_k\partial_\mu\psi_i = (\partial_\mu\varphi_k)\psi_i = \frac{1}{2}\partial_\mu(\varphi_k\psi_i) \,, \; k,j = 1,2,3,4 \,; \; \mu = 0,1,2,3. \tag{4.78}$$

This is equivalent, with Eq. (4.62) and Eq. (4.68), to

$$\xi_k(\partial_\mu\zeta_i^*) = (\partial_\mu\xi_k)\zeta_i^* = \frac{1}{2}\partial_\mu(\xi_k\zeta_i^*),$$

$$\xi_k(\partial_\mu\lambda_i^*) = (\partial_\mu\xi_k)\lambda_i^* = \frac{1}{2}\partial_\mu(\xi_k\lambda_i^*),$$

$$\eta_k(\partial_\mu\zeta_i^*) = (\partial_\mu\eta_k)\zeta_i^* = \frac{1}{2}\partial_\mu(\eta_k\zeta_i^*), \tag{4.79}$$

$$\eta_k(\partial_\mu\lambda_i^*) = (\partial_\mu\eta_k)\lambda_i^* = \frac{1}{2}\partial_\mu(\eta_k\lambda_i^*).$$

Wave equations Eq. (4.76) and Eq. (4.77) are form invariant under the Lorentz dilation D defined by Eq. (1.42) and satisfy

$$\phi_1' = M\phi_1 \; ; \quad \phi_2' = M\phi_2. \tag{4.80}$$

4.4.1   *The electromagnetism of the photon*

We start here from the fact seen in Eq. (4.20) that the electromagnetic potential A is a contravariant space-time vector, that is a vector transforming as x: $A' = MAM^\dagger$. We know in addition that Pauli's principle rules that products must be antisymmetric. We also know that the σ_3 term is privileged [6] in the Dirac equation. We must then consider a space-time vector A and an electromagnetic field F_e so defined:

$$A = \phi_1 i\sigma_3\phi_2^\dagger - \phi_2 i\sigma_3\phi_1^\dagger, \tag{4.81}$$

$$F_e = \nabla\widehat{A}. \tag{4.82}$$

The variance of A and the variance of the electromagnetic field F_e under Cl_3^* are expected variances because

$$A' = \phi_1'i\sigma_3\phi_2'^\dagger - \phi_2'i\sigma_3\phi_1'^\dagger = (M\phi_1)i\sigma_3(M\phi_2)^\dagger - (M\phi_2)i\sigma_3(M\phi_1)^\dagger$$

$$= M(\phi_1 i\sigma_3\phi_2^\dagger - \phi_2 i\sigma_3\phi_1^\dagger)M^\dagger = MAM^\dagger, \tag{4.83}$$

$$F_e = \nabla\widehat{A} = \overline{M}\nabla'\widehat{\overline{M}A} = \overline{M}\nabla'\widehat{MAM^\dagger}(\widehat{M^\dagger})^{-1} = \overline{M}(\nabla'\widehat{A'})\overline{M}^{-1}$$

$$= M^{-1}M\overline{M}F_e'\overline{M}^{-1}M^{-1}M = M^{-1}\det(M)F_e'\det(M^{-1})M = M^{-1}F_e'M,$$

$$F_e' = MF_eM^{-1}. \tag{4.84}$$

A is actually a space-time vector because

$$A^\dagger = (\phi_1 i\sigma_3\phi_2^\dagger - \phi_2 i\sigma_3\phi_1^\dagger)^\dagger = \phi_2(-i\sigma_3)\phi_1^\dagger - \phi_1(-i\sigma_3)\phi_2^\dagger = A. \tag{4.85}$$

---

6. We shall develop this in chapter 5.

The calculation of A with Eq. (4.75) and the Pauli matrices in Eq. (1.19) gives

$$\widehat{A} = 2i \begin{pmatrix} \eta_1\lambda_1^* - \xi_2^*\zeta_2 - \lambda_1\eta_1^* + \zeta_2^*\xi_2 & \eta_1\lambda_2^* + \xi_2^*\zeta_1 - \lambda_1\eta_2^* - \zeta_2^*\xi_1 \\ \eta_2\lambda_1^* + \xi_1^*\zeta_2 - \lambda_2\eta_1^* - \zeta_1^*\xi_2 & \eta_2\lambda_2^* - \xi_1^*\zeta_1 - \lambda_2\eta_2^* + \zeta_1^*\xi_1 \end{pmatrix}. \quad (4.86)$$

We then remark that each product is one of the products in Eq. (4.79) and this gives

$$\partial_\mu\widehat{A} = \partial_\mu(\widehat{\phi}_1 i\sigma_3\overline{\phi}_2 - \widehat{\phi}_2 i\sigma_3\overline{\phi}_1) = 2(\partial_\mu\widehat{\phi}_1)i\sigma_3\overline{\phi}_2 - 2(\partial_\mu\widehat{\phi}_2)i\sigma_3\overline{\phi}_1,$$
$$\nabla\widehat{A} = 2[(\nabla\widehat{\phi}_1)i\sigma_3\overline{\phi}_2 - (\nabla\widehat{\phi}_2)i\sigma_3\overline{\phi}_1]. \quad (4.87)$$

The Dirac equations Eq. (4.76) and Eq. (4.77) give then

$$F_e = m\phi_1(-i\sigma_3)i\sigma_3\overline{\phi}_2 - m\phi_2(-i\sigma_3)i\sigma_3\overline{\phi}_1 = m(\phi_1\overline{\phi}_2 - \phi_2\overline{\phi}_1). \quad (4.88)$$

Any element in the Cl_3 algebra as F_e is a sum

$$F_e = s + \vec{E} + i\vec{H} + ip, \quad (4.89)$$

where s is a scalar, $\vec{E}$ is a vector, $i\vec{H}$ is a pseudo-vector and ip is a pseudo-scalar. But we get

$$\overline{F}_e = s - \vec{E} - i\vec{H} + ip = m\overline{(\phi_1\overline{\phi}_2 - \phi_2\overline{\phi}_1)} = m(\phi_2\overline{\phi}_1 - \phi_1\overline{\phi}_2)$$
$$= -m(\phi_1\overline{\phi}_2 - \phi_2\overline{\phi}_1) = -F_e = -s - \vec{E} - i\vec{H} - ip. \quad (4.90)$$

F_e is [7] therefore a pure bivector:

$$s = 0 \; ; \quad p = 0 \; ; \quad F_e = \vec{E} + i\vec{H}. \quad (4.91)$$

This agrees with all we know about electromagnetism and optics. Now Eq. (4.83) reads

$$\vec{E} + i\vec{H} = (\partial_0 - \vec{\partial})(A^0 - \vec{A}) = \partial_0 A^0 - \vec{\partial}A^0 - \partial_0\vec{A} + \vec{\partial}\vec{A}. \quad (4.92)$$

This is equivalent to the system

$$0 = \partial_\mu A^\mu, \quad (4.93)$$
$$\vec{E} = -\vec{\partial}A^0 - \partial_0\vec{A}, \quad (4.94)$$
$$\vec{H} = \vec{\partial} \times \vec{A}. \quad (4.95)$$

We also get

$$\widehat{\nabla}F_e = m\widehat{\nabla}(\phi_1\overline{\phi}_2 - \phi_2\overline{\phi}_1). \quad (4.96)$$

---

7. We have previously supposed that the electromagnetic field F is a pure bivector, without scalar or pseudo-scalar part, for instance in Eq. (4.3). It is necessary to get Maxwell's laws without supplementary non-physical terms. Here we have nothing to suppose, the pure bivector nature of the electromagnetic field is a consequence of the antisymmetric building from two spinors and wave equations.

A detailed calculation of the matrices shows as previously only the products present in Eq. (4.79) and this gives

$$\widehat{\nabla} F_e = 2m[(\widehat{\nabla}\phi_1)\overline{\phi}_2 - (\widehat{\nabla}\phi_2)\overline{\phi}_1]. \qquad (4.97)$$

And we get with wave equations Eq. (4.76) and Eq. (4.77)

$$\widehat{\nabla}\phi_1 = \frac{m}{2}\widehat{\phi}_1\sigma_{21} \ ; \quad \widehat{\nabla}\phi_2 = \frac{m}{2}\widehat{\phi}_2\sigma_{21}, \qquad (4.98)$$

$$\Box\widehat{A} = \widehat{\nabla}\nabla\widehat{A} = \widehat{\nabla} F_e = m^2[\widehat{\phi}_1(-i\sigma_3)\overline{\phi}_2 - \widehat{\phi}_2(-i\sigma_3)\overline{\phi}_1],$$

$$\widehat{\nabla} F_e = -m^2\widehat{A}. \qquad (4.99)$$

So we get the seven laws of the electromagnetism of Maxwell in the vacuum, completed by the terms found by Louis de Broglie containing the very small proper mass $m_0 = \frac{m\hbar}{c}$ of the photon. The seven laws are exactly the same, but the quantities are here only real or with real components. F_e is therefore exactly the electromagnetic field of classical electromagnetism and optics. The definition Eq. (4.81) – Eq. (4.82) allows us to get a theory of a massive photon with a wave which includes real components of an electromagnetic space-time potential vector A, contravariant, and an electromagnetic bivector field F_e. This is an improvement in the theory of light, coming from the use of Cl_3 allowing an antisymmetric building, instead of Dirac matrices. Moreover the potential term is directly linked to the two spinors as much as the field bivector. It is an important difference with classical electromagnetism where potential terms are often considered as non-physical. This difference comes with quantum physics: potential terms are the electromagnetic terms present in the Dirac or Schrödinger wave equations.

The differential laws Eq. (4.82) and Eq. (4.99) are form invariant under the dilations defined by Eq. (1.42). This invariance under Cl_3^* implies that they are invariant under the restricted Lorentz group.

The quantum wave of this photon is actually an electromagnetic wave, with field and potential term with real components.

Potentials and fields were defined from antisymmetric products of spinors. They can disappear as soon as the two spinors are equal. They can appear as soon as the two spinors are not equal.

The result of the conditions Eq. (4.79) of Louis de Broglie is a linearization of the derivation of products which gives linear equations for the bosons built from the fermions. This is how the linear operator ∇ acts both in the Dirac equation and in the Maxwell equations. This linearization gives Maxwell's laws.

Chapter 5

Miscellaneous

We study a first consequence of the two space-time manifolds and of the dilations between these two manifolds: the non-isotropy of the intrinsic manifold. We link this with the existence of three kinds of leptons. We present new possibilities for the wave of systems of identical particles. We study a wave equation without possibility of Lagrangian mechanism. We present the three other Lochak's photons.

The wave of the electron induces, in each space-time point, a geometric transformation from the tangent space-time to an intrinsic manifold linked to the wave, into the usual space-time of restricted relativity. The intrinsic space-time, contrary to the usual space-time, is not isotropic, and we study now this anisotropy.

5.1 Anisotropy

The fact that there exists, in the Dirac theory, a privileged direction was remarked on by Louis de Broglie in his first work on the Dirac equation: [30] p.138 [1] *"Les fonctions ψ_i solutions de ces équations sont donc intimement liées au choix des axes comme dans la théorie de Pauli; elles doivent servir à calculer des probabilités pour lesquelles l'axe des z joue un rôle particulier"*. The solution to this difficulty is that with a rotation it is always possible to bring the z axis into any direction of the space.

The solution uses then a conveniently chosen element of Cl_3^*, which generates a spatial rotation and rotates the third axis onto the chosen direction. There are always two solutions, and then the final space-time, the relative space-time, is isotropic and has no privileged direction. But the

---

1. Translation:*"The ψ_i functions solutions of these equations are then completely linked to the choice of axis as into the Pauli theory; they must serve to calculate probabilities for which the z axis plays a particular role"*.

initial space-time, the intrinsic space-time, on the contrary, remains perfectly non-isotropic: before as after the rotation, it is always σ_3 which is privileged. We have remarked previously that with the Lorentz rotations of the complex formalism the γ_μ matrices are invariant. They are identical before or after the rotation. Whatever formalism is used it is always the third component of the spin that is measured and the square of the spin vector, never the first or the second component of the spin. The reason is evident if we regard the wave equation or the Lagrangian in the Clifford algebra of space. This third direction is present in the wave equation and in the Lagrangian which both contain an $i\sigma_3$.

Now, and this is the first concrete consequence of calculations with the Pauli algebra, it is perfectly possible to write two other Lagrangian densities, two other wave equations similar to the Dirac equation:

$$\nabla\widehat{\phi} + qA\widehat{\phi}\sigma_{23} + me^{-i\beta}\phi\sigma_{23} = 0, \tag{5.1}$$

$$\nabla\widehat{\phi} + qA\widehat{\phi}\sigma_{31} + me^{-i\beta}\phi\sigma_{31} = 0. \tag{5.2}$$

The invariant wave equations obtained by multiplying on the left by $\overline{\phi}$ are

$$\overline{\phi}(\nabla\widehat{\phi})\sigma_{32} + \overline{\phi}qA\widehat{\phi} + m\rho = 0, \tag{5.3}$$

$$\overline{\phi}(\nabla\widehat{\phi})\sigma_{13} + \overline{\phi}qA\widehat{\phi} + m\rho = 0. \tag{5.4}$$

With the wave equation Eq. (5.1) and Eq. (5.3) it is the first axis which is privileged. The conservative space-time vectors are D_0 and D_1. To solve the wave equation Eq. (5.1) for the hydrogen atom, we shall take again the method of separation of variables of Appendix C, making a circular permutation p on indices 1, 2, 3 of matrices σ: $1 \mapsto 2$, $2 \mapsto 3$, $3 \mapsto 1$, and on indices of formula Eq. (C.1). Since it is the only thing that changes, the results will be similar.

With the wave equation Eq. (5.2), it is the second axis which is privileged. The conservative space-time vectors are D_0 and D_2. To solve the wave equation Eq. (5.2) for the hydrogen atom, we shall take again the method of separation of variables of Appendix C, making a circular permutation $p^2 = p^{-1}$ on indices 1, 2, 3 of matrices σ, and on indices of formula Eq. (C.1). Since it is the only thing that changes, the results will be similar.

In all that we know today about experimental physics there is something very similar. Beside electrons there exist also muons and tauons. The three kinds of objects are similar and nevertheless different. Muons have been known for more than 70 years, and until now there has been no simple explanation why they exist, or what distinguishes them from electrons. We shall associate here to each category, that is to say to each of the three generations, one of the three wave equations Eq. (3.1), Eq. (5.3) and Eq. (5.4). The similarity between the wave equations allows us to explain why electrons, muons and tauons have the same properties, behave in the same way in an electromagnetic field, and have the same energy levels in a Coulombian potential. In fact, to see a difference between these three equations, it is necessary to go past the wave equation of a single particle and to enter the question of a system made of different kinds [2] of particles.

The third direction **or** the first direction may be put after a rotation in any direction **but** a rotation cannot turn both the third direction **and** the first direction into a given direction. So in this direction it is impossible to measure both the spin of an electron following Eq. (3.2) **and** the spin of a muon following Eq. (5.3).

In addition we know that a muon, even though it is a particle with spin $\frac{1}{2}$ like the electron, cannot spontaneously disintegrate into a lone electron. Its disintegration gives an electron plus a muonic neutrino and an electronic antineutrino. This may be understood in the following way: The wave of the muonic neutrino, like the wave of the muon, has a measurable spin in the first direction and takes away the muon's spin. The spin of the electron which is measurable in the third direction is brought by the antineutrino with a spin opposed to the spin of the electron.

We have supposed arbitrarily that the electron follows Eq. (3.1) and that the muon follows Eq. (5.3). One or the other could also follow Eq. (5.4), nothing allows us to say. On the other hand, the choice made by Nature of one or another equation justifies the fact that physical space is oriented: Consider in the intrinsic space three space vectors having respectively the third direction and the wavelength of the electron, the first direction and the wavelength of the muon, and the second direction and the wavelength of a tauon. These three vectors form a basis of the intrinsic space. If we exchange now the second and the third vectors, we get another basis, with another orientation.

Equation (3.1), Eq. (5.3) and Eq. (5.4) are equivalent only if the mass terms are equal in the different equations. But experiment shows that these masses are completely different from one generation to another. This difference, of unknown origin, differentiates the three generations of leptons.

---

2. We know for instance that a muon within the electronic cloud of an atom does not respect the Pauli exclusion principle. This is rather easy to understand if that exclusion principle is linked to the spin of the different particles, because the spin of an electron following Eq. (3.2) is always measured in the third direction and cannot be added or subtracted to the spin of a muon following Eq. (5.1), which is always measured in the first direction.

We have calculated the affine connection of the intrinsic manifold [15]. In the case of plane waves studied in chapters 2 and 3 only two terms are not zero and give a torsion. These torsion terms are linked to the proper mass of the particle.

5.2   Systems of electrons

The non-relativistic Schrödinger equation for a particle system uses in the case of a system of two particles without spin a wave $\psi = \psi_1 \psi_2$ which is the product of the two waves of each particle, when it is possible to neglect the interaction between these particles. We cannot transpose $\psi_1 \psi_2$ into $\phi_1 \phi_2$ which should transform into $M\phi_1 M\phi_2$ under the dilation R defined in Eq. (1.42), because M does not commute with ϕ. Another product is suggested by Eq. (4.14) because if $\phi_{12} = \phi_1 \phi_2^{-1}$ we get

$$\phi'_{12} = \phi'_1 \phi'^{-1}_2 = M\phi_1 \phi_2^{-1} M^{-1} = M\phi_{12} M^{-1}, \tag{5.5}$$

and ϕ_{12} transforms under a dilation as the electromagnetic field. But the factor $e^{-i\frac{E}{\hbar}t}$ of non-relativistic quantum mechanics becomes in the case of the electron $e^{-\frac{E}{\hbar c}x^0 \sigma_{12}}$ with the Cl_3 algebra, and with $\phi_1 \phi_2^{-1}$ the energies are not added but subtracted. To get the addition of energies we can consider terms as $\phi_1 \sigma_1 \phi_2^{-1}$ or $\phi_1 \sigma_2 \phi_2^{-1}$ because σ_1 and σ_2 anti-commute with σ_{12} and

$$\sigma_1 e^{-\frac{E}{\hbar c}x^0 \sigma_{12}} = e^{\frac{E}{\hbar c}x^0 \sigma_{12}} \sigma_1. \tag{5.6}$$

Since we have

$$\sigma_2 = \sigma_1 \sigma_{12} = \sigma_1 e^{\frac{\pi}{2}\sigma_{12}}, \tag{5.7}$$

σ_1 and σ_2 differ only by a constant gauge factor and we can choose σ_1. Since we know that two electrons are identical we can consider only terms such as $\phi_1 \sigma_1 \phi_2^{-1} \pm \phi_2 \sigma_1 \phi_1^{-1}$. The Pauli principle invites us to consider for the wave of a system of two electrons

$$\phi_{12} = \phi_1 \sigma_1 \phi_2^{-1} - \phi_2 \sigma_1 \phi_1^{-1}, \tag{5.8}$$

which is antisymmetric:

$$\phi_{21} = -\phi_{12}, \tag{5.9}$$

and transforms under a dilation R of dilator M as F [14]

$$\phi'_{12} = M\phi_{12} M^{-1}. \tag{5.10}$$

For a system of three electrons whose respective waves are ϕ_1, ϕ_2, ϕ_3 we consider

$$\phi_{123} = \phi_{12}\phi_3 + \phi_{23}\phi_1 + \phi_{31}\phi_2, \tag{5.11}$$

which satisfies

$$\phi_{123} = \phi_{231} = \phi_{312} = -\phi_{132} = -\phi_{321} = -\phi_{132}, \tag{5.12}$$

$$\phi'_{123} = M\phi_{123}. \tag{5.13}$$

The Pauli principle is satisfied and ϕ_{123} transforms [3] as a unique electronic wave. Then for four electrons we consider

$$\phi_{1234} = \phi_{12}\phi_{34} + \phi_{23}\phi_{14} + \phi_{31}\phi_{24} + \phi_{34}\phi_{12} + \phi_{14}\phi_{23} + \phi_{31}\phi_{24}, \tag{5.14}$$

which is antisymmetric, and transforms also as the electromagnetic field

$$\phi'_{1234} = M\phi_{1234}M^{-1}. \tag{5.15}$$

We can easily generalize to n electrons. We get $n + 1$ wave equations, one for each electronic wave

$$\nabla\widehat{\phi}_k + qA_k\widehat{\phi}_k\sigma_{12} + me^{-i\beta_k}\phi_k\sigma_{12} = 0, \tag{5.16}$$

where A_k is the sum of the exterior potential A and the potential created by the $n - 1$ other electrons and β_k is the Yvon–Takabayasi angle of the kth electron. The wave of the system is antisymmetric. The wave equation of this wave is determined by the n wave equations of each particle. If n is even $\phi_{12...n}$ transforms under a dilation as the electromagnetic field F. The wave of an even system appears as a boson wave. Even systems compose greater systems symmetrically as in Eq. (5.14). This is the source of Bose–Einstein statistics. If n is odd $\phi_{12...n}$ transforms under a dilation as a spinor ϕ. The wave of an odd system of electrons transforms under a dilation as the wave of a unique electron.

The wave of a system propagates, like the waves of each electron, in the usual space-time. It is not necessary to use configuration spaces. Difficulties arising from the difference between a unique time and several spaces disappear. Space is, like time, unique in this model. The wave of a system is not very different from the waves of its individual parts; they continue to exist and to propagate.

This model can also explain why a muon in an electronic cloud does not follow the Pauli's exclusion principle: with Eq. (5.1) for instance the phase contains not a σ_{12} factor, but instead a σ_{23} factor, and the muon cannot add its impulse–energy and so cannot enter the process of construction of the wave of a system described here.

---

3. If a similar construction is possible for quarks, this could explain why a proton or a neutron containing three quarks is seen also as a unique spinor, transforming under a Lorentz rotation as the wave of a unique electron.

5.3 Equation without Lagrangian formalism

We have seen in Sec. 2.4 that the Lagrangian density of the Dirac wave is exactly the scalar part of the invariant wave equation. The Lagrangian formalism is a consequence, not the cause of the Dirac equation. Therefore, if we modify the wave equation without changing its scalar part, we shall get a wave equation which cannot result from a Lagrangian mechanism, since the scalar part gives the Dirac equation without change [20]. We consider the invariant wave equation

$$\overline{\phi}(\nabla\widehat{\phi})\sigma_{21} + \overline{\phi}qA\widehat{\phi} + m\overline{\phi}\phi(1 + \epsilon\sigma_3) = 0, \tag{5.17}$$

where ϵ is a very small real constant. Only the mass term is changed from the invariant Eq. (2.64) which is equivalent to the Dirac equation. Computation of the first terms is unchanged, the mass term is

$$m\overline{\phi}\phi(1 + \epsilon\sigma_3) = m(\Omega_1 + i\Omega_2)(1 + \epsilon\sigma_3)$$
$$= m\Omega_1 + m\epsilon\Omega_1\sigma_3 + m\epsilon\Omega_2 i\sigma_3 + im\Omega_2, \tag{5.18}$$

and the system Eq. (2.75) to Eq. (2.82) becomes

$$0 = w^3 + V^0 + m\Omega_1 \ ; \quad 0 = v^2 + V^1, \tag{5.19}$$
$$0 = -v^1 + V^2 \ ; \quad 0 = w^0 + V^3 + m\epsilon\Omega_1, \tag{5.20}$$
$$0 = -v^3 + m\Omega_2 \ ; \quad 0 = w^2, \tag{5.21}$$
$$0 = -w^1 \ ; \quad 0 = -v^0 + m\epsilon\Omega_2. \tag{5.22}$$

This last equation implies that the current of probability is no longer conservative, so this wave equation is certainly unusual. Now it is easy to avoid the problem of the conservation of probabilities: we start from the homogeneous non-linear Eq. (3.1) and we add the same mass term

$$\overline{\phi}(\nabla\widehat{\phi})\sigma_{21} + \overline{\phi}qA\widehat{\phi} + m\rho(1 + \epsilon\sigma_3) = 0. \tag{5.23}$$

The system Eq. (3.10) to Eq. (3.17) becomes

$$0 = w^3 + V^0 + m\rho, \tag{5.24}$$
$$0 = v^2 + V^1, \tag{5.25}$$
$$0 = -v^1 + V^2, \tag{5.26}$$
$$0 = w^0 + V^3 + m\epsilon\rho, \tag{5.27}$$
$$0 = -v^3, \tag{5.28}$$
$$0 = w^2, \tag{5.29}$$
$$0 = -w^1, \tag{5.30}$$
$$0 = -v^0. \tag{5.31}$$

And as previously we have two conservative currents, $J = D_0$ and $K = D_3$. It is easy to see that Eq. (5.23) is invariant under Cl_3^*, there are two gauge invariances (see Sec. 3.1). The angular momentum operators of the Dirac theory are still available, but there is no Hamiltonian to commute with them. This wave equation cannot come from a Lagrangian density since such a density should modify Eq. (5.24), which gives Eq. (3.1), not Eq. (5.23).

5.3.1 *Plane waves*

We consider a plane wave with a phase φ with the vector v defined in Eq. (3.24). Without an exterior electromagnetic field we get in place of Eq. (3.28)

$$-mv\widehat{\phi} + me^{-i\beta}\phi(1 + \epsilon\sigma_3) = 0. \tag{5.32}$$

This gives

$$\phi(1 + \epsilon\sigma_3) = e^{i\beta}v\widehat{\phi}, \tag{5.33}$$

$$\widehat{\phi}(1 - \epsilon\sigma_3) = e^{-i\beta}\widehat{v}\phi, \tag{5.34}$$

$$\phi(1 + \epsilon\sigma_3)(1 - \epsilon\sigma_3) = e^{i\beta}v\widehat{\phi}(1 - \epsilon\sigma_3),$$

$$\phi(1 - \epsilon^2) = e^{i\beta}ve^{-i\beta}\widehat{v}\phi,$$

$$(1 - \epsilon^2)\phi = v\widehat{v}\phi, \tag{5.35}$$

$$v \cdot v = v\widehat{v} = 1 - \epsilon^2, \tag{5.36}$$

$$||v|| = \sqrt{1 - \epsilon^2}. \tag{5.37}$$

We let then

$$c' = c\sqrt{1 - \epsilon^2} \; ; \quad v = v'\sqrt{1 - \epsilon^2}. \tag{5.38}$$

And we get

$$||v'|| = 1. \tag{5.39}$$

First consequence: c', not c, is the velocity limit of this unusual quantum object. The present study has no known physical application, but this wave equation indicates that the limit [4] speed c is not as general [19] as we thought.

---

4. Furthermore, if ϵ tends to 1 the limit speed tends to 0 and may be very small.

5.4   Three other photons of Lochak

Following Eq. (4.81) seven other space-time vectors should be possible on the model $\phi_1 X \phi_2^\dagger - \phi_2 X \phi_1^\dagger$ since the Cl_3 algebra is 8-dimensional. Only three of these seven choices, $X = -\sigma_3$, $X = i$ and $X = 1$, are compatible [5] with Eq. (4.79) and we have established [15] that this gives the three other photons of G. Lochak [50] [51] [52]. Firstly if $X = -\sigma_3$,

$$iB = \phi_1 \widehat{\sigma}_3 \phi_2^\dagger - \phi_2 \widehat{\sigma}_3 \phi_1^\dagger \; ; \; F_m = \nabla \widehat{iB}, \tag{5.40}$$

gives his magnetic photon. As with the electric photon each quantity is real or has real components. It is possible to consider a total field $F = F_e + F_m$ satisfying

$$F = \nabla(\widehat{A + iB}), \tag{5.41}$$

$$\widehat{\nabla} F = -m^2(\widehat{A + iB}), \tag{5.42}$$

which are laws of the electromagnetism with electric charges and magnetic monopoles, and densities of electric current j and magnetic current k satisfying

$$j = -\frac{c}{4\pi} m^2 A \; ; \; k = -\frac{c}{4\pi} m^2 B. \tag{5.43}$$

j and k are very small since m is very small. Even if A and B are contravariant vectors, the variance of m allows j and k to be covariant vectors under Cl_3^*, varying as ∇, not as x. Now

$$A_{(i)} = \phi_1 i \phi_2^\dagger - \phi_2 i \phi_1^\dagger \; ; \; s = \nabla \widehat{A}_{(i)}, \tag{5.44}$$

defines an invariant scalar field s while

$$iB_{(1)} = \phi_1 \phi_2^\dagger - \phi_2 \phi_1^\dagger \; ; \; ip = \nabla \widehat{iB}_{(1)}, \tag{5.45}$$

defines an invariant pseudo-scalar field ip. We can put together cases $X = i$ and $X = 1$. We let

$$P = A_{(i)} + iB_{(1)} \; ; \; F_0 = \nabla \widehat{P} = s + ip, \tag{5.46}$$

and we get

$$\widehat{\nabla} F_0 = -m^2 \widehat{P}. \tag{5.47}$$

So it is possible to get in the frame of Cl_3 all four photons of the theory of de Broglie enlarged by Lochak and the whole thing is form invariant under Cl_3^*. There are differences in comparison with the construction based on the

---

5. This comes from the non-commutative product in Cl_3. Since σ_{12} is present in the Dirac equation, only terms commuting with σ_{12} work here.

Dirac matrices: Physical quantities are real or have real components and they are obtained by antisymmetric products of spinors. This is very easy to get with the internal multiplication of the Cl_3 algebra and was very difficult to make from the complex uni-column matrices. These two differences are advantageous because vectors and tensors of classical electromagnetism and optics have only real components. And de Broglie had understood very early that antisymmetric products are enough to get the Bose–Einstein statistics for bosons made of an even number of fermions. The scalar field of G. Lochak and the pseudo-scalar field for which de Broglie was cautious are perhaps to be identified with the scalar Higgs boson that physicists are today studying. Since s and p fields are obtained here independently from the field of the electric photon and the magnetic photon, their mass is not necessarily very small and may be huge. Curiously it was de Broglie's first idea about the non-Maxwellian part of his theory. Were the Higgs bosons foreseen as early as 1934?

5.5   Uniqueness of the electromagnetic field

The Dirac equation contains a privileged σ_3 which can be generalized as σ_j, $j = 1,\ 2,\ 3$. We generalize then Eq. (4.81) and Eq. (4.82) if we let

$$A_{(j)} = \phi_1 i\sigma_j\phi_2^\dagger - \phi_2 i\sigma_j\phi_1^\dagger, \qquad (5.48)$$

$$F_e = \nabla\widehat{A}_{(j)}, \qquad (5.49)$$

with the ϕ_1 and ϕ_2 waves following

$$\nabla\widehat{\phi}_1 = \frac{m}{2}\phi_1(-i\sigma_j), \qquad (5.50)$$

$$\nabla\widehat{\phi}_2 = \frac{m}{2}\phi_2(-i\sigma_j). \qquad (5.51)$$

We can also start from fields and get from them potentials. The electromagnetic field is then defined by Eq. (4.88)

$$F_e = m(\phi_1\overline{\phi}_2 - \phi_2\overline{\phi}_1). \qquad (5.52)$$

Potentials terms are linked to this field by Dirac equations Eq. (5.1) and Eq. (5.2), they satisfy

$$F_e = \nabla\widehat{A}_{(j)} \ ; \ \widehat{\nabla}F_e = -m^2\widehat{A}_{(j)}. \qquad (5.53)$$

It is interesting to note that F_e is independent of the index j and the electromagnetic field is then unique.

Chapter 6

Electro-weak interactions: The lepton case

We study the weak interactions of the electron with its neutrino. The covariant gauge derivative also accounts for their charge conjugate waves. It is form invariant under Cl_3^* and gauge invariant under the $U(1) \times SU(2)$ gauge group. We study the geometric transformation linked to the wave. We get a remarkable identity which make the wave often invertible. We get a wave equation with mass term that is form invariant and that is gauge invariant under the gauge group of electro-weak interactions.

6.1 The Weinberg–Salam model for the electron

An extension of the Dirac equation up to electro-weak interactions [61] was tried by D. Hestenes [40] and by R. Boudet [3] [4] in the frame of the Clifford algebra $Cl_{1,3}$ of the space-time. We used in [17] another start which implies the use of the greater frame $Cl_{2,3}$. A greater frame was necessary because we wanted to use no supplementary condition. Now, the study that we shall make in this chapter necessitates that we use the condition Eq. (2.90) or Eq. (2.95) which links the wave of the antiparticle to the wave of the particle in the standard model. Therefore the mathematical frame remains the space-time algebra which has 16 dimensions, enough to accommodate 8 real parameters of the wave of the electron and 8 parameters [1] of its neutrino. We saw in Sec. 3.5 that the condition Eq. (2.95) is compatible with the nonlinear equation and that it solves the puzzle of negative energies.

We begin with the electron case and we follow [38]. We change nothing to the Dirac wave of the electron, denoted as ψ_e in the Dirac formalism and

---

1. We shall see further that only four of them are nonzero.

as ϕ_e with space algebra. We use the same notations as previously for Weyl spinors. The wave of the electronic neutrino is denoted as ψ_n, the wave of the positron as ψ_p and the wave of the electronic anti-neutrino as ψ_a. As previously right spinors are ξ Weyl spinors and left ones are η spinors,

$$\psi_e = \begin{pmatrix} \xi_e \\ \eta_e \end{pmatrix} \; ; \; \psi_n = \begin{pmatrix} \xi_n \\ \eta_n \end{pmatrix} \; ; \; \psi_p = \begin{pmatrix} \xi_p \\ \eta_p \end{pmatrix} \; ; \; \psi_a = \begin{pmatrix} \xi_a \\ \eta_a \end{pmatrix}. \qquad (6.1)$$

We have

$$\phi_e = \sqrt{2} \left(\xi_e \; -i\sigma_2 \eta_e^* \right) \; ; \; \widehat{\phi}_e = \sqrt{2} \left(\eta_e \; -i\sigma_2 \xi_e^* \right), \qquad (6.2)$$

$$\phi_n = \sqrt{2} \left(\xi_n \; -i\sigma_2 \eta_n^* \right) \; ; \; \widehat{\phi}_n = \sqrt{2} \left(\eta_n \; -i\sigma_2 \xi_n^* \right), \qquad (6.3)$$

$$\widehat{\phi}_p = \widehat{\phi}_e \sigma_1 \; ; \; \widehat{\phi}_a = \widehat{\phi}_n \sigma_1, \qquad (6.4)$$

which gives

$$\widehat{\phi}_p = \sqrt{2} \left(\eta_p \; -i\sigma_2 \xi_p^* \right) \; ; \; \phi_p = \sqrt{2} \left(\xi_p \; -i\sigma_2 \eta_p^* \right), \qquad (6.5)$$

$$\widehat{\phi}_a = \sqrt{2} \left(\eta_a \; -i\sigma_2 \xi_a^* \right) \; ; \; \phi_a = \sqrt{2} \left(\xi_a \; -i\sigma_2 \eta_a^* \right), \qquad (6.6)$$

$$\xi_{1p} = \eta_{2e}^*, \;\; \xi_{2p} = -\eta_{1e}^*; \;\; \eta_{1p} = -\xi_{2e}^*; \;\; \eta_{2p} = \xi_{1e}^*$$
$$\xi_{1a} = \eta_{2n}^*, \;\; \xi_{2a} = -\eta_{1n}^*; \;\; \eta_{1a} = -\xi_{2n}^*; \;\; \eta_{2a} = \xi_{1n}^*. \qquad (6.7)$$

We used in [17] a wave Ψ function of the space-time with value into the $Cl_{2,3} = M_4(\mathbb{C})$ algebra. We placed waves of particle on the first line and waves of antiparticle on the second line to get correct transformations of left and right waves under Lorentz dilations. We used a σ_1 factor which was a necessary factor exchanging ξ and η terms. This allows us to get a wave for these four particles [2] of the electronic sector and with the link Eq. (2.95) between the wave of the particle and the wave of the antiparticle we have

$$\Psi = \begin{pmatrix} \phi_e & \phi_n \\ \widehat{\phi}_a \sigma_1 & \widehat{\phi}_p \sigma_1 \end{pmatrix} = \begin{pmatrix} \phi_e & \phi_n \\ \widehat{\phi}_n & \widehat{\phi}_e \end{pmatrix}. \qquad (6.8)$$

Now with Eq. (6.4) and Eq. (6.8) the wave is a function of space-time with value in the Clifford algebra of space-time. The Weinberg–Salam model uses ξ_e, η_e, η_n and supposes $\xi_n = 0$. This hypothesis will be used later but not immediately. To separate ξ_e, η_e and η_n the Weinberg–Salam model

---

2. We could exchange the places of ϕ_e and ϕ_n. With Eq. (6.8) the wave of the electron has value in the even sub-algebra and the neutrino has value in the odd part of the algebra. The other choice is possible if we adapt the definition of projectors in Eq. (6.12) to Eq. (6.16).

uses projectors $\frac{1}{2}(1 \pm \gamma_5)$, which read with our choice Eq. (1.75) of Dirac matrices:

$$\frac{1}{2}(1 - \gamma_5)\psi = \psi_L = \begin{pmatrix} 0 & 0 \\ 0 & I \end{pmatrix} \begin{pmatrix} \xi \\ \eta \end{pmatrix} = \begin{pmatrix} 0 \\ \eta \end{pmatrix}, \tag{6.9}$$

$$\frac{1}{2}(1 + \gamma_5)\psi = \psi_R = \begin{pmatrix} I & 0 \\ 0 & 0 \end{pmatrix} \begin{pmatrix} \xi \\ \eta \end{pmatrix} = \begin{pmatrix} \xi \\ 0 \end{pmatrix}. \tag{6.10}$$

Then for particles left waves are η waves and right waves are ξ waves. This is Cl_3^* invariant, consequently relativistic invariant, since under a Lorentz dilation D defined by $D : x \mapsto x' = MxM^\dagger$ we have Eq. (2.9): $\xi' = M\xi$, $\eta' = \widehat{M}\eta$. The γ_5 matrix is not included [3] in the space-time algebra, but this is not a problem here, because the projectors separating ξ and η are in space algebra $\frac{1}{2}(1 \pm \sigma_3)$:

$$\phi_R = \sqrt{2}\,(\xi\ 0) = \phi \begin{pmatrix} 1 & 0 \\ 0 & 0 \end{pmatrix} = \phi\frac{1}{2}(1 + \sigma_3),$$

$$\phi_L = \sqrt{2}\,(0\ -i\sigma_2\eta^*) = \phi \begin{pmatrix} 0 & 0 \\ 0 & 1 \end{pmatrix} = \phi\frac{1}{2}(1 - \sigma_3), \tag{6.11}$$

$$\widehat{\phi}_L = \sqrt{2}\,(\eta\ 0) = \widehat{\phi}\frac{1}{2}(1 + \sigma_3)\ ;\quad \widehat{\phi}_R = \widehat{\phi}\frac{1}{2}(1 - \sigma_3).$$

We define now two projectors $P_\pm$ and four operators P_0, P_1, P_2, P_3 acting in the space-time algebra as follows

$$P_\pm(\Psi) = \frac{1}{2}(\Psi \pm \mathbf{i}\Psi\gamma_{21})\ ;\quad \mathbf{i} = \gamma_{0123}, \tag{6.12}$$

$$P_0(\Psi) = \Psi\gamma_{21} + \frac{1}{2}\Psi\mathbf{i} + \frac{1}{2}\mathbf{i}\Psi\gamma_{30} = \Psi\gamma_{21} + P_-(\Psi)\mathbf{i}, \tag{6.13}$$

$$P_1(\Psi) = \frac{1}{2}(\mathbf{i}\Psi\gamma_0 + \Psi\gamma_{012}) = P_+(\Psi)\gamma_3\mathbf{i}, \tag{6.14}$$

$$P_2(\Psi) = \frac{1}{2}(\Psi\gamma_3 - \mathbf{i}\Psi\gamma_{123}) = P_+(\Psi)\gamma_3, \tag{6.15}$$

$$P_3(\Psi) = \frac{1}{2}(-\Psi\mathbf{i} + \mathbf{i}\Psi\gamma_{30}) = P_+(\Psi)(-\mathbf{i}). \tag{6.16}$$

Noting $P_\mu P_\nu(\Psi) = P_\mu[P_\nu(\Psi)]$ they satisfy

$$\begin{aligned}
P_1 P_2 &= P_3 = -P_2 P_1, \\
P_2 P_3 &= P_1 = -P_3 P_2, \\
P_3 P_1 &= P_2 = -P_1 P_3, \\
P_1^2 = P_2^2 &= P_3^2 = -P_+, \\
P_0 P_j = P_j P_0 &= -\mathbf{i}P_j\ ,\quad j = 1,\ 2,\ 3.
\end{aligned} \tag{6.17}$$

---

3. This was wrongly considered as a reason to forbid the use of space-time algebra.

The Weinberg–Salam model replaces partial derivatives ∂_μ by covariant derivatives

$$D_\mu = \partial_\mu - ig_1 \frac{Y}{2} B_\mu - ig_2 T_j W_\mu^j, \tag{6.18}$$

with $T_j = \frac{\tau_j}{2}$ for a doublet of left-handed particles and $T_j = 0$ for a singlet of right-handed particle. Y is the weak hypercharge, $Y_L = -1$, $Y_R = -2$ for the electron. To transpose into space-time algebra, we let

$$D = \sigma^\mu D_\mu \ ; \quad \mathbf{D} = \gamma^\mu D_\mu = \begin{pmatrix} 0 & D \\ \widehat{D} & 0 \end{pmatrix}, \tag{6.19}$$

$$B = \sigma^\mu B_\mu \ ; \quad \mathbf{B} = \gamma^\mu B_\mu = \begin{pmatrix} 0 & B \\ \widehat{B} & 0 \end{pmatrix}, \tag{6.20}$$

$$W^j = \sigma^\mu W_\mu^j \ ; \quad \mathbf{W}^j = \gamma^\mu W_\mu^j = \begin{pmatrix} 0 & W^j \\ \widehat{W^j} & 0 \end{pmatrix}. \tag{6.21}$$

We will prove now that Eq. (6.18) comes from

$$\mathbf{D} = \partial + \frac{g_1}{2} \mathbf{B} P_0 + \frac{g_2}{2} (\mathbf{W}^1 P_1 + \mathbf{W}^2 P_2 + \mathbf{W}^3 P_3). \tag{6.22}$$

Firstly we have in space-time algebra (see Sec. 1.4.1)

$$\partial\Psi = \begin{pmatrix} 0 & \nabla \\ \widehat{\nabla} & 0 \end{pmatrix} \begin{pmatrix} \phi_e & \phi_n \\ \widehat{\phi}_a \sigma_1 & \widehat{\phi}_p \sigma_1 \end{pmatrix} = \begin{pmatrix} \nabla\widehat{\phi}_a\sigma_1 & \nabla\widehat{\phi}_p\sigma_1 \\ \widehat{\nabla}\phi_e & \widehat{\nabla}\phi_n \end{pmatrix}, \tag{6.23}$$

while we get with Eq. (6.19)

$$\mathbf{D}\Psi = \begin{pmatrix} 0 & D \\ \widehat{D} & 0 \end{pmatrix} \begin{pmatrix} \phi_e & \phi_n \\ \widehat{\phi}_a \sigma_1 & \widehat{\phi}_p \sigma_1 \end{pmatrix} = \begin{pmatrix} D\widehat{\phi}_a\sigma_1 & D\widehat{\phi}_p\sigma_1 \\ \widehat{D}\phi_e & \widehat{D}\phi_n \end{pmatrix}. \tag{6.24}$$

To compute $P_0(\Psi)$ we use

$$P_0(\Psi) = \begin{pmatrix} p_0(\phi_e) & p_0(\phi_n) \\ p_0(\widehat{\phi}_a)\sigma_1 & p_0(\widehat{\phi}_p)\sigma_1 \end{pmatrix}. \tag{6.25}$$

And we get

$$\Psi\gamma_{21} = i \begin{pmatrix} \phi_e\sigma_3 & \phi_n\sigma_3 \\ -\widehat{\phi}_a\sigma_3\sigma_1 & -\widehat{\phi}_p\sigma_3\sigma_1 \end{pmatrix}, \tag{6.26}$$

$$\frac{1}{2}\Psi\mathbf{i} = \frac{i}{2} \begin{pmatrix} \phi_e & -\phi_n \\ \widehat{\phi}_a\sigma_1 & -\widehat{\phi}_p\sigma_1 \end{pmatrix}, \tag{6.27}$$

$$\frac{1}{2}\mathbf{i}\Psi\gamma_{30} = \frac{i}{2} \begin{pmatrix} \phi_e\sigma_3 & -\phi_n\sigma_3 \\ \widehat{\phi}_a\sigma_3\sigma_1 & -\widehat{\phi}_p\sigma_3\sigma_1 \end{pmatrix}. \tag{6.28}$$

Then we get

$$P_0(\Psi) = i \begin{pmatrix} \phi_e \frac{1+3\sigma_3}{2} & \phi_n \frac{-1+\sigma_3}{2} \\ \widehat{\phi}_a \frac{1-\sigma_3}{2}\sigma_1 & \widehat{\phi}_p \frac{-1-3\sigma_3}{2}\sigma_1 \end{pmatrix}, \qquad (6.29)$$

$$p_0(\phi_e) = i\phi_e \frac{1+3\sigma_3}{2} = i(2\phi_{eR} - \phi_{eL}), \qquad (6.30)$$

$$p_0(\phi_n) = i\phi_n \frac{-1+\sigma_3}{2} = -i\phi_{nL}, \qquad (6.31)$$

$$p_0(\widehat{\phi}_p) = i\widehat{\phi}_p \frac{-1-3\sigma_3}{2} = -i(2\widehat{\phi}_{pL} - \widehat{\phi}_{pR}), \qquad (6.32)$$

$$p_0(\widehat{\phi}_a) = i\widehat{\phi}_a \frac{1-\sigma_3}{2} = i\widehat{\phi}_{aR}, \qquad (6.33)$$

with

$$\phi_{eL} = \phi_e \frac{1-\sigma_3}{2}; \quad \phi_{eR} = \phi_e \frac{1+\sigma_3}{2}, \qquad (6.34)$$

$$\phi_{nL} = \phi_n \frac{1-\sigma_3}{2}; \quad \phi_{nR} = \phi_n \frac{1+\sigma_3}{2}, \qquad (6.35)$$

$$\widehat{\phi}_{pL} = \widehat{\phi}_p \frac{1+\sigma_3}{2}; \quad \widehat{\phi}_{pR} = \widehat{\phi}_p \frac{1-\sigma_3}{2}, \qquad (6.36)$$

$$\widehat{\phi}_{aL} = \widehat{\phi}_a \frac{1+\sigma_3}{2}; \quad \widehat{\phi}_{aR} = \widehat{\phi}_a \frac{1-\sigma_3}{2}, \qquad (6.37)$$

which gives

$$\mathbf{B}P_0(\Psi) = \begin{pmatrix} 0 & B \\ \widehat{B} & 0 \end{pmatrix} \begin{pmatrix} p_0(\phi_e) & p_0(\phi_n) \\ p_0(\widehat{\phi}_a)\sigma_1 & p_0(\widehat{\phi}_p)\sigma_1 \end{pmatrix}$$

$$= i \begin{pmatrix} B\widehat{\phi}_{aR}\sigma_1 & -B(2\widehat{\phi}_{pL} - \widehat{\phi}_{pR})\sigma_1 \\ \widehat{B}(2\phi_{eR} - \phi_{eL}) & -\widehat{B}\phi_{nL} \end{pmatrix}. \qquad (6.38)$$

Next we let

$$P_j(\Psi) = \begin{pmatrix} p_j(\phi_e) & p_j(\phi_n) \\ p_j(\widehat{\phi}_a)\sigma_1 & p_j(\widehat{\phi}_p)\sigma_1 \end{pmatrix}, \quad j = 1, 2, 3. \qquad (6.39)$$

We get for $j = 1$

$$i\Psi\gamma_0 = i \begin{pmatrix} \phi_n & \phi_e \\ -\widehat{\phi}_p\sigma_1 & -\widehat{\phi}_a\sigma_1 \end{pmatrix}; \quad \Psi\gamma_{012} = i \begin{pmatrix} -\phi_n\sigma_3 & -\phi_e\sigma_3 \\ \widehat{\phi}_p\sigma_3\sigma_1 & \widehat{\phi}_a\sigma_3\sigma_1 \end{pmatrix},$$

$$P_1(\Psi) = i \begin{pmatrix} \phi_n \frac{1-\sigma_3}{2} & \phi_e \frac{1-\sigma_3}{2} \\ \widehat{\phi}_p \frac{-1+\sigma_3}{2}\sigma_1 & \widehat{\phi}_a \frac{-1+\sigma_3}{2}\sigma_1 \end{pmatrix} = i \begin{pmatrix} \phi_{nL} & \phi_{eL} \\ -\widehat{\phi}_{pR}\sigma_1 & -\widehat{\phi}_{aR}\sigma_1 \end{pmatrix}, \qquad (6.40)$$

$$p_1(\phi_e) = i\phi_{nL} \; ; \; p_1(\phi_n) = i\phi_{eL},$$

$$p_1(\widehat{\phi}_a) = -i\widehat{\phi}_{pR} \; ; \; p_1(\widehat{\phi}_p) = -i\widehat{\phi}_{aR}. \qquad (6.41)$$

We get for $j = 2$

$$\Psi\gamma_3 = \begin{pmatrix} -\phi_n\sigma_3 & \phi_e\sigma_3 \\ \widehat{\phi}_p\sigma_3\sigma_1 & -\widehat{\phi}_a\sigma_3\sigma_1 \end{pmatrix} \ ; \ -\mathbf{i}\Psi\gamma_{123} = \begin{pmatrix} \phi_n & -\phi_e \\ -\widehat{\phi}_p\sigma_1 & \widehat{\phi}_a\sigma_1 \end{pmatrix},$$

$$P_2(\Psi) = \begin{pmatrix} \phi_n\frac{1-\sigma_3}{2} & \phi_e\frac{-1+\sigma_3}{2} \\ \widehat{\phi}_p\frac{-1+\sigma_3}{2}\sigma_1 & \widehat{\phi}_a\frac{1-\sigma_3}{2}\sigma_1 \end{pmatrix} = \begin{pmatrix} \phi_{nL} & -\phi_{eL} \\ -\widehat{\phi}_{pR}\sigma_1 & \widehat{\phi}_{aR}\sigma_1 \end{pmatrix}, \qquad (6.42)$$

$$p_2(\phi_e) = \phi_{nL} \ ; \ p_2(\phi_n) = -\phi_{eL},$$

$$p_2(\widehat{\phi}_a) = -\widehat{\phi}_{pR} \ ; \ p_2(\widehat{\phi}_p) = \widehat{\phi}_{aR}. \qquad (6.43)$$

We get for $j = 3$

$$-\Psi\mathbf{i} = i\begin{pmatrix} -\phi_e & \phi_n \\ -\widehat{\phi}_a\sigma_1 & \widehat{\phi}_p\sigma_1 \end{pmatrix} \ ; \ \mathbf{i}\Psi\gamma_{30} = i\begin{pmatrix} \phi_e\sigma_3 & -\phi_n\sigma_3 \\ \widehat{\phi}_a\sigma_3\sigma_1 & -\widehat{\phi}_p\sigma_3\sigma_1 \end{pmatrix},$$

$$P_3(\Psi) = i\begin{pmatrix} \phi_e\frac{-1+\sigma_3}{2} & \phi_n\frac{1-\sigma_3}{2} \\ \widehat{\phi}_a\frac{-1+\sigma_3}{2}\sigma_1 & \widehat{\phi}_p\frac{1-\sigma_3}{2}\sigma_1 \end{pmatrix} = i\begin{pmatrix} -\phi_{eL} & \phi_{nL} \\ -\widehat{\phi}_{aR}\sigma_1 & \widehat{\phi}_{pR}\sigma_1 \end{pmatrix}, \qquad (6.44)$$

$$p_3(\phi_e) = -i\phi_{eL} \ ; \ p_3(\phi_n) = i\phi_{nL},$$

$$p_3(\widehat{\phi}_a) = -i\widehat{\phi}_{aR} \ ; \ p_3(\widehat{\phi}_p) = i\widehat{\phi}_{pR}. \qquad (6.45)$$

We also have

$$\mathbf{W}^j P_j(\Psi) = \begin{pmatrix} 0 & W^j \\ \widehat{W}^j & 0 \end{pmatrix} \begin{pmatrix} p_j(\phi_e) & p_j(\phi_n) \\ p_j(\widehat{\phi}_a)\sigma_1 & p_j(\widehat{\phi}_p)\sigma_1 \end{pmatrix}$$

$$= \begin{pmatrix} W^j p_j(\widehat{\phi}_a)\sigma_1 & W^j p_j(\widehat{\phi}_p)\sigma_1 \\ \widehat{W}^j p_j(\phi_e) & \widehat{W}^j p_j(\phi_n) \end{pmatrix}. \qquad (6.46)$$

Therefore Eq. (6.22) gives the system

$$D\widehat{\phi}_a = \nabla\widehat{\phi}_a + \frac{g_1}{2}Bp_0(\widehat{\phi}_a) + \frac{g_2}{2}W^j p_j(\widehat{\phi}_a), \qquad (6.47)$$

$$D\widehat{\phi}_p = \nabla\widehat{\phi}_p + \frac{g_1}{2}Bp_0(\widehat{\phi}_p) + \frac{g_2}{2}W^j p_j(\widehat{\phi}_p), \qquad (6.48)$$

$$\widehat{D}\phi_e = \widehat{\nabla}\phi_e + \frac{g_1}{2}\widehat{B}p_0(\phi_e) + \frac{g_2}{2}\widehat{W}^j p_j(\phi_e), \qquad (6.49)$$

$$\widehat{D}\phi_n = \widehat{\nabla}\phi_n + \frac{g_1}{2}\widehat{B}p_0(\phi_n) + \frac{g_2}{2}\widehat{W}^j p_j(\phi_n). \qquad (6.50)$$

With Eq. (6.30) to Eq. (6.33), Eq. (6.41), Eq. (6.43) and Eq. (6.45) this gives

$$D\widehat{\phi}_a = \nabla\widehat{\phi}_a + i\frac{g_1}{2}B\widehat{\phi}_{aR} + \frac{g_2}{2}[(-iW^1 - W^2)\widehat{\phi}_{pR} - iW^3\widehat{\phi}_{aR}], \qquad (6.51)$$

$$D\widehat{\phi}_p = \nabla\widehat{\phi}_p + i\frac{g_1}{2}B(-2\widehat{\phi}_{pL} + \widehat{\phi}_{pR}) + \frac{g_2}{2}[(-iW^1 + W^2)\widehat{\phi}_{aR} + iW^3\widehat{\phi}_{pR}], \qquad (6.52)$$

$$\widehat{D}\phi_e = \widehat{\nabla}\phi_e + i\frac{g_1}{2}\widehat{B}(2\phi_{eR} - \phi_{eL}) + \frac{g_2}{2}[(i\widehat{W}^1 + \widehat{W}^2)\phi_{nL} - i\widehat{W}^3\phi_{eL}], \qquad (6.53)$$

$$\widehat{D}\phi_n = \widehat{\nabla}\phi_n - i\frac{g_1}{2}\widehat{B}\phi_{nL} + \frac{g_2}{2}[(i\widehat{W}^1 - \widehat{W}^2)\phi_{eL} + i\widehat{W}^3\phi_{nL}]. \qquad (6.54)$$

Using the conjugation $M \mapsto \widehat{M}$ in Eq. (6.53) and Eq. (6.54) gives

$$D\widehat{\phi}_a = \nabla\widehat{\phi}_a + i\frac{g_1}{2}B\widehat{\phi}_{aR} + i\frac{g_2}{2}[(-W^1 + iW^2)\widehat{\phi}_{pR} - W^3\widehat{\phi}_{aR}], \qquad (6.55)$$

$$D\widehat{\phi}_p = \nabla\widehat{\phi}_p + i\frac{g_1}{2}B(-2\widehat{\phi}_{pL} + \widehat{\phi}_{pR}) + i\frac{g_2}{2}[-(W^1 + iW^2)\widehat{\phi}_{aR} + W^3\widehat{\phi}_{pR}], \qquad (6.56)$$

$$D\widehat{\phi}_e = \nabla\widehat{\phi}_e + i\frac{g_1}{2}B(-2\widehat{\phi}_{eR} + \widehat{\phi}_{eL}) + i\frac{g_2}{2}[-(W^1 + iW^2)\widehat{\phi}_{nL} + W^3\widehat{\phi}_{eL}], \qquad (6.57)$$

$$D\widehat{\phi}_n = \nabla\widehat{\phi}_n + i\frac{g_1}{2}B\widehat{\phi}_{nL} + i\frac{g_2}{2}[(-W^1 + iW^2)\widehat{\phi}_{eL} - W^3\widehat{\phi}_{nL}]. \qquad (6.58)$$

We study firstly the case of the electron and its neutrino. We have with Eq. (6.34)

$$\widehat{\phi}_{eL} = \widehat{\phi}_e\frac{1+\sigma_3}{2} \; ; \; \widehat{\phi}_{eL}\sigma_3 = \widehat{\phi}_{eL}, \qquad (6.59)$$

$$\widehat{\phi}_{eR} = \widehat{\phi}_e\frac{1-\sigma_3}{2} \; ; \; \widehat{\phi}_{eR}\sigma_3 = -\widehat{\phi}_{eR}, \qquad (6.60)$$

$$-2\widehat{\phi}_{eR} + 2\widehat{\phi}_{eL} = 2(\widehat{\phi}_{eR} + \widehat{\phi}_{eL})\sigma_3 = 2\widehat{\phi}_e\sigma_3, \qquad (6.61)$$

and we get for Eq. (6.57) and Eq. (6.58)

$$D\widehat{\phi}_e = \nabla\widehat{\phi}_e + g_1B\widehat{\phi}_e i\sigma_3 + \frac{i}{2}(-g_1B + g_2W^3)\widehat{\phi}_{eL} - i\frac{g_2}{2}(W^1 + iW^2)\widehat{\phi}_{nL}, \qquad (6.62)$$

$$D\widehat{\phi}_n = \nabla\widehat{\phi}_n - \frac{i}{2}(-g_1B + g_2W^3)\widehat{\phi}_{nL} + i\frac{g_2}{2}(-W^1 + iW^2)\widehat{\phi}_{eL}. \qquad (6.63)$$

We separate left and right parts of the wave:

$$D\widehat{\phi}_{nR} = \nabla\widehat{\phi}_{nR} \; ; \; \widehat{D}\phi_{nR} = \widehat{\nabla}\phi_{nR}, \qquad (6.64)$$

$$D\widehat{\phi}_{nL} = \nabla\widehat{\phi}_{nL} + \frac{i}{2}(g_1B - g_2W^3)\widehat{\phi}_{nL} + i\frac{g_2}{2}(-W^1 + iW^2)\widehat{\phi}_{eL}, \qquad (6.65)$$

$$D\widehat{\phi}_{eR} = \nabla\widehat{\phi}_{eR} - ig_1B\widehat{\phi}_{eR} \; ; \; \widehat{D}\phi_{eR} = \widehat{\nabla}\phi_{eR} + ig_1\widehat{B}\phi_{eR}, \qquad (6.66)$$

$$D\widehat{\phi}_{eL} = \nabla\widehat{\phi}_{eL} + \frac{i}{2}(g_1B + g_2W^3)\widehat{\phi}_{eL} - i\frac{g_2}{2}(W^1 + iW^2)\widehat{\phi}_{nL}. \qquad (6.67)$$

which is equivalent [4] to

$$D_\mu \xi_n = \partial_\mu \xi_n, \tag{6.68}$$

$$D_\mu \eta_n = \partial_\mu \eta_n + i\frac{g_1}{2}B_\mu \eta_n - i\frac{g_2}{2}[(W_\mu^1 - iW_\mu^2)\eta_e + W_\mu^3 \eta_n], \tag{6.69}$$

$$D_\mu \xi_e = \partial_\mu \xi_e + ig_1 B_\mu \xi_e, \tag{6.70}$$

$$D_\mu \eta_e = \partial_\mu \eta_e + i\frac{g_1}{2}B_\mu \eta_e - i\frac{g_2}{2}[(W_\mu^1 + iW_\mu^2)\eta_n - W_\mu^3 \eta_e]. \tag{6.71}$$

Equation (6.69) and Eq. (6.71) give for the "lepton doublet" $\psi_L = \begin{pmatrix} \eta_n \\ \eta_e \end{pmatrix}$ with weak isospin $Y = -1$:

$$D_\mu \psi_L = \partial_\mu \psi_L - ig_1 \frac{Y}{2}B_\mu \psi_L - i\frac{g_2}{2}W_\mu^j \tau_j \psi_L,$$

$$\tau_1 = \gamma_0 \; ; \; \tau_2 = \gamma_{123} \; ; \; \tau_3 = \gamma_5. \tag{6.72}$$

With Eq. (6.68) we see that the right part of the wave of the neutrino does not interact. Therefore we can suppose $\xi_n = 0$. Equation (6.70) is interpreted as a SU(2) singlet $\psi_R = \xi$ with weak isospin $Y = -2$:

$$D_\mu \psi_R = \partial_\mu \psi_R - ig_1 \frac{Y}{2}B_\mu \psi_R. \tag{6.73}$$

Finally we see here that all features of weak interactions, with a doublet of left waves, a singlet of right wave, a non-interacting right neutrino, and a charge conjugation exchanging right and left waves are obtained here from very simple hypotheses:

1 – The wave of all components of the lepton sector, electron, positron, electronic neutrino and anti-neutrino, is the function Eq. (6.8) of space-time with value into the Clifford algebra of the space-time.

2 – Four operators P_0, P_1, P_2, P_3 are defined by Eq. (6.12) to Eq. (6.16).

3 – A covariant derivative that is defined by Eq. (6.22).

It is now easy to use the system Eq. (6.55) to Eq. (6.58) to get all other features of the Weinberg–Salam model. It considers the "charged currents" W^+ and W^- defined by

$$W_\mu^+ = W_\mu^1 + iW_\mu^2 \; ; \; W_\mu^- = -W_\mu^1 + iW_\mu^2,$$

$$W^+ = W^1 + iW^2 \; ; \; W^- = -W^1 + iW^2, \tag{6.74}$$

---

4. Since $\phi_{eR} = \sqrt{2}(\xi_e \; 0)$ we must use the second equality Eq. (6.66) to get Eq. (6.70).

where $i = \sigma_{123}$ is the generator of the chiral gauge [5] not the i_3 of the electric gauge. We will use Eq. (6.61) and similarly

$$\widehat{\phi}_{pR} = \widehat{\phi}_p \frac{1 - \sigma_3}{2} \ ; \ \widehat{\phi}_{pR}\sigma_3 = -\widehat{\phi}_{pR}, \tag{6.75}$$

$$\widehat{\phi}_{pL} = \widehat{\phi}_p \frac{1 + \sigma_3}{2} \ ; \ \widehat{\phi}_{pL}\sigma_3 = \widehat{\phi}_{pL}, \tag{6.76}$$

$$\widehat{\phi}_{pL} - \widehat{\phi}_{pR} = \widehat{\phi}_{pL}\sigma_3 + \widehat{\phi}_{pR}\sigma_3 = (\widehat{\phi}_{pL} + \widehat{\phi}_{pR})\sigma_3 = \widehat{\phi}_p\sigma_3. \tag{6.77}$$

Then Eq. (6.55) to Eq. (6.58) reads

$$D\widehat{\phi}_a = \nabla\widehat{\phi}_a + \frac{i}{2}(g_1 B - g_2 W^3)\widehat{\phi}_{aR} + i\frac{g_2}{2}W^-\widehat{\phi}_{pR}, \tag{6.78}$$

$$D\widehat{\phi}_p = \nabla\widehat{\phi}_p + ig_1 B(\widehat{\phi}_{pR} - \widehat{\phi}_{pL}) + \frac{i}{2}(-g_1 B + g_2 W^3)\widehat{\phi}_{pR} - i\frac{g_2}{2}W^+\widehat{\phi}_{aR}, \tag{6.79}$$

$$D\widehat{\phi}_n = \nabla\widehat{\phi}_n + \frac{i}{2}(g_1 B - g_2 W^3)\widehat{\phi}_{nL} + \frac{i}{2}g_2 W^-\widehat{\phi}_{eL}, \tag{6.80}$$

$$D\widehat{\phi}_e = \nabla\widehat{\phi}_e + ig_1 B(\widehat{\phi}_{eL} - \widehat{\phi}_{eR}) + \frac{i}{2}(-g_1 B + g_2 W^3)\widehat{\phi}_{eL} - \frac{i}{2}g_2 W^+\widehat{\phi}_{nL}. \tag{6.81}$$

The Weinberg–Salam model uses the electromagnetic potential A, a θ_W angle and a Z^0 term [6] satisfying

$$g_1 = \frac{q}{\cos(\theta_W)} \ ; \ g_2 = \frac{q}{\sin(\theta_W)} \ ; \ q = \frac{e}{\hbar c}, \tag{6.82}$$

$$-g_1 B + g_2 W^3 = \sqrt{g_1^2 + g_2^2}Z^0 = \frac{2q}{\sin(2\theta_W)}Z^0, \tag{6.83}$$

$$B = \cos(\theta_W)A - \sin(\theta_W)Z^0 \ ; \ W^3 = \sin(\theta_W)A + \cos(\theta_W)Z^0, \tag{6.84}$$

$$B + iW^3 = e^{i\theta_W}(A + iZ^0) \ ; \ A + iZ^0 = e^{-i\theta_W}(B + iW^3). \tag{6.85}$$

---

5. This is another sufficient reason to abandon the formalism of Dirac matrices, which uses a unique i. It is therefore unable to discriminate between the different gauges at work.

6. Equation (6.85) indicates that Z^0 is similar to Cabibbo–Ferrari's B of Eq. (4.36).

Using Eq. (6.77) this gives for the system Eq. (6.78) to Eq. (6.81)

$$D\widehat{\phi}_a = \nabla\widehat{\phi}_a - \frac{iq}{\sin(2\theta_W)}Z^0\widehat{\phi}_{aR} + i\frac{g_2}{2}W^-\widehat{\phi}_{pR}, \tag{6.86}$$

$$D\widehat{\phi}_p = \nabla\widehat{\phi}_p - qA\widehat{\phi}_p\sigma_{12}$$
$$+ q\tan(\theta_W)Z^0\widehat{\phi}_p\sigma_{12} + i\frac{q}{\sin(2\theta_W)}Z^0\widehat{\phi}_{pR} - i\frac{g_2}{2}W^+\widehat{\phi}_{aR}, \tag{6.87}$$

$$D\widehat{\phi}_e = \nabla\widehat{\phi}_e + qA\widehat{\phi}_e\sigma_{12}$$
$$- q\tan(\theta_W)Z^0\widehat{\phi}_e\sigma_{12} + i\frac{q}{\sin(2\theta_W)}Z^0\widehat{\phi}_{eL} - i\frac{g_2}{2}W^+\widehat{\phi}_{nL}, \tag{6.88}$$

$$D\widehat{\phi}_n = \nabla\widehat{\phi}_n - \frac{iq}{\sin(2\theta_W)}Z^0\widehat{\phi}_{nL} + i\frac{g_2}{2}W^-\widehat{\phi}_{eL}. \tag{6.89}$$

Equation (6.88) contains the first and second terms $\nabla\widehat{\phi} + qA\widehat{\phi}\sigma_{12}$ of the Dirac equation, giving the electromagnetic interaction of the electron. Equation (6.87) contains the first and second terms $-\nabla\widehat{\phi} + qA\widehat{\phi}\sigma_{12}$ of the Dirac equation for a positron. There is no potential A term in Eq. (6.86) nor Eq. (6.89), since anti-neutrinos and neutrinos have no electromagnetic interaction. Since we have

$$\widehat{\phi}_e\sigma_{12} = i(-\widehat{\phi}_{eR} + \widehat{\phi}_{eL}), \tag{6.90}$$

we can read Eq. (6.89) and Eq. (6.88) as

$$D\widehat{\phi}_{nR} = \nabla\widehat{\phi}_{nR}, \tag{6.91}$$

$$D\widehat{\phi}_{nL} = \nabla\widehat{\phi}_{nL} - i\frac{q}{\sin(2\theta_W)}Z^0\widehat{\phi}_{nL} + i\frac{q}{2\sin(\theta_W)}W^-\widehat{\phi}_{eL}, \tag{6.92}$$

$$D\widehat{\phi}_{eR} = \nabla\widehat{\phi}_{eR} + qA\widehat{\phi}_{eR}\sigma_{12} + iq\tan(\theta_W)Z^0\widehat{\phi}_{eR}, \tag{6.93}$$

$$D\widehat{\phi}_{eL} = \nabla\widehat{\phi}_{eL} + qA\widehat{\phi}_{eL}\sigma_{12},$$
$$+ iq[-\tan(\theta_W) + \frac{1}{\sin(2\theta_W)}]Z^0\widehat{\phi}_{eL} - i\frac{q}{2\sin(\theta_W)}W^+\widehat{\phi}_{nL}. \tag{6.94}$$

Terms containing W^+ and W^- which couple left electrons to left neutrinos generate "charged currents"; terms containing Z^0 generate "neutral currents". The Z^0 boson is linked to ϕ_L, ϕ_{nL} and ϕ_R, not to ϕ_{nR}. Similarly we can read Eq. (6.86) and Eq. (6.87) as

$$D\widehat{\phi}_{aL} = \nabla\widehat{\phi}_{aL}, \tag{6.95}$$

$$D\widehat{\phi}_{aR} = \nabla\widehat{\phi}_{aR} - i\frac{q}{\sin(2\theta_W)}Z^0\widehat{\phi}_{aR} + i\frac{q}{2\sin(\theta_W)}W^-\widehat{\phi}_{pR}, \tag{6.96}$$

$$D\widehat{\phi}_{pL} = \nabla\widehat{\phi}_{pL} - qA\widehat{\phi}_{pL}\sigma_{12} + iq\tan(\theta_W)Z^0\widehat{\phi}_{pL}, \tag{6.97}$$

$$D\widehat{\phi}_{pR} = \nabla\widehat{\phi}_{pR} - qA\widehat{\phi}_{pR}\sigma_{12}$$
$$+ iq[-\tan(\theta_W) + \frac{1}{\sin(2\theta_W)}]Z^0\widehat{\phi}_{pR} - i\frac{q}{2\sin(\theta_W)}W^+\widehat{\phi}_{aR}. \tag{6.98}$$

Equation (6.95) signifies that the left anti-neutrino does not interact by electro-weak forces. The electric charge of the positron is opposite to the charge of the electron. But the comparison with the same relation for the electron shows that, contrary to what is said about charge conjugation, thought as changing the sign of any quantum number, only the exchange between left and right waves plus the multiplication on the right by σ_3 give a change of sign. Other coefficients are conserved when passing from electron to positron or from neutrino to anti-neutrino. Charge conjugation must be seen as a pure quantum transformation acting only on the wave, as described [7] in Sec. 3.5. A similar result was obtained by G. Lochak [48] for the magnetic monopole: charge conjugation does not change the sign of magnetic charges, and there is no polarization of the vacuum resulting from spontaneous creation of pairs. It is the same for neutrinos, there is no creation of pairs of neutrino–anti-neutrino similar to the creation of pairs of particle–antiparticle with opposite [8] electric charges.

6.2 Invariances

As with electromagnetism, we can enlarge the relativistic invariance to the greater group Cl_3^*. With the Lorentz dilation R defined by a M element in Cl_3^* satisfying $x \mapsto x' = MxM^\dagger$ we have

$$\phi'_e = M\phi_e \; ; \; \phi'_n = M\phi_n \; ; \; \widehat{\phi}'_p = \widehat{M}\widehat{\phi}_p \; ; \; \widehat{\phi}'_a = \widehat{M}\widehat{\phi}_a \; ; \; \Psi' = N\Psi$$

$$N = \begin{pmatrix} M & 0 \\ 0 & \widehat{M} \end{pmatrix} \; ; \; \widetilde{N} = \begin{pmatrix} \overline{M} & 0 \\ 0 & M^\dagger \end{pmatrix}. \tag{6.99}$$

We may consider $g_1 B$ and $g_2 W^j$, linked to qA, as covariants vectors:

$$g_1 B = \overline{M} g'_1 B' \widehat{M} \; ; \; g_2 W^j = \overline{M} g'_2 W^{j'} \widehat{M},$$

$$g_1 \mathbf{B} = \widetilde{N} g'_1 \mathbf{B}' N \; ; \; g_2 \mathbf{W}^j = \widetilde{N} g'_2 \mathbf{W}^{j'} N. \tag{6.100}$$

This allows D to be a covariant vector, varying as ∇:

$$D = \overline{M} D' \widehat{M} \; ; \; \nabla = \overline{M} \nabla' \widehat{M}$$

$$\mathbf{D} = \widetilde{N} \mathbf{D}' N. \tag{6.101}$$

That also gives for the Weinberg–Salam angle

$$B' + iW'^3 = e^{i\theta_W}(A' + iZ'^0), \tag{6.102}$$

---

7. Furthermore, if we try to build a charge conjugation by changing other signs, we get instead of Eq. (6.17) relations which do not give a $U(1) \times SU(2)$ gauge invariance.

8. This is also consistent with Eq. (4.76) and Eq. (4.77) where charge conjugation in the neutrino case gives the same wave equation.

which means that the θ_W angle is Cl_3^* invariant and therefore is a relativistic invariant. We get

$$D\widehat{\phi}_e = \overline{M}D'\widehat{\phi}_e' \ ; \quad D\widehat{\phi}_n = \overline{M}D'\widehat{\phi}_n' \tag{6.103}$$

$$D\widehat{\phi}_p = \overline{M}D'\widehat{\phi}_p' \ ; \quad D\widehat{\phi}_a = \overline{M}D'\widehat{\phi}_a' \tag{6.104}$$

$$\mathbf{D}\Psi = \widetilde{N}\mathbf{D}'\Psi', \tag{6.105}$$

and the Cl_3^* invariance of electro-weak interactions is completely similar to the invariance of electromagnetism.

Operators P_0, P_1, P_2 and P_3 are built from projectors and have no inverse. They are not directly elements of a gauge group. Nevertheless we can build a Yang–Mills gauge group by using the exponential function. With four real numbers a^0, a^1, a^2, a^3, we define

$$\exp(a^0 P_0) = \sum_{n=0}^{\infty} \frac{(a^0 P_0)^n}{n!}, \tag{6.106}$$

$$\exp(a^j P_j) = \sum_{n=0}^{\infty} \frac{(a^1 P_1 + a^2 P_2 + a^3 P_3)^n}{n!}. \tag{6.107}$$

We get with Eq. (6.25) and Eq. (6.30) to Eq. (6.33)

$$\exp(a^0 P_0)(\Psi) = \begin{pmatrix} \exp(a^0 p_0)(\phi_e) & \exp(a^0 p_0)(\phi_n), \\ \exp(a^0 p_0)(\widehat{\phi}_a)\sigma_1 & \exp(a^0 p_0)(\widehat{\phi}_p)\sigma_1 \end{pmatrix}, \tag{6.108}$$

$$\exp(a^0 p_0)(\phi_{nL}) = e^{-ia^0}\phi_{nL} \ ; \quad \exp(a^0 p_0)(\phi_{nR}) = \phi_{nR}, \tag{6.109}$$

$$\exp(a^0 p_0)(\phi_{eL}) = e^{-ia^0}\phi_{eL} \ ; \quad \exp(a^0 p_0)(\phi_{eR}) = e^{2ia^0}\phi_{eR}, \tag{6.110}$$

$$\exp(a^0 p_0)(\phi_{aR}) = e^{ia^0}\phi_{aR} \ ; \quad \exp(a^0 p_0)(\phi_{aL}) = \phi_{aL}, \tag{6.111}$$

$$\exp(a^0 p_0)(\phi_{pR}) = e^{ia^0}\phi_{pR} \ ; \quad \exp(a_0 p_0)(\phi_{pL}) = e^{-2ia^0}\phi_{pL}, \tag{6.112}$$

$$\exp(-a^0 P_0) = [\exp(a^0 P_0)]^{-1}. \tag{6.113}$$

Next we let

$$a = \sqrt{(a^1)^2 + (a^2)^2 + (a^3)^2} \ ; \quad S = a^j P_j, \tag{6.114}$$

and we get

$$[\exp(S)](\Psi) = \Psi + [-1 + \cos(a)]P_+(\Psi) + \frac{\sin(a)}{a}S(\Psi), \tag{6.115}$$

$$[\exp(-S)](\Psi) = \Psi + [-1 + \cos(a)]P_+(\Psi) - \frac{\sin(a)}{a}S(\Psi), \tag{6.116}$$

which gives

$$\exp(-S) = [\exp(S)]^{-1}. \tag{6.117}$$

Since P_0 commutes with S (see Eq. (6.16)) we get

$$\exp(a^0 P_0 + S) = \exp(a^0 P_0) \exp(S) = \exp(S) \exp(a^0 P_0). \tag{6.118}$$

The set of the operators $\exp(a^0 P_0 + S)$ is a $U(1) \times SU(2)$ Lie group. The local gauge invariance under this group comes from the derivation of products. If we use

$$\Psi' = [\exp(a^0 P_0 + S)](\Psi) \; ; \; \mathbf{D} = \gamma^\mu D_\mu, \tag{6.119}$$

then $D_\mu \Psi$ is replaced by $D'_\mu \Psi'$ where

$$D'_\mu \Psi' = \exp(a^0 P_0 + S) D_\mu \Psi, \tag{6.120}$$

$$B'_\mu = B_\mu - \frac{2}{g_1} \partial_\mu a^0, \tag{6.121}$$

$$W'^j_\mu P_j = \left[\exp(S) W^j_\mu P_j - \frac{2}{g_2} \partial_\mu [\exp(S)] \right] \exp(-S). \tag{6.122}$$

6.3 Geometry linked to the wave in space-time algebra

We saw in Sec. 3.3 that the wave of the electron defines at each point of space-time a geometric transformation Eq. (3.44) from the tangent space-time of an intrinsic manifold into the tangent space-time to our space-time manifold. What does this transformation become when we consider the wave Ψ_l of an electron-neutrino pair, or the complete wave Ψ of the first generation? Any element M in Cl_3 is sum of a scalar s, a vector $\vec{v}$, a bivector $i\vec{w}$ and a pseudo-scalar ip. We have

$$M = s + \vec{v} + i\vec{w} + ip; \; \widehat{M} = s - \vec{v} + i\vec{w} - ip,$$
$$M^\dagger = s + \vec{v} - i\vec{w} - ip; \; \overline{M} = s - \vec{v} - i\vec{w} + ip. \tag{6.123}$$

With the matrix representation of the space-time algebra studied in Sec. 1.4.1 and the N in Eq. (1.80) we associate to $x = x^\mu \sigma_\mu$ in Cl_3 the space-time vector

$$\mathbf{x} = x^\mu \gamma_\mu = \begin{pmatrix} 0 & x \\ \hat{x} & 0 \end{pmatrix}. \tag{6.124}$$

Then the dilation R defined by Eq. (1.42) associates to the space-time vector $\mathbf{x}$ the space-time vector $\mathbf{x}'$ satisfying

$$\mathbf{x}' = N\mathbf{x}\widetilde{N}, \tag{6.125}$$

while the differential operator $\boldsymbol{\partial} = \gamma^\mu \partial_\mu$ satisfies

$$\boldsymbol{\partial} = \widetilde{N}\boldsymbol{\partial}' N. \tag{6.126}$$

Next the dilation D defined by Eq. (3.44) associates to the space-time vector $\mathbf{y}$, element of the tangent space-time to the intrinsic manifold linked to the wave, a space-time vector $\mathbf{x}$ in the usual space-time, satisfying

$$\mathbf{x} = \Psi \mathbf{y} \widetilde{\Psi}; \ \Psi = \begin{pmatrix} \phi & 0 \\ 0 & \widehat{\phi} \end{pmatrix}; \ \widetilde{\Psi} = \begin{pmatrix} \overline{\phi} & 0 \\ 0 & \phi^\dagger \end{pmatrix}; \ \mathbf{y} = y^\mu \gamma_\mu = \begin{pmatrix} 0 & y \\ \widehat{y} & 0 \end{pmatrix}. \quad (6.127)$$

Now we consider the wave of the lepton case Ψ_l which reads

$$\Psi_l = \begin{pmatrix} \phi_e & \phi_n \\ \widehat{\phi}_n & \widehat{\phi}_e \end{pmatrix}; \ \widetilde{\Psi}_l = \begin{pmatrix} \overline{\phi}_e & \phi_n^\dagger \\ \overline{\phi}_n & \phi_e^\dagger \end{pmatrix}. \quad (6.128)$$

The generalization of Eq. (6.127) is

$$\mathbf{x} = \Psi_l \mathbf{y} \widetilde{\Psi}_l. \quad (6.129)$$

But, since

$$\widetilde{\mathbf{x}} = \Psi_l \mathbf{y} \widetilde{\Psi}_l = \mathbf{x}, \quad (6.130)$$

then $\mathbf{x}$ is the sum [9] of a scalar, a vector and a pseudo-scalar. To get only a vector, we must separate the vector part. Denoting the vector part of the multivector M as $\langle M \rangle_1$, we then let instead of Eq. (6.129)

$$\mathbf{x} = \langle \Psi_l \mathbf{y} \widetilde{\Psi}_l \rangle_1. \quad (6.131)$$

We have

$$\Psi_l \mathbf{y} \widetilde{\Psi}_l = \begin{pmatrix} \phi_e & \phi_n \\ \widehat{\phi}_n & \widehat{\phi}_e \end{pmatrix} \begin{pmatrix} 0 & y \\ \widehat{y} & 0 \end{pmatrix} \begin{pmatrix} \overline{\phi}_e & \phi_n^\dagger \\ \overline{\phi}_n & \phi_e^\dagger \end{pmatrix}$$

$$= \begin{pmatrix} \phi_n \widehat{y} \overline{\phi}_e + \phi_e y \overline{\phi}_n & \phi_e y \phi_e^\dagger + \phi_n \widehat{y} \phi_n^\dagger \\ \widehat{\phi}_e \widehat{y} \overline{\phi}_e + \widehat{\phi}_n y \overline{\phi}_n & \widehat{\phi}_n y \phi_e^\dagger + \widehat{\phi}_e \widehat{y} \phi_n^\dagger \end{pmatrix}, \quad (6.132)$$

which gives

$$\mathbf{x} = \langle \Psi_l \mathbf{y} \widetilde{\Psi}_l \rangle_1 = \begin{pmatrix} 0 & x \\ \widehat{x} & 0 \end{pmatrix}; \ x = \phi_e y \phi_e^\dagger + \phi_n \widehat{y} \phi_n^\dagger. \quad (6.133)$$

We let

$$D = D_e + D_n \ ; \ D_e(y) = \phi_e y \phi_e^\dagger \ ; \ D_n(y) = \phi_n \widehat{y} \phi_n^\dagger. \quad (6.134)$$

D_e is a direct dilation, conserving the orientation of time and space. D_n is an inverse dilation, conserving the orientation of time and changing the orientation of space. The geometric transformation $D : y \mapsto x$ is the sum of these two dilations.

---

9. The same property in Cl_3 proves that x is the sum of a scalar and a vector and this is exact for a space-time vector.

The element **y** is independent of the relative observer: if M is any element of Cl_3^* and N is given by Eq. (1.80), the dilation R defined in Eq. (1.42) satisfies

$$x' = MxM^\dagger \; ; \; \phi'_e = M\phi_e \; ; \; \phi'_n = M\phi_n, \tag{6.135}$$

$$\Psi'_l = \begin{pmatrix} \phi'_e & \phi'_n \\ \widehat{\phi}'_n & \widehat{\phi}'_e \end{pmatrix} = \begin{pmatrix} M\phi_e & M\phi_n \\ \widehat{M\phi_n} & \widehat{M\phi_e} \end{pmatrix} = N\Psi_l, \tag{6.136}$$

$$x' = \begin{pmatrix} 0 & x' \\ \widehat{x}' & 0 \end{pmatrix} = Nx\widetilde{N} = N\langle \Psi_l \mathbf{y} \widetilde{\Psi}_l \rangle_1 \widetilde{N} = \langle N\Psi_l \mathbf{y} \widetilde{\Psi}_l \widetilde{N} \rangle_1$$

$$= \langle \Psi'_l \mathbf{y} \widetilde{\Psi}'_l \rangle_1. \tag{6.137}$$

The non trivial equalities in Eq. (6.137) come from the decomposition of the sum of a vector and a pseudo-vector in space time which is conserved when we multiply by N and $\widetilde{N}$. We then have

$$\mathbf{x} = \langle \Psi_l \mathbf{y} \widetilde{\Psi}_l \rangle_1 \; ; \; \mathbf{x}' = \langle \Psi'_l \mathbf{y} \widetilde{\Psi}'_l \rangle_1 \tag{6.138}$$

with the same **y** for the observer of **x** as for the observer of **x**′.

6.4   Existence of the inverse

Our study in Sec. 5.2 of systems of electrons has introduced the inverse ϕ^{-1} which is defined only where $\det \phi \neq 0$. We can see that this condition is satisfied everywhere for each bound state of the H atom (see Appendix C). We saw previously that the wave of the electron is a part of the wave Ψ_l with value in $Cl_{1,3}$ which must be also invertible. We must then get $\det(\Psi_l) \neq 0$.

We have not yet used one of the features of the standard model, because it was not useful until now: the right part of the neutrino wave does not interact, and the standard model can do anything without ξ_n. We can then suppose

$$\xi_{1n} = \xi_{2n} = 0. \tag{6.139}$$

We then have with Eq. (6.2), Eq. (6.3) and Eq. (6.8):

$$\Psi_l = \sqrt{2} \begin{pmatrix} \xi_{1e} & -\eta_{2e}^* & 0 & -\eta_{2n}^* \\ \xi_{2e} & \eta_{1e}^* & 0 & \eta_{1n}^* \\ \eta_{1n} & 0 & \eta_{1e} & -\xi_{2e}^* \\ \eta_{2n} & 0 & \eta_{2e} & \xi_{1e}^* \end{pmatrix}. \tag{6.140}$$

We let

$$\rho_e = |\det(\phi_e)| = 2|\xi_{1e}\eta_{1e}^* + \xi_{2e}\eta_{2e}^*|, \tag{6.141}$$

$$\rho_L = |\det(\phi_L)| \; ; \; \phi_L = \sqrt{2}\begin{pmatrix} \eta_{1n} & \eta_{1e} \\ \eta_{2n} & \eta_{2e} \end{pmatrix}, \tag{6.142}$$

$$\rho_l = [\det(\Psi_l)]^{1/2}. \tag{6.143}$$

The calculation of the determinant of the matrix Eq. (6.140) gives the remarkable result:

$$\rho_l = \sqrt{\rho_e^2 + \rho_L^2}. \tag{6.144}$$

It is then very easy to get an invertible Ψ_l: it happens as soon as ϕ_e is invertible (for instance everywhere for each bound state of the H atom), or as soon as η_e and η_n are linearly independent. This is a very interesting use of the condition $\xi_n = 0$, and means that all features of the standard model are important. It also means that the true mathematical frame is the Clifford algebras and that the existence of an inverse wave at each point is physically useful.

6.5   Wave equations

The mass term of the Dirac equation links the right wave to the left wave, we can read this in Eq. (2.12) and Eq. (2.13). W^1 and W^2 terms in the electro-weak theory link left waves η_e η_n of the electron and its neutrino, while B and W^3 terms work separately with left and right waves. The Weinberg–Salam model takes advantage of the very small mass of the electron to neglect [10] its mass term. Mass is then missing in Sec. 6.1 to Sec. 6.4.

When we consider the three spinors, one right and two left ones, necessary to get the gauge group of electro-weak interactions, we have many more tensorial densities: from $8 \times 9/2 = 36$ they are now $12 \times 13/2 = 78$. Amongst these tensorial densities 6 form 3 complex quantities **a**, **b**, **c** in the place of $\mathbf{a} = \Omega_1 + i\Omega_2$. The identity Eq. (6.144) uses two of these three terms that, in the case electron+neutrino, replace the unique density ρ in Eq. (2.33):

$$a_1 = \mathbf{a} = \det(\phi_e) \; ; \; -\mathbf{b} = \det(\phi_L) \; ; \; a_2 = -\mathbf{b}^*. \tag{6.145}$$

---

10. This approximation is *a posteriori* satisfied by the huge mass of the Z^0 which is 180,000 times the mass of the electron. This approximation was inevitable because the mass term of the linear Dirac equation cannot be compatible with the electro-weak gauge.

The third term $a_3 = \mathbf{c} = 2(\xi_{1e}\eta_{1n}^* + \xi_{2e}\eta_{2n}^*)$ allows us to build the density

$$\rho = \sqrt{a_1 a_1^* + a_2 a_2^* + a_3 a_3^*}. \qquad (6.146)$$

This allows a mass term for a wave equation that is both form invariant and gauge invariant for the Ψ_l wave of the electron and its neutrino [21]:

$$\widetilde{\Psi}_l(\mathbf{D}\Psi_l)\gamma_{012} + m\rho\widetilde{\Psi}_l\chi_l = 0, \qquad (6.147)$$

where χ_l is a term depending on Ψ_l, defined in Eq. (B.91). For the form invariance (therefore also for the relativistic invariance) we establish in Eq. (B.41) the invariance of $m\rho$ and in Eq. (B.104) and Eq. (B.105) the invariance of the mass term. The wave equation Eq. (6.147) is then invariant under the R transformation defined by M in Eq. (1.42) and N in Eq. (1.80). The wave equation becomes [21]:

$$\widetilde{\Psi}_l'(\mathbf{D}'\Psi_l')\gamma_{012} + m'\rho'\widetilde{\Psi}_l'\chi_l' = 0 \; ; \quad m\rho = m'\rho' \; ; \quad \widetilde{\Psi}_l'\chi_l' = \widetilde{\Psi}_l\chi_l. \qquad (6.148)$$

This wave equation cannot be obtained from the linear Dirac equation because the relation Eq. (B.91) linking χ_l to Ψ_l destroys the linearity of the equation. Consequently the homogeneous nonlinear equation of chapter 3 is an obligatory intermediate allowing us to link our extended wave equation to the Dirac equation: The wave equation Eq. (3.1) is exactly what remains in Eq. (6.147) when the wave of the neutrino is canceled. And the Dirac equation is the linear approximation of our Eq. (3.1).

In any domain of the space-time where the wave of the electron is null or negligible $\rho = 0$ then the wave equation of the neutrino is reduced to $\nabla\eta_n = 0$ which is the wave equation of the usual neutrino, that moves with the velocity of light. This wave is without interaction, since the neutrino interacts only with the electron and its ϕ_e wave.

Under the gauge transformation defined by Eq. (6.119) to Eq. (6.122) we get (a detailed calculation is in B.3):

$$\widetilde{\Psi}_l'(\mathbf{D}'\Psi_l')\gamma_{012} + m\rho\widetilde{\Psi}_l'\chi_l' = 0. \qquad (6.149)$$

A detailed calculation of Eq. (6.147) shows that two of the 16 numeric equations are cancelled, and only two of the 14 numeric equations are simple: the real part is $\mathcal{L} = 0$ where $\mathcal{L}$ is the Lagrangian density giving Eq. (6.147). Then a double link exists, like in the case of the electron alone, between the wave equation and the Lagrangian density: this density is the real scalar part of the wave equation, and this wave equation may be obtained from this density by the variational calculus. This double link may be considered as the true reason for the presence in quantum physics of the variational calculus. The other simple equation is the conservation of the current [21]:

$$\partial_\mu J^\mu = 0 \; ; \quad J = D_0 + D_n \; ; \quad D_0 = \phi_e\phi_e^\dagger \; ; \quad D_n = \phi_n\phi_n^\dagger. \qquad (6.150)$$

The wave of the pair electron+neutrino may then be normalized. But the interpretation of this density of probability as a probability of presence of the particle is impossible since the density uses both the wave of the electron and the wave of the neutrino. We shall see in chapter 9 where the density of probability comes from.

Chapter 7

Electro-weak and strong interactions

We extend the gauge invariance to the quark sector, using $Cl_{1,5}$. We present in this frame the $SU(3)$ group of chromodynamics. We study the geometric transformation generated by the complete wave. We get another remarkable identity which makes the complete wave invertible. We get a wave equation with mass term that is form invariant and gauge invariant under the gauge group of the standard model.

7.1 Electro-weak interactions: the quark sector

For the first generation of fundamental fermions the standard model includes 16 fermions, 8 particles and their antiparticles. We studied previously the case of the electron, its neutrino, its antiparticle the positron and its anti-neutrino. We put these waves into a unique wave Ψ_l. Each generation includes also two quarks with three states, so we get six waves similar to ϕ_e or ϕ_n. Quarks of the first generation are named u and d and the couple d–u is similar to n–e for electro-weak interactions but with differences since the electric charge of u is $\frac{2}{3}|e|$ while the charge of d is $-\frac{1}{3}|e|$. Similarly to the lepton sector, the electric charges of antiparticles are opposite to charges of particles. Three states of "color" are named r, g, b (red, green, blue). So we build a wave with all fermions of the first generation as

$$\Psi = \begin{pmatrix} \Psi_l & \Psi_r \\ \Psi_g & \Psi_b \end{pmatrix}, \tag{7.1}$$

where Ψ_l is defined by Eq. (6.8) and Ψ_r, Ψ_g, Ψ_b are defined on the same model:

$$\Psi_r = \begin{pmatrix} \phi_{dr} & \phi_{ur} \\ \widehat{\phi_{\overline{u}r}}\sigma_1 & \widehat{\phi_{\overline{d}r}}\sigma_1 \end{pmatrix} = \begin{pmatrix} \phi_{dr} & \phi_{ur} \\ \widehat{\phi}_{ur} & \widehat{\phi}_{dr} \end{pmatrix}, \tag{7.2}$$

$$\Psi_g = \begin{pmatrix} \phi_{dg} & \phi_{ug} \\ \widehat{\phi_{\overline{u}g}}\sigma_1 & \widehat{\phi_{\overline{d}g}}\sigma_1 \end{pmatrix} = \begin{pmatrix} \phi_{dg} & \phi_{ug} \\ \widehat{\phi_{ug}} & \widehat{\phi_{dg}} \end{pmatrix}, \tag{7.3}$$

$$\Psi_b = \begin{pmatrix} \phi_{db} & \phi_{ub} \\ \widehat{\phi_{\overline{u}b}}\sigma_1 & \widehat{\phi_{\overline{d}b}}\sigma_1 \end{pmatrix} = \begin{pmatrix} \phi_{db} & \phi_{ub} \\ \widehat{\phi_{ub}} & \widehat{\phi_{db}} \end{pmatrix}. \tag{7.4}$$

The wave is a function of space-time with value into $Cl_{1,5}$ which is a sub-algebra of $Cl_{5,2} = M_8(\mathbb{C})$ (see Sec. 1.5). As previously, electro-weak inter-actions are obtained by replacing partial derivatives with covariant deriva-tives. Now we use the notation of Sec. 1.5 and let

$$\underline{W}^j = L^\mu W^j_\mu, \ j = 1,2,3 \ ; \ \underline{D} = L^\mu D_\mu \ ; \ L^0 = L_0 \ ; \ L^j = -L_j, \tag{7.5}$$

for $j = 1,2,3$. The covariant derivative reads now

$$\underline{D}(\Psi) = \underline{\partial}(\Psi) + \frac{g_1}{2}\underline{B}\,\underline{P}_0(\Psi) + \frac{g_2}{2}\underline{W}^j\underline{P}_j(\Psi). \tag{7.6}$$

We use two projectors $\underline{P}_\pm$ satisfying

$$\underline{P}_\pm(\Psi) = \frac{1}{2}(\Psi \pm \mathbf{i}\Psi L_{21}) \ ; \ \mathbf{i} = L_{0123}. \tag{7.7}$$

Three operators act on the quark sector as on the lepton sector :

$$\underline{P}_1(\Psi) = \underline{P}_+(\Psi)L_{35}, \tag{7.8}$$

$$\underline{P}_2(\Psi) = \underline{P}_+(\Psi)L_{5012}, \tag{7.9}$$

$$\underline{P}_3(\Psi) = \underline{P}_+(\Psi)L_{0132}. \tag{7.10}$$

The fourth operator acts differently on the lepton wave and on the quarks:

$$\underline{P}_0(\Psi) = \begin{pmatrix} P_0(\Psi_l) & P'_0(\Psi_r) \\ P'_0(\Psi_g) & P'_0(\Psi_b) \end{pmatrix}, \tag{7.11}$$

$$P_0(\Psi_l) = \Psi_l\gamma_{21} + P_-(\Psi_l)\mathbf{i} = \Psi_l\gamma_{21} + \frac{1}{2}(\Psi_l\mathbf{i} + \mathbf{i}\Psi_l\gamma_{30}), \tag{7.12}$$

$$P'_0(\Psi_r) = -\frac{1}{3}\Psi_r\gamma_{21} + P_-(\Psi_r)\mathbf{i} = -\frac{1}{3}\Psi_r\gamma_{21} + \frac{1}{2}(\Psi_r\mathbf{i} + \mathbf{i}\Psi_r\gamma_{30}). \tag{7.13}$$

This is very important: first the value $-1/3$ shall give the four correct values of the charges of quarks and antiquarks. Next if all four operators were identical we should get four states and an SU(4) group for chromodynamics and the electron should be sensitive to strong interactions. Since only three parts of the wave are similar, we will get in the next paragraph an SU(3) group for chromodynamics. Now we get two identical formulas by replacing the r index by g and b. We can abbreviate and we remove indices r, g, b to study the electro-weak covariant derivative. We let

$$P'_0(\Psi) = \begin{pmatrix} p'_0(\phi_d) & p'_0(\phi_u) \\ p'_0(\widehat{\phi_{\overline{u}}})\sigma_1 & p'_0(\widehat{\phi_{\overline{d}}})\sigma_1 \end{pmatrix}, \tag{7.14}$$

which gives with Eq. (7.13):

$$P'_0(\Psi) = -\frac{i}{3}\begin{pmatrix} \phi_d\sigma_3 & \phi_u\sigma_3 \\ -\widehat{\phi_{\bar{u}}}\sigma_3\sigma_1 & -\widehat{\phi_{\bar{d}}}\sigma_3\sigma_1 \end{pmatrix}$$
$$+ \frac{i}{2}\begin{pmatrix} \phi_d & -\phi_u \\ \widehat{\phi_{\bar{u}}}\sigma_1 & -\widehat{\phi_{\bar{d}}}\sigma_1 \end{pmatrix} + \frac{i}{2}\begin{pmatrix} \phi_d\sigma_3 & -\phi_u\sigma_3 \\ \widehat{\phi_{\bar{u}}}\sigma_3\sigma_1 & -\widehat{\phi_{\bar{d}}}\sigma_3\sigma_1 \end{pmatrix}. \qquad (7.15)$$

We then get the system :

$$p'_0(\phi_d) = -\frac{i}{3}\phi_d\sigma_3 + \frac{i}{2}\phi_d + \frac{i}{2}\phi_d\sigma_3 = \frac{i}{3}(2\phi_{dR} + \phi_{dL}),$$

$$p'_0(\phi_u) = -\frac{i}{3}\phi_u\sigma_3 - \frac{i}{2}\phi_u - \frac{i}{2}\phi_u\sigma_3 = \frac{i}{3}(-4\phi_{uR} + \phi_{uL}),$$

$$p'_0(\widehat{\phi_{\bar{u}}}) = \frac{i}{3}\widehat{\phi_{\bar{u}}}\sigma_3 + \frac{i}{2}\widehat{\phi_{\bar{u}}} + \frac{i}{2}\widehat{\phi_{\bar{u}}}\sigma_3 = \frac{i}{3}(4\widehat{\phi_{\bar{u}L}} - \widehat{\phi_{\bar{u}R}}), \qquad (7.16)$$

$$p'_0(\widehat{\phi_{\bar{d}}}) = \frac{i}{3}\widehat{\phi_{\bar{d}}}\sigma_3 - \frac{i}{2}\widehat{\phi_{\bar{d}}} - \frac{i}{2}\widehat{\phi_{\bar{d}}}\sigma_3 = \frac{i}{3}(-2\widehat{\phi_{\bar{d}L}} - \widehat{\phi_{\bar{d}R}}).$$

Since P_1, P_2 and P_3 are unchanged in the quark sector, we get from Eq. (6.41), Eq. (6.43) and Eq. (6.45)

$$p_1(\phi_d) = i\phi_{uL}; \; p_1(\phi_u) = i\phi_{dL}; \; p_1(\widehat{\phi_{\bar{u}}}) = -i\widehat{\phi_{\bar{d}R}}; \; p_1(\widehat{\phi_{\bar{d}}}) = -i\widehat{\phi_{\bar{u}R}}, \quad (7.17)$$

$$p_2(\phi_d) = \phi_{uL}; \; p_2(\phi_u) = -\phi_{dL}; \; p_2(\widehat{\phi_{\bar{u}}}) = -\widehat{\phi_{\bar{d}R}}; \; p_2(\widehat{\phi_{\bar{d}}}) = \widehat{\phi_{\bar{u}R}}, \qquad (7.18)$$

$$p_3(\phi_d) = -i\phi_{dL}; \; p_3(\phi_u) = i\phi_{uL}; \; p_3(\widehat{\phi_{\bar{u}}}) = -i\widehat{\phi_{\bar{u}R}}; \; p_3(\widehat{\phi_{\bar{d}}}) = i\widehat{\phi_{\bar{d}R}}. \quad (7.19)$$

Now Eq. (7.6) gives

$$\mathbf{D}\Psi_r = \boldsymbol{\partial}\Psi_r + \frac{g_1}{2}\mathbf{B}P'_0(\Psi_r) + \frac{g_2}{2}\mathbf{W}^j P_j(\Psi_r), \qquad (7.20)$$

and we get, similarly to Eq. (6.47) to Eq. (6.50)

$$D\widehat{\phi_{\bar{u}}} = \nabla\widehat{\phi_{\bar{u}}} + \frac{g_1}{2}Bp'_0(\widehat{\phi_{\bar{u}}}) + \frac{g_2}{2}W^j p_j(\widehat{\phi_{\bar{u}}}), \qquad (7.21)$$

$$D\widehat{\phi_{\bar{d}}} = \nabla\widehat{\phi_{\bar{d}}} + \frac{g_1}{2}Bp'_0(\widehat{\phi_{\bar{d}}}) + \frac{g_2}{2}W^j p_j(\widehat{\phi_{\bar{d}}}), \qquad (7.22)$$

$$\widehat{D}\phi_d = \widehat{\nabla}\phi_d + \frac{g_1}{2}\widehat{B}p'_0(\phi_d) + \frac{g_2}{2}\widehat{W}^j p_j(\phi_d), \qquad (7.23)$$

$$\widehat{D}\phi_u - \widehat{\nabla}\phi_u + \frac{g_1}{2}\widehat{B}p'_0(\phi_u) + \frac{g_2}{2}\widehat{W}^j p_j(\phi_u). \qquad (7.24)$$

With Eq. (7.16) to Eq. (7.19) this gives

$$D\widehat{\phi_{\bar{u}}} = \nabla\widehat{\phi_{\bar{u}}} + \frac{g_1}{2}B\frac{i}{3}(4\widehat{\phi_{\bar{u}L}} - \widehat{\phi_{\bar{u}R}})$$
$$+ \frac{g_2}{2}[W^1(-i\widehat{\phi_{\bar{d}R}}) + W^2(-\widehat{\phi_{\bar{d}R}}) + W^3(-i\widehat{\phi_{\bar{u}R}})], \qquad (7.25)$$

$$D\widehat{\phi_{\bar{d}}} = \nabla\widehat{\phi_{\bar{d}}} + \frac{g_1}{2}B\frac{i}{3}(-2\widehat{\phi_{\bar{d}L}} - \widehat{\phi_{\bar{d}R}})$$
$$+ \frac{g_2}{2}[W^1(-i\widehat{\phi_{\bar{u}R}}) + W^2\widehat{\phi_{\bar{u}R}}) + W^3 i\widehat{\phi_{\bar{d}R}}], \qquad (7.26)$$

$$\widehat{D}\phi_d = \widehat{\nabla}\phi_d + \frac{g_1}{2}\widehat{B}\frac{i}{3}(2\phi_{dR} + \phi_{dL})$$
$$+ \frac{g_2}{2}[\widehat{W}^1(i\phi_{uL}) + \widehat{W}^2(\phi_{uL}) + \widehat{W}^3(-i\phi_{dL})], \qquad (7.27)$$

$$\widehat{D}\phi_u = \widehat{\nabla}\phi_u + \frac{g_1}{2}\widehat{B}\frac{i}{3}(-4\phi_{uR} + \phi_{uL})$$
$$+ \frac{g_2}{2}[\widehat{W}^1(i\phi_{dL}) + \widehat{W}^2(-\phi_{dL}) + \widehat{W}^3(i\phi_{uL})]. \qquad (7.28)$$

We separate right and left waves, this gives

$$D\widehat{\phi_{\bar{u}L}} = \nabla\widehat{\phi_{\bar{u}L}} - i(-\tfrac{2}{3})g_1 B\widehat{\phi_{\bar{u}L}}; \quad D_\mu \eta_{\bar{u}} = \partial_\mu \eta_{\bar{u}} - i(-\tfrac{2}{3})g_1 B_\mu \eta_{\bar{u}}, \qquad (7.29)$$

$$D\widehat{\phi_{\bar{d}L}} = \nabla\widehat{\phi_{\bar{d}L}} - i(+\tfrac{1}{3})g_1 B\widehat{\phi_{\bar{d}L}}; \quad D_\mu \eta_{\bar{d}} = \partial_\mu \eta_{\bar{d}} - i(+\tfrac{1}{3})g_1 B_\mu \eta_{\bar{d}}, \qquad (7.30)$$

$$\widehat{D}\phi_{dR} = \widehat{\nabla}\phi_{dR} - i(-\tfrac{1}{3})g_1 \widehat{B}\phi_{dR}; \quad D_\mu \xi_d = \partial_\mu \xi_d - i(-\tfrac{1}{3})g_1 B_\mu \xi_d, \qquad (7.31)$$

$$\widehat{D}\phi_{uR} = \widehat{\nabla}\phi_{uR} - i(+\tfrac{2}{3})g_1 \widehat{B}\phi_{uR}; \quad D_\mu \xi_u = \partial_\mu \xi_u - i(+\tfrac{2}{3})g_1 B_\mu \xi_u. \qquad (7.32)$$

Comparison with Sec. 6.1 shows that quarks and anti-quarks have prede-
termined electric charges: $-\tfrac{2}{3}|e|$ for anti-quark $\bar{u}$, $+\tfrac{1}{3}|e|$ for anti-quark $\bar{d}$,
$+\tfrac{2}{3}|e|$ for the u quark and $-\tfrac{1}{3}|e|$ for [1] the d quark. Separation of right and
left waves from Eq. (7.25) to Eq. (7.28) gives also

$$D\widehat{\phi_{\bar{u}R}} = \nabla\widehat{\phi_{\bar{u}R}} - i\frac{g_1}{6}B\widehat{\phi_{\bar{u}R}} + \frac{g_2}{2}[-iW^1\widehat{\phi_{\bar{d}R}}W^2\widehat{\phi_{\bar{d}R}} - iW^3\widehat{\phi_{\bar{u}R}}], \qquad (7.33)$$

$$D\widehat{\phi_{\bar{d}R}} = \nabla\widehat{\phi_{\bar{d}R}} - i\frac{g_1}{6}B\widehat{\phi_{\bar{d}R}} + \frac{g_2}{2}[-iW^1\widehat{\phi_{\bar{u}R}} + W^2\widehat{\phi_{\bar{u}R}}) + iW^3\widehat{\phi_{\bar{d}R}}], \qquad (7.34)$$

$$\widehat{D}\phi_{dL} = \widehat{\nabla}\phi_{dL} + i\frac{g_1}{6}\widehat{B}\phi_{dL} + \frac{g_2}{2}[i\widehat{W}^1\phi_{uL} + \widehat{W}^2\phi_{uL} - i\widehat{W}^3\phi_{dL}], \qquad (7.35)$$

$$\widehat{D}\phi_{uL} = \widehat{\nabla}\phi_{uL} + i\frac{g_1}{6}\widehat{B}\phi_{uL} + \frac{g_2}{2}[i\widehat{W}^1\phi_{dL} - \widehat{W}^2\phi_{dL} + i\widehat{W}^3\phi_{uL}]. \qquad (7.36)$$

Using the conjugation $\phi \mapsto \widehat{\phi}$ we get

$$\widehat{D}\phi_{\bar{u}R} = \widehat{\nabla}\phi_{\bar{u}R} + i\frac{g_1}{6}\widehat{B}\phi_{\bar{u}R} + \frac{g_2}{2}[+i\widehat{W}^1\phi_{\bar{d}R} - \widehat{W}^2\phi_{\bar{d}R} + i\widehat{W}^3\phi_{\bar{u}R}], \qquad (7.37)$$

$$\widehat{D}\phi_{\bar{d}R} = \widehat{\nabla}\phi_{\bar{d}R} + i\frac{g_1}{6}\widehat{B}\phi_{\bar{d}R} + \frac{g_2}{2}[+i\widehat{W}^1\phi_{\bar{u}R} + \widehat{W}^2\phi_{\bar{u}R}) - i\widehat{W}^3\phi_{\bar{d}R}], \qquad (7.38)$$

$$D\widehat{\phi_{dL}} = \nabla\widehat{\phi_{dL}} - i\frac{g_1}{6}B\widehat{\phi_{dL}} + \frac{g_2}{2}[-iW^1\widehat{\phi_{uL}} + W^2\widehat{\phi_{uL}} + iW^3\widehat{\phi_{dL}}], \qquad (7.39)$$

$$D\widehat{\phi_{uL}} = \nabla\widehat{\phi_{uL}} - i\frac{g_1}{6}B\widehat{\phi_{uL}} + \frac{g_2}{2}[-iW^1\widehat{\phi_{dL}} - W^2\widehat{\phi_{dL}} - iW^3\widehat{\phi_{uL}}]. \qquad (7.40)$$

---

1. Another mechanism giving the $\pm\frac{e}{3}$ and $\pm\frac{2e}{3}$ charges of quarks was proposed in
Sec. 5.3 of [15].

This gives a left doublet of particles and a right doublet of antiparticles. With Eq. (6.72) and

$$\psi_L = \begin{pmatrix} \eta_u \\ \eta_d \end{pmatrix} \; ; \quad \psi_R = \begin{pmatrix} \xi_{\bar u} \\ \xi_{\bar d} \end{pmatrix}, \tag{7.41}$$

we get

$$D_\mu \psi_L = \partial_\mu \psi_L - i\frac{g_1}{6} B_\mu \psi_L - i\frac{g_2}{2}(W\mu^1 \tau_1 + W\mu^2 \tau_2 + W\mu^3 \tau_3)\psi_L, \tag{7.42}$$

$$D_\mu \psi_R = \partial_\mu \psi_R + i\frac{g_1}{6} B_\mu \psi_R - i\frac{g_2}{2}(W\mu^1 \tau_1 - W\mu^2 \tau_2 + W\mu^3 \tau_3)\psi_R. \tag{7.43}$$

We can then say that charge conjugation not only changes the signs of electric charges, but also the right and the left waves. It also changes the orientation of the space of the τ_j, where a direct basis (τ_1, τ_2, τ_3), is replaced by an inverse basis $(\tau_1, -\tau_2, \tau_3)$. We encounter this basis both here and in the wave of an antiparticle Eq. (4.64) used by de Broglie.

7.2 Chromodynamics

We start from generators of the $SU(3)$ gauge group of chromodynamics

$$\lambda_1 = \begin{pmatrix} 0&1&0 \\ 1&0&0 \\ 0&0&0 \end{pmatrix}, \lambda_2 = \begin{pmatrix} 0&-i&0 \\ i&0&0 \\ 0&0&0 \end{pmatrix}, \lambda_3 = \begin{pmatrix} 1&0&0 \\ 0&-1&0 \\ 0&0&0 \end{pmatrix},$$

$$\lambda_4 = \begin{pmatrix} 0&0&1 \\ 0&0&0 \\ 1&0&0 \end{pmatrix}, \lambda_5 = \begin{pmatrix} 0&0&-i \\ 0&0&0 \\ i&0&0 \end{pmatrix}, \lambda_6 = \begin{pmatrix} 0&0&0 \\ 0&0&1 \\ 0&1&0 \end{pmatrix},$$

$$\lambda_7 = \begin{pmatrix} 0&0&0 \\ 0&0&-i \\ 0&i&0 \end{pmatrix}, \lambda_8 = \frac{1}{\sqrt{3}}\begin{pmatrix} 1&0&0 \\ 0&1&0 \\ 0&0&-2 \end{pmatrix}. \tag{7.44}$$

To simplify notations we use now l, r, g, b instead of $\Psi_l, \Psi_r, \Psi_g, \Psi_b$. So we have $\Psi = \begin{pmatrix} l & r \\ g & b \end{pmatrix}$. Then this gives

$$\lambda_1 \begin{pmatrix} r \\ g \\ b \end{pmatrix} = \begin{pmatrix} g \\ r \\ 0 \end{pmatrix}, \lambda_2 \begin{pmatrix} r \\ g \\ b \end{pmatrix} = \begin{pmatrix} -ig \\ ir \\ 0 \end{pmatrix}, \lambda_3 \begin{pmatrix} r \\ g \\ b \end{pmatrix} = \begin{pmatrix} r \\ -g \\ 0 \end{pmatrix},$$

$$\lambda_4 \begin{pmatrix} r \\ g \\ b \end{pmatrix} = \begin{pmatrix} b \\ 0 \\ r \end{pmatrix}, \lambda_5 \begin{pmatrix} r \\ g \\ b \end{pmatrix} = \begin{pmatrix} -ib \\ 0 \\ ir \end{pmatrix}, \lambda_6 \begin{pmatrix} r \\ g \\ b \end{pmatrix} = \begin{pmatrix} 0 \\ b \\ g \end{pmatrix}, \tag{7.45}$$

$$\lambda_7 \begin{pmatrix} r \\ g \\ b \end{pmatrix} = \begin{pmatrix} 0 \\ -ib \\ ig \end{pmatrix}, \lambda_8 \begin{pmatrix} r \\ g \\ b \end{pmatrix} = \frac{1}{\sqrt{3}}\begin{pmatrix} r \\ g \\ -2b \end{pmatrix}.$$

We name Γ_k the operators corresponding to λ_k acting on Ψ. We get with projectors Eq. (1.93):

$$\Gamma_1(\Psi) = \frac{1}{2}(L_4\Psi L_4 + L_{01235}\Psi L_{01235}) = \begin{pmatrix} 0 & g \\ r & 0 \end{pmatrix}, \qquad (7.46)$$

$$\Gamma_2(\Psi) = \frac{1}{2}(L_5\Psi L_4 - L_{01234}\Psi L_{01235}) = \begin{pmatrix} 0 & -ig \\ ir & 0 \end{pmatrix}, \qquad (7.47)$$

$$\Gamma_3(\Psi) = P^+\Psi P^- - P^-\Psi P^+ = \begin{pmatrix} 0 & r \\ -g & 0 \end{pmatrix}, \qquad (7.48)$$

$$\Gamma_4(\Psi) = L_{01253}\Psi P^- = \begin{pmatrix} 0 & b \\ 0 & r \end{pmatrix} ; \ \Gamma_5(\Psi) = L_{01234}\Psi P^- = \begin{pmatrix} 0 & -ib \\ 0 & ir \end{pmatrix}, \quad (7.49)$$

$$\Gamma_6(\Psi) = P^-\Psi L_{01253} = \begin{pmatrix} 0 & 0 \\ b & g \end{pmatrix} ; \ \Gamma_7(\Psi) = -\underline{i}P^-\Psi L_4 = \begin{pmatrix} 0 & 0 \\ -ib & ig \end{pmatrix}, \quad (7.50)$$

$$\Gamma_8(\Psi) = \frac{1}{\sqrt{3}}(P^-\Psi L_{012345} + L_{012345}\Psi P^-) = \frac{1}{\sqrt{3}}\begin{pmatrix} 0 & r \\ g & -2b \end{pmatrix}. \qquad (7.51)$$

Everywhere the upper left term is 0, so all Γ_k project the wave Ψ on its quark sector.

We can extend the covariant derivative of electro-weak interactions Eq. (7.6):

$$\underline{D}(\Psi) = \underline{\partial}(\Psi) + \frac{g_1}{2}\underline{B}\,\underline{P}_0(\Psi) + \frac{g_2}{2}\underline{W}^j\underline{P}_j(\Psi) + \frac{g_3}{2}\underline{G}^k\underline{i}\Gamma_k(\Psi), \qquad (7.52)$$

where g_3 is another constant and $\underline{G}^k$ are eight terms called "gluons". Since I_4 commute with any element of $Cl_{1,3}$ and since $P_j(\underline{i}\Psi_{ind}) = \underline{i}P_j(\Psi_{ind})$ for $j = 0, 1, 2, 3$ and $ind = l, r, g, b$ each operator $\underline{i}\Gamma_k$ commutes with all operators $\underline{P}_j$.

Now we use 12 real numbers a^0, a^j, $j = 1, 2, 3$ and b^k, $k = 1, 2, ..., 8$. We let

$$S_1 = \sum_{j=1}^{j=3} a^j \underline{P}_j \ ; \ S_2 = \sum_{k=1}^{k=8} b^k \underline{i}\Gamma_k, \qquad (7.53)$$

and we get, using exponentiation (see Sec. 6.2)

$$\exp(a^0\underline{P}_0 + S_1 + S_2) = \exp(a^0\underline{P}_0)\exp(S_1)\exp(S_2) \qquad (7.54)$$

The set of these operators is a $U(1) \times SU(2) \times SU(3)$ Lie group. The only difference with the standard model is that the structure of this group is not postulated but calculated. The invariance under Cl_3^* (and in particular the relativistic invariance) of this covariant derivative is similar to Eq. (6.105) with underlined terms. The gauge invariance reads with

$$\Psi' = [\exp(a^0\underline{P}_0 + S_1 + S_2)](\Psi) \ ; \ \underline{D} = L^\mu\underline{D}_\mu \ ; \ \underline{D}' = L^\mu\underline{D}'_\mu, \qquad (7.55)$$

$$\underline{D}'_\mu \Psi' = \exp(a^0 \underline{P}_0 + S_1 + S_2) \underline{D}_\mu \Psi, \tag{7.56}$$

$$B'_\mu = B_\mu - \frac{2}{g_1} \partial_\mu a^0, \tag{7.57}$$

$$W'^j_\mu \underline{P}_j = \left[\exp(S_1) W^j_\mu \underline{P}_j - \frac{2}{g_2} \partial_\mu [\exp(S_1)] \right] \exp(-S_1), \tag{7.58}$$

$$\underline{G}'^k_\mu i \Gamma_k = \left[\exp(S_2) \underline{G}^k_\mu i \Gamma_k - \frac{2}{g_3} \partial_\mu [\exp(S_2)] \right] \exp(-S_2). \tag{7.59}$$

The $SU(3)$ group generated by operators projecting on the quark sector acts only on this sector of the wave:

$$P^+ [\exp(b^k i \Gamma_k)(\Psi) P^+ = P^+ \Psi P^+ = \begin{pmatrix} \Psi_l & 0 \\ 0 & 0 \end{pmatrix}. \tag{7.60}$$

We get then a $U(1) \times SU(2) \times SU(3)$ gauge group for a wave including all fermions of the first generation. This group acts on the lepton sector only by its $U(1) \times SU(2)$ part. The physical translation is: leptons do not strongly interact, they have only electromagnetic and weak interactions. This is fully satisfied in experiments. The novelty here is that this comes from the structure itself of the quantum wave. Since it is independent of the energy scale, we understand why great unified theories do not work.

7.3 Three generations, four neutrinos

The aim of theoretical physics is to understand experimental facts. Today we have to understand both why we get only three kinds of leptons and quarks and a fourth neutrino, without electro-weak interactions. Actual experiments show both the limitation to three kinds of light leptons from the study of the Z^0 and the possible existence of a fourth neutrino without electro-weak interactions. We explained the existence of three kinds of leptons in chapter 5. This is easily generalized to the three generations of the standard model. Two other generations are gotten by replacing the privileged third direction σ_3 by σ_1 or σ_2, everywhere this direction is used. The passage from one to another generation must be seen as a circular permutation of indices $1 \mapsto 2 \mapsto 3 \mapsto 1$ or $1 \mapsto 3 \mapsto 2 \mapsto 1$ for the other. For instance the σ_3 in Eq. (6.11) which defines left and right projectors must be replaced by σ_1 or σ_2. The σ_1 in Eq. (6.8) which links the wave of the particle to the wave of the antiparticle must be replaced by σ_2 or σ_3. These changes imply that each generation should be treated separately and it is the reason for this separate treatment in the standard model. Now for a fourth generation we have no other similar possibility since the Cl_3 algebra is based on the 3-dimensional physical space. We cannot get a fourth set of operators similar to the P_μ.

But the existence of a fourth neutrino [20] is possible because Cl_3 has four generators with square -1. The wave equation of the electron includes one of these four generators, $i\sigma_3 = \sigma_{12}$. Now $i\sigma_1 = \sigma_{23}$ and $i\sigma_2 = \sigma_{31}$ explain why two other kinds of leptons exist. We can also build an invariant wave equation with the fourth generator, $i = \sigma_{123}$:

$$\overline{\phi}(\nabla\widehat{\phi})\sigma_{123} + m\rho = 0. \tag{7.61}$$

Multiplying on the left by $\overline{\phi}^{\,-1}$ we get with $\rho = e^{-i\beta}\overline{\phi}\phi$ the equivalent equation

$$\nabla\widehat{\phi}i + me^{-i\beta}\phi = 0 \ ; \ \ \nabla\widehat{\phi} = ime^{-i\beta}\phi. \tag{7.62}$$

Contrary to our homogeneous non-linear wave equation Eq. (3.1) which has the Dirac equation as linear approximation, this wave equation cannot come from linear quantum theory: it has no linear approximation because the β angle is not small, it is now the angle of the phase [2] of the wave. We can nevertheless get plane waves. We search now solutions satisfying

$$\phi = e^{-i\varphi}\phi_0 \ ; \ \ \varphi = mv_\mu x^\mu \ ; \ \ v = \sigma^\mu v_\mu, \tag{7.63}$$

where v is a fixed reduced speed and ϕ_0 is also a fixed term. We get:

$$\nabla\widehat{\phi} = \sigma^\mu \partial_\mu(e^{i\varphi}\widehat{\phi}_0) = imve^{i\varphi}\widehat{\phi}_0. \tag{7.64}$$

And we have

$$\phi\overline{\phi} = e^{-i\varphi}\phi_0 e^{-i\varphi}\overline{\phi}_0 = e^{-2i\varphi}\phi_0\overline{\phi}_0. \tag{7.65}$$

Then if we let

$$\phi_0\overline{\phi}_0 = \rho_0 e^{i\beta_0}, \tag{7.66}$$

we get

$$\beta = \beta_0 - 2\varphi \ ; \ \ e^{-i\beta}\phi = e^{-i(\beta_0 - 2\varphi)}e^{-i\varphi}\phi_0 = e^{-i(\beta_0 - \varphi)}\phi_0. \tag{7.67}$$

Then Eq. (7.61) is equivalent to

$$imve^{i\varphi}\widehat{\phi}_0 = ime^{-i(\beta_0 - \varphi)}\phi_0 \tag{7.68}$$

$$v\widehat{\phi}_0 = e^{-i\beta_0}\phi_0$$

$$e^{i\beta_0}v\widehat{\phi}_0 = \phi_0. \tag{7.69}$$

---

2. This is another reason to think that the homogeneous non-linear equation is better than its linear approximation.

Conjugating, we get

$$e^{-i\beta_0}\widehat{v}\widehat{\phi}_0 = \widehat{\phi}_0. \tag{7.70}$$

So we get

$$\phi_0 = e^{i\beta_0}v\widehat{\phi}_0 = e^{i\beta_0}v[e^{-i\beta_0}\widehat{v}\widehat{\phi}_0] = v\widehat{v}\phi_0. \tag{7.71}$$

Then if $\phi_0 \neq 0$ we get

$$1 = v\widehat{v} \tag{7.72}$$

which gives Eq. (3.30). Since Eq. (7.70) implies Eq. (3.33) we get the same results as with our non-linear wave equation: existence of plane waves with only positive energy. Developing Eq. (7.61) we get a system of eight equations similar to the system Eq. (2.75) to Eq. (2.82) and four of these equations are the conservation of the D_μ currents ($\partial_\nu D_\mu^\nu = 0$) [20]. Then the density of probability is conservative and there is no possible disintegration of such a particle. Without a set of operators P_μ there are no electro-weak forces. Therefore only gravitational interactions remain possible. Such an object could be a part [3] of the dark [4] matter.

7.4   Geometric transformation linked to the complete wave

The complete wave Ψ containing the wave of leptons and quarks of the first generation defined in Eq. (7.1) satisfies (the proof is in Appendix A)

$$\widetilde{\Psi} = \begin{pmatrix} \widetilde{\Psi}_b & \widetilde{\Psi}_r \\ \widetilde{\Psi}_g & \widetilde{\Psi}_l \end{pmatrix}. \tag{7.73}$$

The wave has value in the Clifford algebra $Cl_{1,5}$. Each element reads

$$\Psi = \sum_{n=0}^{n=6} \Psi_n, \tag{7.74}$$

where $\langle\Psi\rangle_n = \Psi_n$ is named an n vector. The reverse satisfies

$$\widetilde{\Psi} = \Psi_0 + \Psi_1 - \Psi_2 - \Psi_3 + \Psi_4 + \Psi_5 - \Psi_6. \tag{7.75}$$

We define the v-part A_v of any multivector A as the sum of the vectorial part and of the pseudo-vectorial part in the complete space-time:

$$A_v = A_1 + A_5, \tag{7.76}$$

---

3. The fourth neutrino, insensitive to weak interactions, is not forbidden by the disintegrating Z^0, that gives a maximum of three weakly interacting light neutrinos.

4. The fourth neutrino is its own anti-particle, because the charge conjugation Eq. (2.96) does not change the differential term.

because it is this vectorial part which replaces in $Cl_{1,5}$ the vectorial part $\langle M \rangle_1$ of the space-time algebra. It is linked to the fact that vectors of $Cl_{1,5}$ are 5-vectors of $Cl_{5,1}$ and vice-versa. We should then get the same definition of the vectorial part by using $Cl_{5,1}$. We use

$$x_v = x^n L_n + x_5^n L_n L_{012345} \; ; \quad y_v = y^n L_n + y_5^n L_n L_{012345}, \qquad (7.77)$$

$$M_v = M_1 + M_5 \qquad (7.78)$$

$$x_v = 2(\Psi y_v \widetilde{\Psi})_v \; ; \quad x = x^\mu \sigma_\mu \; ; \quad \mathbf{x} = x^\mu \gamma_\mu. \qquad (7.79)$$

The transformation linked to the wave reads

$$f : y_v \mapsto x_v = 2(\Psi y_v \widetilde{\Psi})_v. \qquad (7.80)$$

Contrary to the R dilation obtained in the case electron+neutrino , where the usual space-time acts alone, the f transformation is a transformation from the sub-space of M_v into the same subspace of the usual manifold. We get

$$x_v = \begin{pmatrix} 0 & x_2 \\ x_1 & 0 \end{pmatrix} ; \quad y_v = \begin{pmatrix} 0 & y_2 \\ y_1 & 0 \end{pmatrix} \qquad (7.81)$$

$$x_1 = 2(\Psi_b y_1 \widetilde{\Psi}_b + \Psi_g y_2 \widetilde{\Psi}_g) \qquad (7.82)$$

$$x_2 = 2(\Psi_r y_1 \widetilde{\Psi}_r + \Psi_l y_2 \widetilde{\Psi}_l). \qquad (7.83)$$

Next we let

$$x_1 = \mathbf{x} + \mathbf{x}_5 + x^4 + x_5^4 + (x^5 + x_5^5)\mathbf{i} \qquad (7.84)$$

$$x_2 = \mathbf{x} - \mathbf{x}_5 - x^4 + x_5^4 + (x^5 - x_5^5)\mathbf{i} \qquad (7.85)$$

$$y_1 = \mathbf{y} + \mathbf{y}_5 + y^4 + y_5^4 + (y^5 + y_5^5)\mathbf{i} \qquad (7.86)$$

$$y_2 = \mathbf{y} - \mathbf{y}_5 - y^4 + y_5^4 + (y^5 - y_5^5)\mathbf{i} \qquad (7.87)$$

that finally gives

$$
\begin{aligned}
x = {}& \phi_e(y - y_5)\phi_e^\dagger + \phi_n(\widehat{y} - \widehat{y}_5)\phi_n^\dagger + 2\Re\{\phi_e\phi_n^\dagger[-y^4 + y_5^4 + (y^5 - y_5^5)i]\} \\
& + \phi_{dr}(y + y_5)\phi_{dr}^\dagger + \phi_{ur}(\widehat{y} + \widehat{y}_5)\phi_{ur}^\dagger + 2\Re\{\phi_{dr}\phi_{ur}^\dagger[y^4 + y_5^4 + (y^5 + y_5^5)i]\} \\
& + \phi_{dg}(y - y_5)\phi_{dg}^\dagger + \phi_{ug}(\widehat{y} - \widehat{y}_5)\phi_{ug}^\dagger + 2\Re\{\phi_{dg}\phi_{ug}^\dagger[-y^4 + y_5^4 + (y^5 - y_5^5)i]\} \\
& + \phi_{db}(y + y_5)\phi_{db}^\dagger + \phi_{ub}(\widehat{y} + \widehat{y}_5)\phi_{ub}^\dagger + 2\Re\{\phi_{db}\phi_{ub}^\dagger[y^4 + y_5^4 + (y^5 + y_5^5)i]\}.
\end{aligned}
$$
$$\qquad (7.88)$$

This equality is a generalization of Eq. (6.133) obtained in the case electron+neutrino. We may remark that the supplementary dimensions are mixed with the ordinary dimensions. We may also remark that the transformation is linear in y.

7.4.1 Invariance

The dilation R induced by any element M of Cl_3 satisfying Eq. (1.42) reads in space-time algebra, with the N of Eq. (1.80):

$$\mathbf{x}' = N\mathbf{x}\widetilde{N} \; ; \; \mathbf{x} = \sum_{\mu=0}^{\mu=3} x^\mu \gamma_\mu \; ; \; \mathbf{x}' = \sum_{\mu=0}^{\mu=3} x'^\mu \gamma_\mu \; ; \; \Psi_l' = N\Psi_l, \tag{7.89}$$

and we need the same transformation for the waves of quarks:

$$\Psi_r' = N\Psi_r \; ; \; \Psi_g' = N\Psi_g \; ; \; \Psi_b' = N\Psi_b. \tag{7.90}$$

We let

$$\mathbf{N} = \begin{pmatrix} N & 0 \\ 0 & N \end{pmatrix} = \begin{pmatrix} M & 0 & 0 & 0 \\ 0 & \widehat{M} & 0 & 0 \\ 0 & 0 & M & 0 \\ 0 & 0 & 0 & \widehat{M} \end{pmatrix}. \tag{7.91}$$

We then get

$$\widetilde{\mathbf{N}} = \begin{pmatrix} \widetilde{N} & 0 \\ 0 & \widetilde{N} \end{pmatrix} = \begin{pmatrix} \overline{M} & 0 & 0 & 0 \\ 0 & M^\dagger & 0 & 0 \\ 0 & 0 & \overline{M} & 0 \\ 0 & 0 & 0 & M^\dagger \end{pmatrix}. \tag{7.92}$$

With

$$x_v' = \begin{pmatrix} 0 & x_2' \\ x_1' & 0 \end{pmatrix} \; ; \; \mathbf{x}' = x'^\mu \gamma_\mu \; ; \; \mathbf{x}_5' = x'^\mu_5 \gamma_\mu, \tag{7.93}$$

$$x_1' = \mathbf{x}' + \mathbf{x}_5' + x'^4 + x'^4_5 + (x'^5 + x'^5_5)\mathbf{i}, \tag{7.94}$$

$$x_2' = \mathbf{x}' - \mathbf{x}_5' - x'^4 + x'^4_5 + (x'^5 - x'^5_5)\mathbf{i}, \tag{7.95}$$

the generalization of Eq. (6.135) and Eq. (6.136) in $Cl_{1,5}$ reads

$$x_v' = \mathbf{N} x_v \widetilde{\mathbf{N}} \; ; \; \Psi' = \mathbf{N}\Psi, \tag{7.96}$$

which gives

$$\widetilde{\Psi}' = \widetilde{\mathbf{N}\Psi} = \widetilde{\Psi}\widetilde{\mathbf{N}}. \tag{7.97}$$

Then the first equality Eq. (7.96) is equivalent to the system:

$$\mathbf{x}' + x'^4_5 + x'^5\mathbf{i} = N(\mathbf{x} + x^4_5 + x^5\mathbf{i})\widetilde{N}, \tag{7.98}$$

$$\mathbf{x}_5' + x'^4 + x'^5_5\mathbf{i} = N(\mathbf{x}_5 + x^4 + x^5_5\mathbf{i})\widetilde{N}. \tag{7.99}$$

And since we can separate the different multivector parts, this is equivalent
to the system:

$$\mathbf{x'} = N\mathbf{x}\widetilde{N}, \tag{7.100}$$

$$\mathbf{x}_5' = N\mathbf{x}_5\widetilde{N}, \tag{7.101}$$

$$x_5'^4 + x'^5\mathbf{i} = (x_5^4 + x^5\mathbf{i})N\widetilde{N}, \tag{7.102}$$

$$x'^4 + x'^5_5\mathbf{i} = (x^4 + x^5_5\mathbf{i})N\widetilde{N}. \tag{7.103}$$

With Eq. (1.43) these two last equalities read

$$x_5'^4 + ix'^5 = re^{i\theta}(x_5^4 + ix^5), \tag{7.104}$$

$$x'^4 + ix'^5_5 = re^{i\theta}(x^4 + ix^5_5). \tag{7.105}$$

This separation between the different components of the global space-time
explains why we usually see only the real components of the 4-dimensional
space-time vector $\mathbf{x}$. Only the usual space-time has real components. Equation (7.104) and Eq. (7.105) indicates both that these two supplementary
dimensions act as complex dimensions and that they separate completely
the usual space-time in the global space-time. A space-time with one or two
supplementary conditions has been used as early as [59]. The problem was
always to explain why classical physics does not see these supplementary
dimensions. Here this problem is automatically solved by the difference
coming from the invariance group of physical laws.

The form invariance of the geometric transformation f results from
Eq. (7.96) and Eq. (7.97) which give

$$f: y_v \mapsto x_v = (\Psi y_v \widetilde{\Psi})_v \tag{7.106}$$

$$f: y_v \mapsto x'_v = (\Psi' y_v \widetilde{\Psi}')_v = (N\Psi y_v \widetilde{\Psi}\widetilde{N})_v$$

$$= N(\Psi y_v \widetilde{\Psi})_v \widetilde{N} = N x_v \widetilde{N}. \tag{7.107}$$

Similarly to what we said in Sec. 3.3, y_v is independent of the observer and
intrinsic to the wave.

7.5   Existence of the inverse

To extend the complete wave to a system we need the inverse Ψ^{-1}, and
we are not in a field, only in an algebra, where the inverse does not always
exist. The standard model uses only left waves for the quarks. We get then
for the color r:

$$\Psi_r = \sqrt{2}\begin{pmatrix} 0 & -\eta_{2dr}^* & 0 & -\eta_{2ur}^* \\ 0 & \eta_{1dr}^* & 0 & \eta_{1ur}^* \\ \eta_{1ur} & 0 & \eta_{1dr} & 0 \\ \eta_{2ur} & 0 & \eta_{2dr} & 0 \end{pmatrix}, \tag{7.108}$$

and two similar equalities for colors g and b. Now we consider two matrices:

$$L = \sqrt{2}\begin{pmatrix} \eta_{1e} & \eta_{1n} & \eta_{1dr} & \eta_{1ur} \\ \eta_{2e} & \eta_{2n} & \eta_{2dr} & \eta_{2ur} \\ \eta_{1dg} & \eta_{1ug} & \eta_{1db} & \eta_{1ub} \\ \eta_{2dg} & \eta_{2ug} & \eta_{2db} & \eta_{2ub} \end{pmatrix} ; M = \sqrt{2}\begin{pmatrix} -\xi_{2e}^* & \eta_{1e} & \eta_{1dr} & \eta_{1ur} \\ \xi_{1e}^* & \eta_{2e} & \eta_{2dr} & \eta_{2ur} \\ 0 & \eta_{1dg} & \eta_{1db} & \eta_{1ub} \\ 0 & \eta_{2dg} & \eta_{2db} & \eta_{2ub} \end{pmatrix}. \quad (7.109)$$

We get the remarkable identity

$$\det(\Psi) = |\det(L)|^2 + |\det(M)|^2. \quad (7.110)$$

We can then see the waves of the standard model as having the maximum number of degrees of freedom compatible with the existence of an inverse wave Ψ^{-1}.

In the M matrix in Eq. (7.109) the g color is less present than the other colors b and g. This seems abnormal. Technically the reason is simple: since the only right term ξ_e is on the same column as the u_g wave, when we suppress all terms of one column or all terms of one line in the calculation of a determinant the u_g term necessarily disappears.

Here the main mathematical tool is Clifford algebra and not complex matrix algebra. The space-time algebra is not identical to the algebra of 4×4 complex matrices. The $Cl_{1,5}$ algebra of the 6-dimensional space-time is not identical to the algebra of 8×8 complex matrices. We encounter only sub-algebras, moreover not as complex algebras but as real algebras. The use of complex matrices is somewhere based on a kind of mathematical accident, a fortuitous coincidence: the identification between the Clifford algebra of the physical space and the algebra of 2×2 complex matrices, even if it is only as algebras on the real field.

For instance a consequence is that Ψ matrices are not at all symmetric. Zero terms are in columns, not in lines. Lines 1, 2 and 5, 6 of the Ψ matrix are multiplied by M while line 3,4 and 7, 8 are multiplied on the contrary by $\widetilde{M}$ when we get the form invariance. We may also remark that all terms of L and M matrices in Eq. (7.109) are left terms, multiplied by $\widehat{M}$ in the form invariance. When we look at operators in the electro-weak gauge group we see that they operate on columns of matrices, not on lines.

Another consequence: to take the adjoint is not an important transformation. This should be the case if the theory of hermitian and unitary matrices is fundamental. The main transformation, that we use again and again all over this book, is the reversion ($A \mapsto \widetilde{A}$). This reversion is defined in any Clifford algebra. It is this reversion and nothing else that appears in calculations. It happens that the reversion is equivalent to take the adjoint in space algebra, and only in space algebra. Both in the space-time algebra and in the algebra of 6-dimensional space-time the reversion is not equivalent to taking the adjoint and it uses also an exchange between the matrix bloc up-left and the matrix bloc down-right while the two other blocs remain at their places because they are twice exchanged. If the $SU(3)$ that exchanges the r, g, b color states was fundamentally unitary it should be effectively necessary for the green color to play exactly the same role as the red and blue colors.

And since this is not exactly true we see that the unitarity of the $U(1) \times SU(2) \times SU(3)$ structure is largely accidental from the mathematical point of view, which means that it does not imply the necessity of unitary groups. The main structure is the $Cl_{1,5}$ algebra and its multiplicative left and right automorphisms.

7.6   Wave equation with mass term

The wave equation [22]

$$0 = (\underline{D}\Psi)L_{012} + \mathbf{M}, \tag{7.111}$$

has for mass term

$$\mathbf{M} = \begin{pmatrix} m_2\rho_2\chi_b & m_2\rho_2\chi_g \\ m_2\rho_2\chi_r & m_1\rho_1\chi_l \end{pmatrix}, \tag{7.112}$$

with the a_j defined from the Ψ_l and s_j in Eq. (B.168) to Eq. (B.182) and

$$\rho_1^2 = a_1 a_1^* + a_2 a_2^* + a_3 a_3^* \; ; \quad \rho_2^2 = \sum_{j=1}^{j=15} s_j s_j^*. \tag{7.113}$$

Since only the $U(1) \times SU(2)$ part of the gauge group acts on the electron+neutrino wave, the wave equation acts separately in a lepton part and a quark part:

$$0 = (\underline{D}\Psi^l)L_{012} + m_1\rho_1 \begin{pmatrix} 0 & 0 \\ 0 & \chi_l \end{pmatrix} ; \quad \Psi^l = \begin{pmatrix} \Psi_l & 0 \\ 0 & 0 \end{pmatrix}, \tag{7.114}$$

$$0 = (\underline{D}\Psi^c)L_{012} + m_2\rho_2\chi^c; \quad \chi^c = \begin{pmatrix} \chi_b & \chi_g \\ \chi_r & 0 \end{pmatrix} ; \quad \Psi^c = \begin{pmatrix} 0 & \Psi_r \\ \Psi_g & \Psi_b \end{pmatrix}. \tag{7.115}$$

The χ_c, $c = r, g, b$ are defined in Eq. (B.184) to Eq. (B.186). The wave equation Eq. (7.114) is equivalent to

$$\mathbf{D}\Psi_l\gamma_{012} + m_1\rho_1\chi_l = 0 \; ; \quad \gamma_{012} = \gamma_0\gamma_1\gamma_2, \tag{7.116}$$

which is the equation that we have previously studied, with $m_1 = m$, $\rho_1 = \rho$. This wave equation is equivalent to the invariant form:

$$\widetilde{\Psi}_l(\mathbf{D}\Psi_l)\gamma_{012} + m_1\rho_1\widetilde{\Psi}_l\chi_l = 0 \; ; \quad \widetilde{\Psi}_l = \begin{pmatrix} \overline{\phi}_e & \phi_n^\dagger \\ \overline{\phi}_n & \phi_e^\dagger \end{pmatrix}. \tag{7.117}$$

We want to establish the double link between the lagrangian density and the wave equation for the complete wave equation Eq. (7.117). We shall use only this: the real part of the invariant equation is the sum of the lepton term previously studied and the corresponding term for the quark part of the wave equation. This one is equivalent to the invariant form:

$$0 = \widetilde{\Psi}^c(\underline{D}\Psi^c)L_{012} + m_2\rho_2\widetilde{\Psi}^c\chi^c, \tag{7.118}$$

$$\widetilde{\Psi}^c = \begin{pmatrix} \widetilde{\Psi}_b & \widetilde{\Psi}_r \\ \widetilde{\Psi}_g & 0 \end{pmatrix}; \quad \chi^c = \begin{pmatrix} \chi_b & \chi_g \\ \chi_r & 0 \end{pmatrix}. \tag{7.119}$$

From the covariant derivative Eq. (7.52), with the $\underline{P}_j$ operators in Eq. (7.7) to Eq. (7.13), the Γ_k in Eq. (7.46) to Eq. (7.51) and with Ψ^c in Eq. (7.115) we get

$$\underline{D}\Psi^c = \begin{pmatrix} A_g & A_b \\ 0 & A_r \end{pmatrix}, \tag{7.120}$$

$$A_g = \partial\Psi_g - \frac{g_1}{6}\mathbf{B}\Psi_g\gamma_{21} + \frac{g_2}{2}(\mathbf{W}^1\Psi_g\gamma_3\mathbf{i} + \mathbf{W}^2\Psi_g\gamma_3 - \mathbf{W}^3\Psi_g\mathbf{i})$$
$$+ \frac{g_3}{2}(\mathbf{G}^1\mathbf{i}\Psi_r - \mathbf{G}^2\Psi_r - \mathbf{G}^3\mathbf{i}\Psi_g + \mathbf{G}^6\mathbf{i}\Psi_b + \mathbf{G}^7\Psi_b + \frac{1}{\sqrt{3}}\mathbf{G}^8\mathbf{i}\Psi_g), \tag{7.121}$$

$$A_b = \partial\Psi_b - \frac{g_1}{6}\mathbf{B}\Psi_b\gamma_{21} + \frac{g_2}{2}(\mathbf{W}^1\Psi_b\gamma_3\mathbf{i} + \mathbf{W}^2\Psi_b\gamma_3 - \mathbf{W}^3\Psi_b\mathbf{i})$$
$$+ \frac{g_3}{2}(\mathbf{G}^4\mathbf{i}\Psi_r - \mathbf{G}^5\Psi_r + \mathbf{G}^6\mathbf{i}\Psi_g - \mathbf{G}^7\Psi_g - \frac{2}{\sqrt{3}}\mathbf{G}^8\mathbf{i}\Psi_b), \tag{7.122}$$

$$A_r = \partial\Psi_r - \frac{g_1}{6}\mathbf{B}\Psi_r\gamma_{21} + \frac{g_2}{2}(\mathbf{W}^1\Psi_r\gamma_3\mathbf{i} + \mathbf{W}^2\Psi_r\gamma_3 - \mathbf{W}^3\Psi_r\mathbf{i})$$
$$+ \frac{g_3}{2}(\mathbf{G}^1\mathbf{i}\Psi_g + \mathbf{G}^2\Psi_g + \mathbf{G}^3\mathbf{i}\Psi_r + \mathbf{G}^4\mathbf{i}\Psi_b + \mathbf{G}^5\Psi_b + \frac{1}{\sqrt{3}}\mathbf{G}^8\mathbf{i}\Psi_r). \tag{7.123}$$

Next we get

$$\widetilde{\Psi}^c(\underline{D}\Psi^c)L_{012} + m_2\rho_2\widetilde{\Psi}^c\chi^c$$
$$= \begin{pmatrix} \widetilde{\Psi}_b(A_b\gamma_{012} + m_2\rho_2\chi_b) + \widetilde{\Psi}_r(A_r\gamma_{012} + m_2\rho_2\chi_r) & \widetilde{\Psi}_b(A_g\gamma_{012} + m_2\rho_2\chi_g) \\ \widetilde{\Psi}_g(A_b\gamma_{012} + m_2\rho_2\chi_b) & \widetilde{\Psi}_g(A_g\gamma_{012} + m_2\rho_2\chi_g) \end{pmatrix} \tag{7.124}$$

The calculation of the Lagrangian density for the complete equation is similar to the calculation in the electron+neutrino case. We have

$$\mathcal{L} = \mathcal{L}_l + \mathcal{L}_c \tag{7.125}$$

$$\mathcal{L}_c = \sum_{c=r,g,b} \mathcal{L}_{0c} + g_1 \sum_{c=r,g,b} \mathcal{L}_{1c} + g_2 \sum_{c=r,g,b} \mathcal{L}_{2c} + g_3\mathcal{L}_3 + m_2\rho_2. \tag{7.126}$$

The calculation of $\mathcal{L}_{jc}$, for $j = 0, 1, 2$, replaces the e-n pair by the dc-uc pair and suppress the right ξ term. Then we get, on the model of Eq. (B.72) to Eq. (B.74):

$$\mathcal{L}_{0c} = \Re[-i(\eta_{dc}^\dagger \sigma^\mu \partial_\mu \eta_{dc} + \eta_{uc}^\dagger \sigma^\mu \partial_\mu \eta_{uc})], \tag{7.127}$$

$$\mathcal{L}_{1c} = -\frac{B_\mu}{6}(\eta_{dc}^\dagger \sigma^\mu \eta_{dc} + \eta_{uc}^\dagger \sigma^\mu \eta_{uc}), \tag{7.128}$$

$$\mathcal{L}_{2c} = -\Re[(W_\mu^1 + iW_\mu^2)\eta_{dc}^\dagger \sigma^\mu \eta_{uc}] + \frac{W_\mu^3}{2}(\eta_{dc}^\dagger \sigma^\mu \eta_{dc} - \eta_{uc}^\dagger \sigma^\mu \eta_{uc}). \tag{7.129}$$

Since three $SU(2)$ groups are contained in $SU(3)$ the calculation of $\mathcal{L}_3$ has similarities to the calculation of $\mathcal{L}_2$ and we get

$$\begin{aligned}
\mathcal{L}_3 = &- \Re[(G_\mu^1 + iG_\mu^2)(\eta_{dr}^\dagger \sigma^\mu \eta_{dg} + \eta_{ur}^\dagger \sigma^\mu \eta_{ug})] \\
&- \Re[(G_\mu^4 + iG_\mu^5)(\eta_{dr}^\dagger \sigma^\mu \eta_{db} + \eta_{ur}^\dagger \sigma^\mu \eta_{ub})] \\
&- \Re[(G_\mu^6 + iG_\mu^7)(\eta_{dg}^\dagger \sigma^\mu \eta_{db} + \eta_{ug}^\dagger \sigma^\mu \eta_{ub})] \\
&+ \frac{G_\mu^3}{2}(-\eta_{dr}^\dagger \sigma^\mu \eta_{dr} - \eta_{ur}^\dagger \sigma^\mu \eta_{ur} + \eta_{dg}^\dagger \sigma^\mu \eta_{dg} + \eta_{ug}^\dagger \sigma^\mu \eta_{ug}) \\
&+ \frac{G_\mu^8}{2\sqrt{3}}(-\eta_{dr}^\dagger \sigma^\mu \eta_{dr} - \eta_{ur}^\dagger \sigma^\mu \eta_{ur} + 2\eta_{db}^\dagger \sigma^\mu \eta_{db} \\
&+ 2\eta_{ub}^\dagger \sigma^\mu \eta_{ub} - \eta_{dg}^\dagger \sigma^\mu \eta_{dg} - \eta_{ug}^\dagger \sigma^\mu \eta_{ug}). \tag{7.130}
\end{aligned}$$

Like in the lepton case, the real part of the wave equation is simply the equality

$$\mathcal{L} = 0. \tag{7.131}$$

This link between the wave equation and the Lagrangian density is very strong from the mathematical point of view, since it comes from an algebraic calculation, similar to taking the real part of a complex number. The calculation going from the Lagrangian density, by the variational calculus and an integration by parts, is very dubious from the physical point of view for propagating waves. This method is nevertheless always available on the mathematical point of view. It is in this way that we got the wave equation Eq. (7.111) and the χ_c. Similarly to Eq. (6.150) only one numeric equation coming from Eq. (7.117) is simple, the law of conservation of the total current:

$$\partial_\mu J_t^\mu = 0 \tag{7.132}$$

$$J_t = \phi_{dr}\phi_{dr}^\dagger + \phi_{ur}\phi_{ur}^\dagger + \phi_{dg}\phi_{dg}^\dagger + \phi_{ug}\phi_{ug}^\dagger + \phi_{db}\phi_{db}^\dagger + \phi_{ub}\phi_{ub}^\dagger. \tag{7.133}$$

7.6.1 *Form invariance of the wave equation*

Under the Lorentz dilation R induced by an invertible M satisfying

$$x' = MxM^\dagger \; ; \quad \det(M) = re^{i\theta} \; ; \quad x = x^\mu \sigma_\mu \; ; \quad x' = x'^\mu \sigma_\mu, \qquad (7.134)$$

$$\eta'_{uc} = \widehat{M}\eta_{uc} \; ; \quad \eta'_{dc} = \widehat{M}\eta_{dc} \; ; \quad \phi'_{dc} = M\phi_{dc} \; ; \quad \phi'_{uc} = M\phi_{uc}, \qquad (7.135)$$

$$\Psi'_c = \begin{pmatrix} \phi'_{dc} & \phi'_{uc} \\ \widehat{\phi}'_{uc} & \widehat{\phi}'_{dc} \end{pmatrix} = \begin{pmatrix} M & 0 \\ 0 & \widehat{M} \end{pmatrix} \begin{pmatrix} \phi_{dc} & \phi_{uc} \\ \widehat{\phi}_{uc} & \widehat{\phi}_{dc} \end{pmatrix} = N\Psi_c \; ; \quad c = r, g, b, \qquad (7.136)$$

we then get

$$\underline{N} = \begin{pmatrix} N & 0 \\ 0 & N \end{pmatrix} \; ; \quad \underline{\partial} = L^\mu \partial_\mu = \begin{pmatrix} 0 & \partial \\ \partial & 0 \end{pmatrix}, \qquad (7.137)$$

which implies

$$\Psi'^c = \underline{N}\Psi^c; \quad \widetilde{\Psi}'^c = \widetilde{\Psi}^c \underline{\widetilde{N}}; \quad \underline{\widetilde{N}} = \begin{pmatrix} \widetilde{N} & 0 \\ 0 & \widetilde{N} \end{pmatrix}; \quad \underline{D} = \underline{\widetilde{N}}\, \underline{D'}\underline{N}. \qquad (7.138)$$

We then get

$$\widetilde{\Psi}^c (\underline{D}\Psi^c) L_{012} = \widetilde{\Psi}^c \underline{\widetilde{N}}\, \underline{D'}\underline{N}\Psi^c L_{012} = \widetilde{\Psi}'^c (\underline{D'}\Psi'^c) L_{012}. \qquad (7.139)$$

It remains to study the mass term. All s_j are determinants of terms similar to ϕ, which implies:

$$s'_j = \det(\phi') = \det(M\phi) = \det(M)\det(\phi) = re^{i\theta} s_j, \qquad (7.140)$$

$$s'^*_j = re^{-i\theta} s^*_j; \quad \rho'_2 = r\rho_2. \qquad (7.141)$$

This gives

$$\chi'^c = \begin{pmatrix} \chi'_b & \chi'_g \\ \chi'_r & 0 \end{pmatrix}; \quad r^2 \rho_2^2 \chi'^c = \rho'^2_2 \chi'^c = \begin{pmatrix} re^{-i\theta} M & 0 \\ 0 & re^{i\theta} \widehat{M} \end{pmatrix} \rho_2^2 \chi^c, \qquad (7.142)$$

$$\chi'^c = \begin{pmatrix} r^{-1}e^{-i\theta} M & 0 \\ 0 & r^{-1}e^{i\theta} \widehat{M} \end{pmatrix} \chi_c = \widetilde{N}^{-1}\chi^c, \qquad (7.143)$$

$$\widetilde{\Psi}'^c \chi'^c = \widetilde{\Psi}^c \widetilde{N}\widetilde{N}^{-1}\chi^c = \widetilde{\Psi}^c \chi^c. \qquad (7.144)$$

Therefore the form invariance of the wave equation is equivalent to the following condition on the mass term:

$$m'_2 \rho'_2 = m_2 \rho_2 \; ; \quad m'_2 r = m_2. \qquad (7.145)$$

And we saw in chapter 3 the link between this relation and the existence of the Planck constant [18]. A detailed study of the gauge invariance is in appendix B. Like in the case of the lone electron or with the electron-neutrino pair, the wave equation Eq. (7.111), that describe all objects of the first generation, particles and antiparticles, is a wave equation with mass term, form invariant then relativistic invariant, and gauge invariant under the $U(1) \times SU(2) \times SU(3)$ group.

7.7　Charge of quarks

We saw how the fractional charge of quarks is determined by the only change of the P_0 operator in Eq. (6.13) into the similar P_0' operator in Eq. (7.13). Since the right part of the quark wave is canceled this operator is reduced to

$$P_0'(\Psi_c) = k\Psi_c\gamma_{21}, \quad c = r, g, b; \quad k = -\frac{1}{3}. \tag{7.146}$$

It is easy to establish that the k factor must be the same for all quarks, because if it was not the same we should not be able to get the gauge invariance under the $U(1) \times SU(2) \times SU(3)$ group. But where does the $k = -\frac{1}{3}$ value come from? To see this, we separate in the complete wave its left part L and its right part:

$$L = \frac{1}{2}(\Psi + i\Psi L_{21}) = \begin{pmatrix} \Psi_{lL} & \Psi_r \\ \Psi_b & \Psi_c \end{pmatrix}. \tag{7.147}$$

We use S such as

$$S = L_{012345}; \quad P^+ = \frac{1}{2}(1 + S); \quad P^- = \frac{1}{2}(1 - S). \tag{7.148}$$

We then get

$$\underline{P}_0(L) = \begin{pmatrix} \Psi_{lL} & k\Psi_r \\ k\Psi_b & k\Psi_c \end{pmatrix} L_{21}$$

$$= (P^+LP^+ + kP^+LP^- + kP^-LP^+ + kP^-LP^-)L_{21} \tag{7.149}$$

$$= [\frac{1+3k}{4}L + \frac{1-k}{4}(SL + LS + SLS)]L_{21}. \tag{7.150}$$

The choice $k = -\frac{1}{3}$ that we made in Sec. 6.3 and that gives the value of each charge of quark, value coming from the model of quarks, is then the choice allowing to simplify the calculations: the $\underline{P}_0$ operator has only one term, and this term is symmetric on S:

$$\underline{P}_0(L) = \frac{1}{3}(SL + LS + SLS)L_{21}. \tag{7.151}$$

The study of the boson part of the standard model will allow us to see what is interesting in this simplification. The fractional charge of quarks is then not at all arbitrary; the choice $k = -\frac{1}{3}$ is necessary to get the simplified form of the $\underline{P}_0$ operator.

Chapter 8

Magnetic monopoles

We present here the recent experimental works on magnetic monopoles. Next we apply to the magnetic monopole our study of electro-weak interactions.

8.1 Russian experimental works

Recent experimental work on magnetic monopoles began with V.F. Mikhailov [60]. He was continuing work made fifty years ago by F. Ehrenhaft. An electric arc produces ferromagnetic dusts that are conducted by Ar gas into a chamber where a laser lights them up. Into the chamber the ferromagnetic particles are moved by a magnetic field and an electric field orthogonal to the magnetic field. The direction of the fields may be reversed. Movements are observed, under the light of the laser, with an optical microscope.

The measurement of the magnetic charge of these particles took advantage of the fact that some of them have also an electric charge and the movement of an electric charge in an electric field is well known. Mikhailov observed an elementary magnetic charge $g = n\alpha\dfrac{e}{6}$. The fine structure constant α is small ($\alpha \approx \frac{1}{137}$).

But the expected value is completely different [38]. A calculation made by Dirac, obtained again in a very smart way by G. Lochak from his theory of the monopole [48] gives for the elementary magnetic charge

$$\frac{eg}{\hbar c} = \frac{n}{2}, \tag{8.1}$$

where n is an integer. The elementary magnetic charge observed by Mikhailov is much smaller than the theoretical charge. We may ask if there is a reason to refute the theoretical calculation, or if there exists an experimental reason for this divergence. Both things are possible: each derivation of Eq. (8.1) includes a calculation of the potentials created by charges, and we can doubt its validity. The magnetic charges observed by Mikhailov were visible only during the illumination by the strong laser light, and there may be second order effects coming from this illumination. Mikhailov realized also an experiment where the ferromagnetic particles were included into water droplets, with spherical symmetry. Then he measured magnetic charges compatible with the elementary magnetic charge calculated by Dirac. The value of such a charge is then a question that must be solved experimentally.

The experimental work of L. Urutskoev has in common with Mikhailov's work only the use of an electric arc. To shatter concrete, little holes are made and filled with water, an electric wire is put into each hole and an electric condenser is discharged into the wires. The discharge produces an explosion and this explosion shatters the concrete. The first astonishing fact was the great speed of the pieces of concrete smashed by the explosion, this induced a need to better study what was going there.

The continuation of experiments was to shoot into pure water, without concrete. An intense glow was found to appear above the device. The duration of this phenomenon, about 5ms, was much greater than the duration of the discharge, 0.15ms. A spectral analysis of the emitted light was performed. Spectral lines of nitrogen or oxygen were very weak, while the glow was emitted into the air. The strongest spectral lines showed the presence of Ti, Fe, Cu, Zn, Cr, Ni, Ca and Na. The presence of Cu and Zn could come from the electric wires, the presence of Ti signified that the Ti foils used in discharges spread above the device, in spite of the cover. The presence of the other elements was enigmatic. This induced Urutskoev to analyze more finely the metal powder resulting from the explosion of the Ti foil in water. Observations made were still stranger. While the foil was made of 99.7% Ti the ratio of Ti in the powder may go down to 92%. The amount of titanium disappearing corresponded to the amount of new elements appearing, Fe, Si, Al, Ca, Na, Cu and Zn, principally. In addition, an isotopic analysis shows that the isotopic composition of Ti had changed, with a significant decline in the ratio of ^{48}Ti. The experiments were repeated many times, with all necessary precautions. Other metals were used, in particular zirconium. The ratios of different outside elements change depending on the composition of the exploded foil. For instance there is much more Cr with zirconium than with titanium, and much less Si and Al.

Since the transformation from an element to another is usually associated with radioactivity, an intensive search of radioactive emission was made. There were no X or γ rays detected, in spite of 10^{19}–10^{20} transformed atoms each experiment. Detection of neutrons was also performed. Scintillator detectors indicated a pulse that allowed to estimate the speed of the radiation as $20-40$ m/s. Such a low speed cannot match a neutron flux, because neutrons should be ultra cold. An attempt to detect the radiation with photo emulsions was made, for lack of better means. We will come back later on what was seen with these films. Urutskoev saw next that the presence of a strong magnetic field changed aspects of these traces, and he deduced that the radiation coming out of his shots had magnetic properties. He then led experiments to trap the radiation with strong magnets and he used the Moessbauer effect to prove the reality of these captures.

Urutskoev noted also that the transformations come principally from even-even kernels, that is to say from kernels with an even number of protons and an even number of neutrons. He noticed that the mean binding energy of produced kernels is very few different from the mean binding energy of initial kernels: there is no nuclear energy emitted or absorbed in significant amounts. And all the produced kernels are in the ground state; there is no radioactivity.

Experiments by N. G. Ivoilov [43] indicate that it is possible to get similar traces on photographic films with much less energy: he discharged an electric arc into water, with a current not exceeding 40 A with an 80 V tension. He got traces that agreed with properties of magnetic monopoles predicted by the G. Lochak's theory.

8.2 Works at E.C.N.

Research performed at the Ecole Centrale de Nantes, in the laboratory of Guillaume Racineux by Didier Priem and Claude Daviau [54] with the help of Henri Lehn † and the Fondation Louis de Broglie, aimed to satisfy and to continue Urutskoev's work. This seemed necessary in view of the extraordinary nature of obtained results. The experimental device is dependent on the available equipment at the E.C.N. and is different, even if it is as few ways as possible, from that used by Urutskoev. The generator is an American one, Maxwell type, maximum power 12 kJ at 8.4 kV, capacity 360 μF and a vessel (Figure 1). The first containment vessel was made of aluminum. It was replaced by a second vessel to allow the gas produced during an experiment to be collected. Experiments by Urutskoev allow him to conclude that the gas is almost totally hydrogen. This second vessel was made of stainless steel and it contained a tank with an internal diameter of 20 mm covered by polyurethane. The internal diameter was then reduced to 16 mm which improved the yield. A third vessel was made when the second was worn out. The current coming from the generator is distributed into two electrodes, one up and one down. They are linked by a fuse made of Ti40 titanium.

Figure 1: Vessel

After an experiment, the gas is collected and its volume is measured. Powders are collected with the liquid which contains them, and are placed for 24 hours under a photographic plate exposed to the radiation coming out of powders. This photographic film is then developed and examined with the optical microscope. Powders are then dessicated and examined with the electronic microscope of the E.C.N. This allows us to get three kinds of results, about powders, gas and traces on the photographic films.

8.2.1 *Results about powder and gas*

Our observations confirm the results obtained by L. Urutskoev, even if our ratios of production are lower than those he got. The energy of the discharge being lower than that of Urutskoev, and the discharge being shorter, this is not astonishing. But aside from this, the strange elements which we got spectrograms using the electronic microscope have a composition very near that obtained by Urutskoev. At the same time our observations make the things still stranger: if we noticed the presence of one per cent of iron in our powders, this iron is not dispersed a little everywhere. On the contrary what we notice is: one per cent of the particles are made of so much iron than titanium is quasi-missing. It is often iron which is dominating but there are experiments where we find more copper than iron. The particles made of copper have different scales, some are numerous and have length of about one micrometer, others (much rarer) are bigger and even visible to the naked eye. Those particles contain very little titanium. The composition of the exotic particles may be more complicated: we observe particles of iron-chromium, and of copper-zinc. Iron is rarely alone, it is most of time with chromium, a little nickel, sometimes 1% manganese, and with carbon and oxygen. The composition of particles is often not homogeneous, a particle may have not transformed titanium at places while at another place nearly all the titanium may have been replaced.

Figure 2 shows a particle with an evident continuity, which has dark places and one light place, in addition to many holes. On the left and above, titanium remains intact. At the center, the spectral analysis indicates the following mass composition: Fe 69.8%, Ti 10.81%, Ni 7.28%, Cr 4.33%, O 3.98%, C 3.8%. Holes are also significant, because they indicate gas production just before the solidification caused by the intense cooling in water.

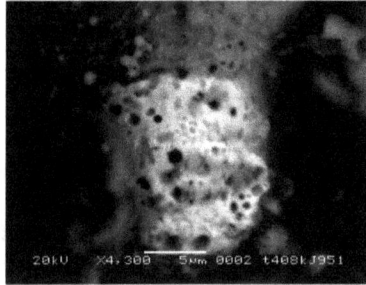

Figure 2: Particle with an iron place

The fact that iron is rarely alone, and that it appears with chromium and nickel has much complicated our work, because the stainless steel of our tank is made of those three metals, and it was possible that the stainless steel of our tank contaminated the powders. Stainless steel was therefore removed from the inside of the tank, which now contained only titanium and polyurethane. The suppression of the stainless steel changed nothing actually, there was still iron in the powders when the only metal inside the tank was titanium. This was predictable since the composition indicated above is not that of the stainless steel of our vessel. We can also easily verify that the titanium used to make our fuse does not contain the ratio of iron, copper and other materials found in the powders.

Extraordinary results obtained by Urutskoev are therefore real. The one who thinks they are impossible only has to reproduce the experiment. If he is honest, he will be obliged to see that something really happens.

But nothing should happen: the conditions of the experiment generate an energy measured in kJ, this provides only a hundred of eV at most to each concerned atom of titanium. This is ridiculously small in comparison with nuclear binding energies. In addition the interactions known then work in a completely different way. For instance weak interactions allow one proton of a kernel to transform into a neutron, or vice versa, and that is subject to the general laws of quantum mechanics, where randomness plays an obligatory and permanent role. If the kernel of a titanium atom was transformed by weak interaction, it could give a kernel of scandium or vanadium. Neither of those metals was seen. We saw vanadium rays not once, and vanadium is obligatory if you want to go, with weak interactions, from titanium to iron or copper. And if weak interactions were happening, transformed kernels should arrive at random, in time and in space, not in macroscopic bundles.

We must not forget that our experiment is an explosion and an explosion is not exactly the best way to assemble some dispersed atoms into a packet. It is on the other hand a very good way to disperse concentrated matter. Since we see particles made of iron, or of copper, or of nickel, or of iron-chromium, with very little titanium, these elements were produced together. We do not understand how it is possible, but that changes nothing about the reality of the phenomenon.

In addition there are energy constraints. The mass of the elements found in our powders that should not be there is 10^{10} times greater than the mass of the energy brought by the electric discharge. To excellent precision we can then say that the total energy of the produced atoms is equal to the total energy of the destroyed atoms. This conservation of the total energy restricts considerably the possibilities of reaction. We cannot get for instance vanadium. The isotopes of vanadium are heavier than those of titanium, which allows to ^{48}V to be β^+ radioactive and to disintegrate into ^{48}Ti. And as we have detected no radioactivity linked to these transformations, it is necessary that the total number of electrons, protons and neutrons are also conserved. So strange as it may be, all these conditions of conservation do not forbid the observed transformations. As Urutskoev said, it is as if for instance 100 kernels of ^{48}Ti go together for some reason to form a big "kernel", then reallocate their nucleons to form at the same time lighter and heavier kernels. Doing so they also respect the conservation laws of energy, electric charge, baryonic charge, leptonic charge... And in addition this magical transformation is accompanied by no significant radioactivity!

Some gas was always produced during the metallurgical works made at the Ecole Centrale. The presence of this gas was considered as a nuisance, limiting the repetitiveness of the experiments. The device that we use allows us to easily measure the quantity of produced gas. This gas is quasi totally made of hydrogen. As titanium heated to a very high temperature is a reducing agent, this is not surprising. It is also difficult to estimate the quantity of oxygen going into the powders as oxide or dioxide of titanium, or dissolving into water. We have estimates indicating that a part of the hydrogen does not come from the dissociation of water. To prove this an experiment in heavy water has been done with success by L. Urutskoev. He got not only D_2, but also HD and H_2. And this hydrogen cannot come from the water. Transformations of titanium can leave isolated protons and electrons which form hydrogen atoms. This hydrogen, either of chemical origin or not, is formed inside particles, which are often so spongy that they float on the water in which we collect the powders.

8.2.2 *Stains*

After each trial, the titanium powders from the fusible were collected along with the water contained in the trial chamber and are placed under a photographic plate. The traces are produced, not immediately in the electric arc, but by what is in the water and powders, and only several hours after the electric arc. Sometimes, some things emerge from the water, it is not only the things that make the traces, but also a part of the powders on the surface of the water. They emerge from the water, despite the gravity and the surface tension of water, and are glued on the wrapping paper of the photographic plate (experiments 103, 62, 79)

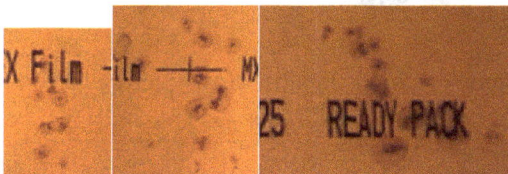

Figure 3: Stained paper.

8.2.3   *Traces*

Powders collected after an experiment and the photographic plate, following both indications of Urutskoev and Lochak, are placed between two metallic plates forming a plane condenser, under a low 10V tension. The movement of a magnetic monopole in a fixed uniform electric field is analogous to the movement of an electric charge in a fixed uniform magnetic field. The Laplace force is

$$\vec{F} = g(\vec{H} - \frac{\vec{v}}{c} \times \vec{E}),\tag{8.2}$$

where g is the charge of the magnetic monopole. In a constant electric field orthogonal to the plane of the plate, a monopole must have a circular trajectory. We expect rotations into the plane of the photographic plate, and it is what happens rather often, as figures 4 to 8 show.

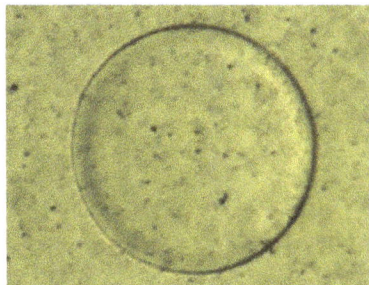

Figure 4: Circle. Diameter: 0.2 mm

Figure 5: Circle. Length of the picture: 2.6 mm

Figure 6: Circle. Length of the picture: 2.47 mm

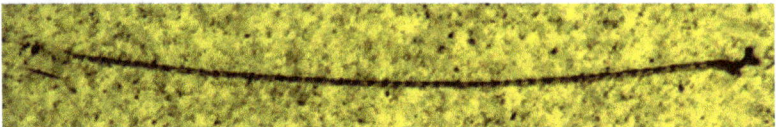

Figure 7: Circle. Length of the picture: 0.95 mm

Figure 8: Circle. Length of the picture: 1.45 mm

We must not expect all traces to be circular, because the presence of glass dishes between plates induces a non-uniform electric field. We must also notice we do not know *a priori* what we seek, we see probably only a little part of traces, in the absence of knowing completely the dynamics of magnetic monopoles. We also do not know how monopoles interact with the photographic plate. It is easier to see the very long and stark traces, more difficult to see the short and weak traces. Circles are not the only curved traces, we obtain also horseshoes:

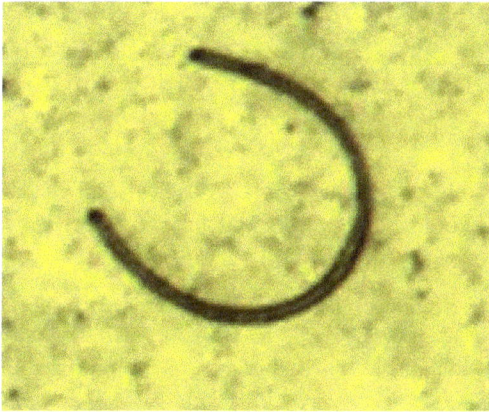

Figure 9: Horseshoe. Length of the picture: 0.19 mm

We must expect imperfect circles, notably because the loss of energy gives a smaller radius. This is visible in the following pictures

Figure 10: Braking. Length of the picture: 1.78 mm

Figure 11: Braking. Length of the picture: 1.9 mm

Figure 12: Braking. Length of the picture: 0.57 mm

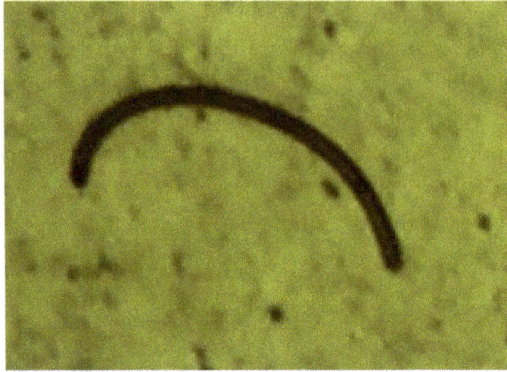

Figure 13: Braking. Length of the picture: 0.2 mm

Large traces, as in figures 9 or 13, are actually double traces. This doubling of traces is more visible when the two traces are well separated:

Figure 14: Double trace. Length of the picture: 2.67 mm

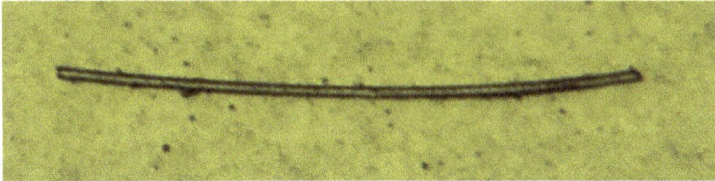

Figure 15: Double trace. Length of the picture: 0.58 mm

The magnetic monopole of G. Lochak is a chiral object, built from an angle which is pseudo-scalar. The simplest object of our everyday world explaining what is chirality is a screw. There are left screws and right screws. This property is satisfied for several observed [1] traces. We can see spirals, often with difficulty. Sometimes the spiral is very visible, as on this trace:

---

1. This is at the moment the best proof of the predictive power of Lochak's idea of a Dirac wave for the magnetic monopole.

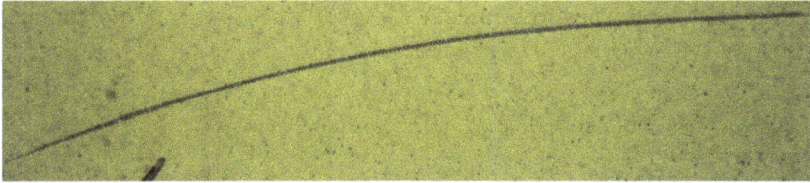

Figure 16: Spiral trace. Length 2mm

and its magnifications:

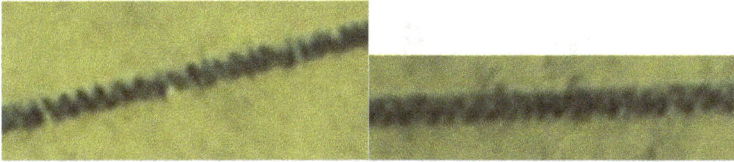

Figures 17 and 18: Magnifications of figure 15

Undulations are often seen on magnifications of our pictures:

Figure 19: Wave. Length of the picture: 2.67 mm

A wavelength is directly measurable in this picture, where we count 30 wavelengths. This gives a 89μm wavelength. A wavelength is also directly measurable in the following picture:

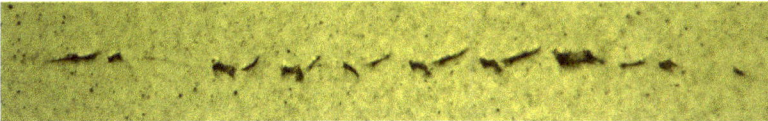

Figure 20: Wave. Length of the picture: 1.54 mm

Considering the four undulations in the middle we can estimate the wavelength: 130μm. Moreover a second thing is visible on this picture, a double pattern with alternately rising and descending traces.

The Lochak's theory of the magnetic monopole can account for this double pattern: the wave is a Dirac spinor made of two Weyl spinors, a right one and a left one. If the proper mass of the monopole is null these two Weyl spinors are independent and may move one without the other. If the proper mass is not null the two Weyl spinors are coupled by the mass term. Perhaps what we see in figure 20 is exactly that, a left wave and a right wave, of which we see only pieces. They are superposed at the ends and successively seen in the middle. A double pattern is rather common, as we can see in the following figures:

Figure 21: Double pattern. Length of the picture: 2 mm

The wavelength is estimated 143μm.

Figure 22: Double pattern. Length of the picture: 1.64 mm

Here the wavelength is estimated 65μm.

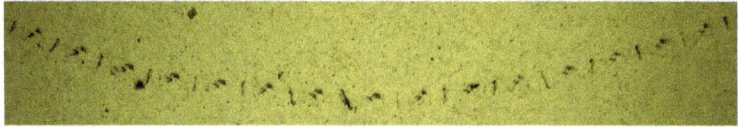

Figure 23: Double pattern. Length of the picture: 2.65 mm

The wavelength is estimated 177μm. If the wavelength is the de Broglie's wavelength, not[2] an artifact, it is possible to calculate the impulse:

$$p = mv = \frac{h}{\lambda} \qquad (8.3)$$

For the wave of figure 22 where the wavelength is the shortest the impulse is about 10^{-29} kgm/s. The big question is then the velocity of the magnetic monopole. If it is the speed of light the energy is very small. Can a wave with only 0.02 eVc^{-2} make the visible effects in figure 22? This is dubious. The only experimental velocity was given by Urutskoev and it is very low: 20–40 m/s. A velocity of 20 m/s gives then a mass: $5 \cdot 10^{-31}$ kg, similar to the proper mass of the electron. A velocity still lower is possible since it is perhaps at the end of the braking that we saw this trace. Another theoretical possibility is given by Eq. (5.23) where the limit speed has a null limit when ϵ is near 1.

2. G. Lochak thinks that what we see is not the de Broglie's wavelength, but only a scale corresponding to the response of the plate to the movement of the wave. But then why two patterns?

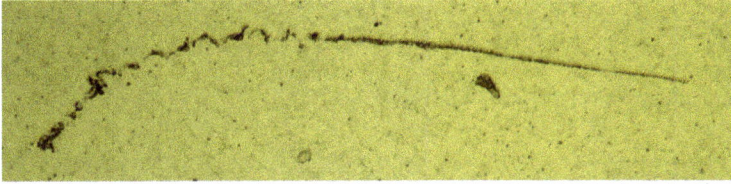

Figure 24: Continuous-broken trace. Length of the picture: 1.47 mm.

Continuity of many traces is only an artifact arising from a blurred picture. We can see this on the next picture, where a numeric enlargement allows us to count grains and to estimate the distance between two grains: 8μm.

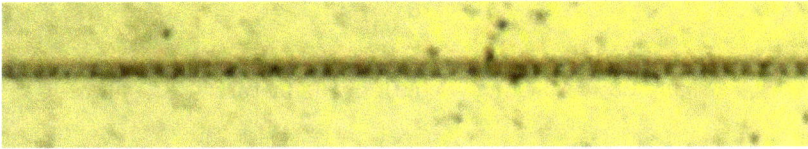

Figure 25: Enlarged trace. Length of the picture: 0.38 mm.

Another frequent aspect of our traces is the quasi-parallelism of very long traces, as in the following figure:

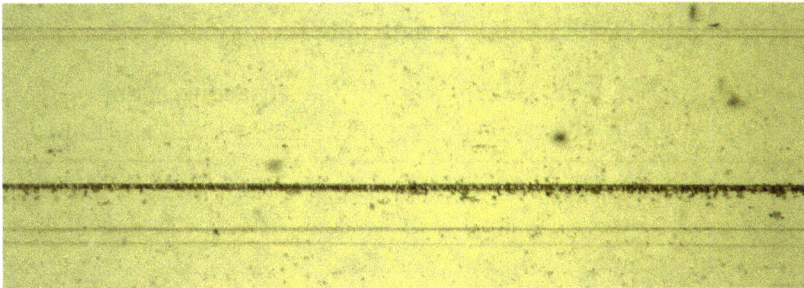

Figure 26: Multiple traces. Length of the picture: 1.97 mm.

Figure 26 shows only a part of each trace which extends on the two sides of the picture. We see five traces nearly parallel and we guess two other ones. We can suppose the double character of these traces is linked to the double character of the wave, with a left and a right part. Following this hypothesis we can conclude that single traces are due to superimposed left and right parts. The parallelism of some traces can come from a weak separation of divergent traces, as in the following figure:

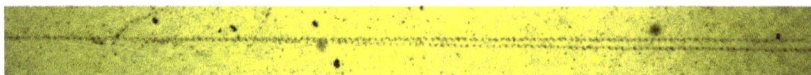

Figure 27: Divergent traces. Length of the picture: 2.67 mm.

These traces present obviously a granular structure, with very similar distances between grains: this argues strongly for the hypothesis of a unique wave, left and right. The wavelength is estimated as 19.6μm. A few branchings between traces may be seen:

Figure 28: Branching. Length of the picture: 1.87 mm.

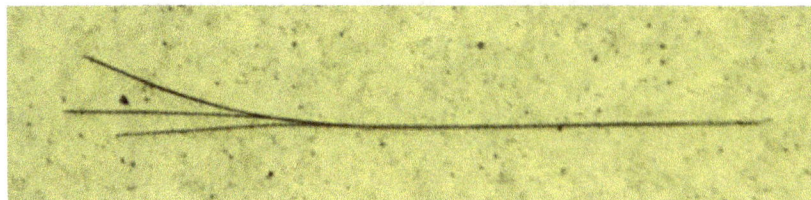

Figure 29: Branching. Length of the picture: 0.77 mm.

One trace favoring best the hypothesis of the left and right spinors is the following, with a magnification of the upper trace and another of the down trace:

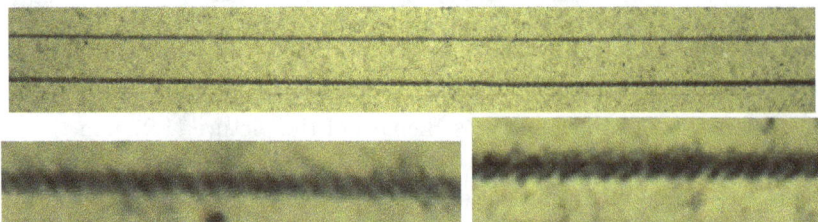

Figure 30: Double spiral. Length of the picture: 1.94 mm.

The two magnifications are similar to two screws turning in the opposite direction.

All these traces show stark differences from the physics of particles with an electric charge. To see the left or right nature of a trace will necessitate a three-dimensional observation of these traces. Such observations show that monopoles make depressions on the surface of the plate [26].

8.3 Electrons and monopoles

The invariant wave equation Eq. (3.1) of the electron was obtained from the wave equation of Lochak's magnetic monopole Eq. (3.3) in the particular case Eq. (3.4) where the wave equation is homogeneous. To do this we replaced the local chiral gauge by the local electric gauge. We shall then get the invariant wave equation of the magnetic monopole by using the inverse transformation, replacing the electric gauge by the chiral gauge. We read this gauge in space algebra as:

$$\phi' = e^{ia}\phi \; ; \; QB' = QB - \nabla a \; ; \; Q = \frac{g}{\hbar c}, \tag{8.4}$$

where a is a real number and where g is the charge of the magnetic monopole. iB is the pseudo-vector of space-time magnetic potential, which is also the Cabibbo–Ferrari's potential of the theory of the monopole as well as the potential term that is multiplied by the projector P_0 in Eq. (6.22). The invariant wave equation of the magnetic monopole reads then :

$$\overline{\phi}(\nabla\widehat{\phi})\sigma_{21} + \overline{\phi}QiB\widehat{\phi}\sigma_{21} + m\rho = 0. \tag{8.5}$$

First difference with the case of the electron: this wave equation has no linear approximation. It is not allowed to add an $e^{-i\beta}$ term into the mass term because β is not chiral gauge invariant.

To get the 8 numeric equations of this invariant wave equation we use a space-time vector U satisfying

$$\overline{\phi}QB\widehat{\phi} = U^{\mu}\sigma_{\mu}, \tag{8.6}$$

and we get in the place of Eq. (3.10) to Eq. (3.17) the system

$$0 = w^3 - U^3 + m\rho, \tag{8.7}$$

$$0 = v^2, \tag{8.8}$$

$$0 = -v^1, \tag{8.9}$$

$$0 = w^0 - U^0, \tag{8.10}$$

$$0 = -v^3, \tag{8.11}$$

$$0 = w^2 + U^1, \tag{8.12}$$

$$0 = -w^1 - U^2, \tag{8.13}$$

$$0 = -v^0. \tag{8.14}$$

As with the electron the scalar part of the invariant wave Eq. (8.7) is the Lagrangian density. Lochak immediately remarked that in this Lagrangian density the current $J = D_0$ is replaced by $K = D_3$. From there comes in Eq. (8.7) the 3 index instead of a 0 index. We can say that the invariant wave equation is somewhere simpler than the invariant wave of the electron: all four v^μ terms are zero. This means that the four D_μ vectors are conservative. We recall that the density D_0^0 gives in the case of the electron what quantum theory sees, from the Schrödinger equation, as a probability density. Lochak has proved that $K = D_3$ is the conservative current linked to the invariance of the Lagrangian density Eq. (8.7) under the chiral gauge Eq. (8.4). Vectors D_1 and D_2, equally conservative, are unknown in the formalism of Dirac matrices. We have seen in Eq. (2.86) and Eq. (2.87) that the electric gauge gives a rotation in the (D_1, D_2) plane. With the chiral gauge all four D_μ are invariant. They are with Eq. (2.39) the elements of an orthogonal basis, and their components are the elements of the matrix of the dilatation D in Eq. (3.44).

8.3.1   *Charge conjugation*

We use again the link between the wave of the particle and the wave of the antiparticle. We note the wave of the antimonopole ϕ_a:

$$\widehat{\phi} = \widehat{\phi}_a \sigma_1 \; ; \; \phi_a = -\phi\sigma_1 \; ; \; \overline{\phi} = \sigma_1 \overline{\phi}_a. \tag{8.15}$$

The invariant wave equation is then read as

$$\sigma_1 \overline{\phi}_a (\nabla\widehat{\phi}_a)\sigma_1\sigma_{21} + \sigma_1\overline{\phi}QiB\widehat{\phi}_a\sigma_1\sigma_{21} + m\rho = 0. \tag{8.16}$$

Multiplying on the right and on the left by σ_1 we get

$$-\overline{\phi}_a(\nabla\widehat{\phi}_a)\sigma_{21} - \overline{\phi}_a QiB\widehat{\phi}_a\sigma_{21} + m\rho = 0. \tag{8.17}$$

This is usually simplified into

$$\overline{\phi}_a(\nabla\widehat{\phi}_a)\sigma_{21} + \overline{\phi}_a QiB\widehat{\phi}_a\sigma_{21} - m\rho = 0. \tag{8.18}$$

Therefore Lochak remarked immediately that the charge conjugation does not change the sign of the magnetic charge, contrary to the case of the electric charge. Then there is no polarization of the vacuum from magnetic charges [46] [47] [48]. But the form invariance of the wave equation indicates that the true wave equation is Eq. (8.17), not Eq. (8.18). It should then be more correct to say that, contrary to the case of the electron, the charge conjugation changes here not only the differential term, but also the charge, therefore it does not change the gauge nor the sign of the mass–energy.

8.3.2 *The interaction electron-monopole*

The space-time vector B is, like the vector electromagnetic potential A, a contravariant vector. This is correct because O. Costa de Beauregard explained [28] why potential terms are moving with sources that are electric and magnetic charges. The QB vector, similar to the qA vector, is a covariant vector (see chapter 4). This allows the interaction by gauge invariance. We have seen in chapter 6 that the Weinberg–Salam θ_W angle is invariant under the group of dilations. An electric charge creating a potential A creates then also, with Eq. (6.85), a potential :

$$B = \cos(\theta_W)A. \tag{8.19}$$

Since this B potential is present in the wave equation of the magnetic monopole, it is able to interact with the electric charge. This interaction was detailed by Lochak. The basis of his calculation is the continuity of the wave function under the group of rotations. The continuity of the wave being comforted by the continuity of the potential, it is not necessary to review the calculation and we can use [46] [48]. The B potential used there was questionable because it is not continuous in each point of the z axis. It is why the result, even if the physical reasoning was perfect, is a little too short. In the case of a potential created by an electric charge we have

$$A^0 = -\frac{e}{r} \; ; \; B^0 = \cos(\theta_W)A^0 = \cos(\theta_W)(-\frac{e}{r}) = -\frac{e\cos(\theta_W)}{r} = -\frac{e'}{r}, \tag{8.20}$$

where $e' = e\cos(\theta_W)$. The calculation of the solutions of the wave equation for electron+neutrino in the hydrogen case implies [25] simply

$$B = B^0 = -\frac{e}{r}. \tag{8.21}$$

The Dirac formula giving the magnetic charge that Lochak obtained from only the condition of continuity of the wave under the group of rotations becomes then

$$\frac{eg}{\hbar c} = \frac{n}{2} \tag{8.22}$$

where n is an integer. This gives a magnetic charge which is a multiple of:

$$g = \frac{\hbar c}{2e}. \tag{8.23}$$

We get then the charge calculated by the Dirac formula. This charge has been gotten by numerous ways, for instance from the angular momentum of the electromagnetic field. Poincaré's equation giving the trajectory of an electron under the influence of a magnetic monopole [42] is unchanged, as is the cone that he introduced. Lochak proved that this cone is the Poinsot cone of a quantum top [49].

The presence of a σ_{21} term in the invariant wave equation implies, similarly to the electron case, the existence of two other wave equations obtained by a circular permutation of indices 1, 2, 3 in Pauli matrices (see chapter 5). A fourth kind of magnetic monopole comes from the wave equation of a fourth neutrino Eq. (7.61) by adding a gauge term. We can then think that four kinds of magnetic monopoles may exist, three of them similarly to the fact that there are electrons, muons and tauons. These three generations must be treated separately in the electro-weak interactions that we look at now.

8.3.3 *Electro-weak interactions with monopoles*

We want to get an identity similar to Eq. (6.144) allowing to Ψ^{-1} to exist everywhere, we suppose then that the wave of the monopole interacting is

$$\Psi = \begin{pmatrix} \phi_L & \phi_n \\ \widehat{\phi_n} & \widehat{\phi_L} \end{pmatrix} ; \quad \phi_n = \phi_{nL} + \phi_{nR} \qquad (8.24)$$

where ϕ_n is the wave of the magnetic monopole. We use here Lochak's idea of the monopole as an excited state of the neutrino, and we place the wave of the monopole where the place of the neutrino was. The supplementary left spinor ϕ_L may be seen as a part of an electric wave. We conserve the form Eq. (6.22) of the covariant derivative. Since only P_0 was changed when we went from the lepton case to the quark case, we shall use the same projectors $P_\pm$ of Eq. (6.12) and we use again projectors P_j in Eq. (6.14) to Eq. (6.16). In place of Eq. (6.13) we let

$$P_0(\Psi) = a\Psi\gamma_{21} + bP_-(\Psi)\mathbf{i}, \qquad (8.25)$$

where a and b are real numbers. We get the same commutation relations as in Eq. (6.17), except the last equality which must be replaced by:

$$P_0 P_j = P_j P_0 = -a\mathbf{i}P_j. \qquad (8.26)$$

Therefore the gauge group has the same structure $U(1) \times SU(2)$. We get with Eq. (8.24)

$$P_+(\Psi) = \begin{pmatrix} \phi_L & \phi_{nL} \\ \widehat{\phi_{nL}} & \widehat{\phi_L} \end{pmatrix} ; \quad P_-(\Psi) = \begin{pmatrix} 0 & \phi_{nR} \\ \widehat{\phi_{nR}} & 0 \end{pmatrix} . \qquad (8.27)$$

We recall that

$$\gamma_{21} = \begin{pmatrix} i\sigma_3 & 0 \\ 0 & i\sigma_3 \end{pmatrix} \;;\quad \phi_R \sigma_3 = \phi_R \;;\; \phi_L \sigma_3 = -\phi_L. \tag{8.28}$$

We then have

$$\Psi\gamma_{21} = i\begin{pmatrix} -\phi_L & \phi_{nR} - \phi_{nL} \\ -\widehat{\phi}_{nR} + \widehat{\phi}_{nL} & \widehat{\phi}_L \end{pmatrix} \;;\quad P_-(\Psi) = i\begin{pmatrix} 0 & -\phi_{nR} \\ \widehat{\phi}_{nR} & 0 \end{pmatrix},$$

$$P_0(\Psi) = i\begin{pmatrix} -a\phi_L & (a-b)\phi_{nR} - a\phi_{nL} \\ (-a+b)\widehat{\phi}_{nR} + a\widehat{\phi}_{nL} & a\widehat{\phi}_L \end{pmatrix}, \tag{8.29}$$

$$\mathbf{B}P_0(\Psi) = i\begin{pmatrix} (-a+b)B\widehat{\phi}_{nR} + aB\widehat{\phi}_{nL} & aB\widehat{\phi}_L \\ -a\widehat{B}\phi_L & (a-b)\widehat{B}\phi_{nR} - a\widehat{B}\phi_{nL} \end{pmatrix}. \tag{8.30}$$

We use Eq. (8.28) and Eq. (6.15), then we get

$$P_2(\Psi) = \begin{pmatrix} \phi_{nL} & -\phi_L \\ -\widehat{\phi}_L & \widehat{\phi}_{nL} \end{pmatrix}, \tag{8.31}$$

$$P_1(\Psi) = P_2(\Psi)\mathbf{i} = i\begin{pmatrix} \phi_{nL} & \phi_L \\ -\widehat{\phi}_L & -\widehat{\phi}_{nL} \end{pmatrix}, \tag{8.32}$$

$$(\mathbf{W}^1 P_1 + \mathbf{W}^2 P_2)(\Psi) = i\begin{pmatrix} (-W^1 + iW^2)\widehat{\phi}_L & (-W^1 - iW^2)\widehat{\phi}_{nL} \\ (\widehat{W}^1 - i\widehat{W}^2)\phi_{nL} & (\widehat{W}^1 + i\widehat{W}^2)\phi_L \end{pmatrix}. \tag{8.33}$$

Using W^+ and W^- defined in Eq. (6.74) we get

$$(\mathbf{W}^1 P_1 + \mathbf{W}^2 P_2)(\Psi) = i\begin{pmatrix} W^-\widehat{\phi}_L & -W^+\widehat{\phi}_{nL} \\ \widehat{W}^+\phi_{nL} & -\widehat{W}^-\phi_L \end{pmatrix}. \tag{8.34}$$

We have also :

$$P_3(\Psi) = P_+(\Psi)(-\mathbf{i}) = i\begin{pmatrix} -\phi_L & \phi_{nL} \\ -\widehat{\phi}_{nL} & \widehat{\phi}_L \end{pmatrix}, \tag{8.35}$$

$$\mathbf{W}^3 P_3(\Psi) = i\begin{pmatrix} -W^3\widehat{\phi}_{nL} & W^3\widehat{\phi}_L \\ -\widehat{W}^3\phi_L & \widehat{W}^3\phi_{nL} \end{pmatrix}. \tag{8.36}$$

The gauge derivative Eq. (6.22) is then equivalent to the system

$$D\widehat{\phi}_n = \nabla\widehat{\phi}_n + i\frac{g_1}{2}[aB\widehat{\phi}_{nL} + (-a+b)B\widehat{\phi}_{nR}] + i\frac{g_2}{2}(W^-\widehat{\phi}_L - W^3\widehat{\phi}_{nL}),$$

$$D\widehat{\phi}_L = \nabla\widehat{\phi}_L + i\frac{g_1}{2}aB\widehat{\phi}_L + i\frac{g_2}{2}(-W^+\widehat{\phi}_{nL} + W^3\widehat{\phi}_L). \tag{8.37}$$

With Eq. (6.83) we get

$$g_2 W^3 = \sqrt{g_1^2 + g_2^2}\, Z^0 + g_1 B. \tag{8.38}$$

Then we get

$$D\widehat{\phi}_n = \nabla\widehat{\phi}_n + i\frac{g_1}{2}[(a-1)B\widehat{\phi}_{nL} + (-a+b)B\widehat{\phi}_{nR}]$$

$$+ i\frac{g_2}{2}W^-\widehat{\phi}_L - \frac{i}{2}\sqrt{g_1^2 + g_2^2}Z^0\widehat{\phi}_{nL}. \qquad (8.39)$$

We want to get $\nabla\widehat{\phi}_n + iQB\widehat{\phi}_n$ so we must have

$$g_1(a-1) = g_1(b-a) = 2Q. \qquad (8.40)$$

The first equality gives $b = 2a - 1$ and if this condition is satisfied we get

$$g_1 = \frac{e}{\hbar c},$$

$$\frac{e(a-1)}{\hbar c} = 2Q = \frac{2g}{\hbar c},$$

$$e(a-1) = 2g = \frac{\hbar c}{e},$$

$$a - 1 = \frac{\hbar c}{e^2} = \frac{1}{\alpha}, \qquad (8.41)$$

where α is the fine structure constant, and we must take

$$a = 1 + \frac{1}{\alpha} ; \quad b = 1 + \frac{2}{\alpha}. \qquad (8.42)$$

This gives in Eq. (8.37)

$$D\widehat{\phi}_L = \nabla\widehat{\phi}_L + i\frac{g_1}{2}(1 + \frac{1}{\alpha})B\widehat{\phi}_L + i\frac{g_2}{2}(-W^+\widehat{\phi}_{nL} + W^3\widehat{\phi}_L), \qquad (8.43)$$

which is not the derivative term of an electron and remains to be interpreted. Since terms containing a are much bigger than other terms the magnetic charge term seems dominant in Eq. (8.39) and Eq. (8.43).

8.3.4   *Gauge invariant wave equation*

Since the pair: (wave of the monopole-ϕ_L wave) is the analog of the electron-neutrino pair, the wave equation for the Ψ in Eq. (8.24) is identical to Eq. (6.147):

$$\widetilde{\Psi}(\mathbf{D}\Psi)\gamma_{012} + m\boldsymbol{\rho}_m\widetilde{\Psi}\chi_m = 0 ; \quad \chi_m = \chi_l\gamma_0. \qquad (8.44)$$

The $\boldsymbol{\rho}$ term and χ_l are obtained by replacing Ψ_l by Ψ_m in the formulas of Appendix B. This wave equation is form invariant under the transformation R in Eq. (1.42) induced by M because we have Eq. (6.105) and:

$$\Psi' = N\Psi ; \quad N = \begin{pmatrix} M & 0 \\ 0 & \widehat{M} \end{pmatrix}, \qquad (8.45)$$

$$\widetilde{\Psi}' = \widetilde{\Psi}\widetilde{N}, \qquad (8.46)$$

$$\boldsymbol{\rho}' = r\boldsymbol{\rho}, \qquad (8.47)$$

$$m'\boldsymbol{\rho}' = m'r\boldsymbol{\rho} = m\boldsymbol{\rho}. \qquad (8.48)$$

The existence of only one mass term, and hence the existence of only one impulse-energy, implies the existence of a single wavelength for all three spinors. This is visible in figure 30. The wave equation Eq. (8.44) is also gauge invariant under the gauge transformation defined by Eq. (6.119) to Eq. (6.122), because P_0 has the general form studied in Appendix B and we get

$$\widetilde{\Psi}'(\mathbf{D}'\Psi')\gamma_{012} + m\boldsymbol{\rho}\widetilde{\Psi}'\chi'_m = 0. \tag{8.49}$$

The mechanism of the spontaneously broken gauge symmetry is then not necessary, neither for the electron nor for the magnetic monopole since the wave equations are simply gauge invariant.

Chapter 9

Inertia and gravitation

Using the preceding wave equations with mass terms, we introduce the inertial terms corresponding to the use of variable M matrices. The equality between gravitational and inertial mass is the cause of the existence of the density of probability. The physical necessity of normalization of the wave implies that the momentum–energy tensor of particle physics becomes symmetric, allowing the Einstein–Ricci gravitational tensor to rule our space-time.

9.1 Differential geometry

All differential operators in the preceding chapters are built on the $\nabla = \sigma^\mu \partial_\mu$ operator. It is then this operator that we shall consider now. The invariance group uses M matrices which are independent on $x = x^\mu \sigma_\mu$. Moreover the σ_μ in the R transformation Eq. (1.42) are the same in x or in x': they are invariant. Then the γ^μ of the relativistic quantum theory are also invariant under the Lorentz group. We shall then consider these σ_μ, γ^μ and L^μ as invariant when we allow M to vary. We have three kinds of "variance": the wave ϕ, the contravariant x and the covariant ∇:

$$\phi' = \phi'(x') = M\phi(x) = M\phi \; ; \quad x' = MxM^\dagger \; ; \quad \nabla = M\nabla'\widehat{M}. \qquad (9.1)$$

Since the Cl_3^* invariance group is the set of the M we must start from these terms. We suppose now that $M = M(x)$ is an analytic function with $M(0) = 1$ satisfying

$$M = \begin{pmatrix} 1 + \dfrac{dx^\mu}{2}(a_\mu + ib_\mu + p_\mu + iq_\mu) & \dfrac{dx^\mu}{2}(f_\mu + g_\mu + ih_\mu - il_\mu) \\ \dfrac{dx^\mu}{2}(f_\mu - g_\mu + ih_\mu + il_\mu) & 1 + \dfrac{dx^\mu}{2}(-a_\mu - ib_\mu + p_\mu + iq_\mu) \end{pmatrix},$$

$$\qquad (9.2)$$

where a_μ, b_μ,... q_μ, are 32 numeric functions of x and dx^μ are the infinitesimal increments of x^μ coordinates. This gives

$$M = 1 + \frac{dx^\mu}{2}(p_\mu + f_\mu\sigma_1 + l_\mu\sigma_2 + a_\mu\sigma_3 + h_\mu i\sigma_1 + g_\mu i\sigma_2 + b_\mu i\sigma_3 + iq_\mu),$$
(9.3)

$$M^\dagger = 1 + \frac{dx^\mu}{2}(p_\mu + f_\mu\sigma_1 + l_\mu\sigma_2 + a_\mu\sigma_3 - h_\mu i\sigma_1 - g_\mu i\sigma_2 - b_\mu i\sigma_3 - iq_\mu),$$
(9.4)

$$\widehat{M} = 1 + \frac{dx^\mu}{2}(p_\mu - f_\mu\sigma_1 - l_\mu\sigma_2 - a_\mu\sigma_3 + h_\mu i\sigma_1 + g_\mu i\sigma_2 + b_\mu i\sigma_3 - iq_\mu),$$
(9.5)

$$\overline{M} = 1 + \frac{dx^\mu}{2}(p_\mu - f_\mu\sigma_1 - l_\mu\sigma_2 - a_\mu\sigma_3 - h_\mu i\sigma_1 - g_\mu i\sigma_2 - b_\mu i\sigma_3 + viq_\mu).$$
(9.6)

We get

$$M\overline{M} = \det(M) = 1 + dx^\mu(p_\mu + iq_\mu),$$ (9.7)
$$\det(M^{-1}) = 1 - dx^\mu(p_\mu + iq_\mu),$$ (9.8)
$$\overline{M}^{-1} = M\det(M^{-1}),$$
$$= 1 + \frac{dx^\mu}{2}(-p_\mu + f_\mu\sigma_1 + l_\mu\sigma_2 + a_\mu\sigma_3 + h_\mu i\sigma_1 + g_\mu i\sigma_2 + b_\mu i\sigma_3 - iq_\mu),$$
(9.9)

$$\widehat{M}^{-1} = (\overline{M}^{-1})^\dagger$$
$$= 1 + \frac{dx^\mu}{2}(-p_\mu + f_\mu\sigma_1 + l_\mu\sigma_2 + a_\mu\sigma_3 - h_\mu i\sigma_1 - g_\mu i\sigma_2 - b_\mu i\sigma_3 + iq_\mu).$$
(9.10)

The dilation R defined from M satisfies

$$x' = R(x) = MxM^\dagger.$$ (9.11)

We get

$$x'^0 = x^0 + (p_\mu x^0 + f_\mu x^1 + l_\mu x^2 + a_\mu x^3)dx^\mu,$$ (9.12)
$$x'^1 = x^1 + (f_\mu x^0 + p_\mu x^1 + b_\mu x^2 - g_\mu x^3)dx^\mu,$$ (9.13)
$$x'^2 = x^2 + (l_\mu x^0 - b_\mu x^1 + p_\mu x^2 + h_\mu x^3)dx^\mu,$$ (9.14)
$$x'^3 = x^3 + (a_\mu x^0 + g_\mu x^1 - h_\mu x^2 + p_\mu x^3)dx^\mu.$$ (9.15)

Christoffel's symbols $\Gamma^\alpha_{\beta\gamma}$ being defined as

$$x'^\alpha = x^\alpha + \Gamma^\alpha_{\beta\gamma}x^\beta dx^\gamma,$$ (9.16)

we then get

$$\Gamma^0_{0\mu} = \Gamma^1_{1\mu} = \Gamma^2_{2\mu} = \Gamma^3_{3\mu} = p_\mu, \tag{9.17}$$

$$\Gamma^1_{0\mu} = \Gamma^0_{1\mu} = f_\mu \ ; \ \ \Gamma^2_{0\mu} = \Gamma^0_{2\mu} = l_\mu \ ; \ \ \Gamma^3_{0\mu} = \Gamma^0_{3\mu} = a_\mu, \tag{9.18}$$

$$\Gamma^2_{3\mu} = -\Gamma^3_{2\mu} = h_\mu \ ; \ \ \Gamma^3_{1\mu} = -\Gamma^1_{3\mu} = g_\mu \ ; \ \ \Gamma^1_{2\mu} = -\Gamma^2_{1\mu} = b_\mu. \tag{9.19}$$

Since R is a dilation, product in any order of a Lorentz transformation and a homothety, the Christoffel's symbols have this particular form and we get not 64 but only $28 = 4 \times 7$ functions: the four q_μ present in Eq. (9.2) are not in the geometry, because the kernel of the group homomorphism $M \mapsto R$ (that is at the origin of the spin $1/2$) is the $U(1)$ group generated by i [13] [17]. Since the Christoffel's symbols are not symmetric, a torsion exists. Vectors transforming as Eq. (9.16) are the contravariant ones. Now for covariant vectors we have

$$\nabla = \sigma^\mu \partial_\mu = \overline{M} \sigma^\mu \widehat{M} \partial'_\mu, \tag{9.20}$$

with the same σ^μ. This gives

$$\nabla' = \sigma^\nu \partial'_\nu = \overline{M}^{-1} \sigma^\nu \widehat{M}^{-1} \partial_\nu = \sigma^\nu (\partial_\nu - dx^\mu \Gamma^\rho_{\nu\mu} \partial_\rho). \tag{9.21}$$

Therefore we get for covariant vectors the usual

$$\partial'_\nu = \partial_\nu - dx^\mu \Gamma^\rho_{\nu\mu} \partial_\rho. \tag{9.22}$$

This relation allows the covariant derivative to be commutative with contractions. It leads the covariant derivative back to partial derivative for scalars. The connection Eq. (9.17) to Eq. (9.19) is new, because all preceding attempts have used variable γ^μ, which is incompatible with the relativistic invariance of the quantum wave. A nonvanishing torsion has been used previously by A. Einstein [37] to unify gravitation and electromagnetism. Since his attempt was made very early in the history of quantum mechanics he evidently did not start from the Dirac wave, which was invented 3 years after. We next get

$$\begin{aligned}
\overline{\phi}' \nabla' \widehat{\phi}' &= (\overline{M\phi}) \, \overline{M}^{-1} \sigma^\mu \widehat{M}^{-1} \partial_\mu (\widehat{M\phi}) \\
&= \overline{\phi} \, \overline{M} \, \overline{M}^{-1} \sigma^\mu \widehat{M}^{-1} [(\partial_\mu \widehat{M}) \widehat{\phi} + \widehat{M} (\partial_\mu \widehat{\phi})] \\
&= \overline{\phi} \sigma^\mu \widehat{M}^{-1} (\partial_\mu \widehat{M}) \widehat{\phi} + \overline{\phi} \sigma^\mu \partial_\mu \widehat{\phi} \\
&= \overline{\phi} \sigma^\mu [-(\partial_\mu \widehat{M}^{-1}) \widehat{M}] \widehat{\phi} + \overline{\phi} \nabla \widehat{\phi} \\
&= \overline{\phi} [\nabla - (\nabla \widehat{M}^{-1}) \widehat{M}] \widehat{\phi}, \tag{9.23}
\end{aligned}$$

$$\overline{\phi}' \nabla' \widehat{\phi}' = \overline{\phi} D \widehat{\phi}, \tag{9.24}$$

where we have let

$$D = \nabla - (\nabla \widehat{M}^{-1})\widehat{M} \qquad (9.25)$$

$$= \sigma^\mu [\partial_\mu + \frac{1}{2}(p_\mu - f_\mu \sigma_1 - l_\mu \sigma_2 - a_\mu \sigma_3 + h_\mu i\sigma_1 + g_\mu i\sigma_2 + b_\mu i\sigma_3 - iq_\mu)].$$

This introduces 8 space-time vectors that we name "potentials of inertia":

$$p = \sigma^\mu p_\mu = \sigma^\mu \Gamma^0_{0\mu} \ ; \ \ f = \sigma^\mu f_\mu = \sigma^\mu \Gamma^1_{0\mu} \ ; \ \ l = \sigma^\mu l_\mu = \sigma^\mu \Gamma^2_{0\mu}, \qquad (9.26)$$

$$a = \sigma^\mu a_\mu = \sigma^\mu \Gamma^3_{0\mu} \ ; \ \ h = \sigma^\mu h_\mu = \sigma^\mu \Gamma^2_{3\mu} \ ; \ \ g = \sigma^\mu g_\mu = \sigma^\mu \Gamma^3_{1\mu}, \qquad (9.27)$$

$$b = \sigma^\mu b_\mu = \sigma^\mu \Gamma^1_{2\mu} \ ; \ \ q = \sigma^\mu q_\mu, \qquad (9.28)$$

$$D = \nabla + \frac{1}{2}(p - f\sigma_1 - l\sigma_2 - a\sigma_3 + hi\sigma_1 + gi\sigma_2 + bi\sigma_3 - iq). \qquad (9.29)$$

If we compare this with the attempts made before quantum physics [37], two main differences appear: Unification is made not with fields but with potentials, like in [57]. And the q term does not come from the geometry of space-time. Einstein had very early the intuition that something was lacking in quantum mechanics. Here it is the contrary: q is not lacking in quantum physics but it is lacking in geometry. These eight potentials become under a dilation R induced by a constant M

$$D = \overline{M}D'\widehat{M} \ ; \ \ \nabla = \overline{M}\nabla'\widehat{M} \ ; \ \ p = \overline{M}p'\widehat{M} \ ; \ \ q = \overline{M}q'\widehat{M}. \qquad (9.30)$$

The true covariant terms are not f... b but $f\sigma_1$... $bi\sigma_3$ since

$$f\sigma_1 = \overline{M}f'\sigma_1\widehat{M} \ ; \ \ l\sigma_2 = \overline{M}l'\sigma_2\widehat{M} \ ; \ \ a\sigma_3 = \overline{M}a'\sigma_3\widehat{M},$$

$$hi\sigma_1 = \overline{M}h'i\sigma_1\widehat{M} \ ; \ \ gi\sigma_2 = \overline{M}g'i\sigma_2\widehat{M} \ ; \ \ bi\sigma_3 = \overline{M}b'i\sigma_3\widehat{M}. \qquad (9.31)$$

In space-time algebra we shall need

$$\widehat{D} = \widehat{\nabla} - (\widehat{\nabla}M^{-1})M$$

$$= \widehat{\nabla} + \frac{1}{2}(\widehat{p} + \widehat{f}\sigma_1 + \widehat{l}\sigma_2 + \widehat{a}\sigma_3 + \widehat{h}i\sigma_1 + \widehat{g}i\sigma_2 + \widehat{b}i\sigma_3 + i\widehat{q}), \qquad (9.32)$$

$$\mathbf{D} = \begin{pmatrix} 0 & D \\ \widehat{D} & 0 \end{pmatrix} \ ; \ \ \underline{\mathbf{D}} = \begin{pmatrix} 0 & \mathbf{D} \\ \mathbf{D} & 0 \end{pmatrix}. \qquad (9.33)$$

And the covariant derivative unifying inertia to gauge interactions becomes

$$\underline{D} = \mathbf{D} + \frac{g_1}{2}\underline{B} \, P_0 + \frac{g_2}{2}\underline{W}^j P_j + \frac{g_3}{2}\underline{G}^k i\Gamma_k. \qquad (9.34)$$

Contrary to all other terms that contains projectors, the term of inertia acts on the whole wave. This is why we can name the consequences of the action of $\mathbf{D}$ inertial. Another reason is the fact that it is linked to the geometry of the space-time manifold, even if this link is only partial.

The relation Eq. (9.24) must be seen as reversed: if we exchange the role of x and x' by changing M into M^{-1} we have

$$x = x' + dx' = R^{-1}(x') = M^{-1}x'M^{-1\dagger} \;;\quad dx' = x - x' = -dx, \quad (9.35)$$

$$x^\alpha = x'^\alpha - \Gamma^\alpha_{\beta\gamma} x'^\beta dx'^\gamma, \quad (9.36)$$

$$\partial_\nu = \partial'_\nu + dx'^\mu \Gamma^\rho_{\nu\mu} \partial'_\rho \;;\quad \phi = M^{-1}\phi', \quad (9.37)$$

$$D' = \nabla' - (\nabla'\widehat{M})\widehat{M}^{-1}, \quad (9.38)$$

$$\overline{\phi}\nabla\widehat{\phi} = \overline{\phi}'D'\widehat{\phi}'. \quad (9.39)$$

The difference between the left and right terms in Eq. (9.39) comes from the two kinds of frames: inertial versus non-inertial frames. Newton laws of mechanics and also laws of general relativity separate all frames into two kinds. Inertial frames are the only frames in which laws of movements are simple. From an inertial frame in which an event is labeled as $x = x^\mu \sigma_\mu$ any other inertial frame labels the same event as $x' = MxM^\dagger + a$ where a is a fixed space-time vector and M is any fixed element in Cl_3^*. In all these inertial frames the covariant derivative has the same form Eq. (7.52). The Lagrangian density is the scalar real part of the wave equation and the wave equation is obtained by variational calculus from this Lagrangian density. In all inertial frames the invariance of the Lagrangian density under the translations $x \mapsto x' = x + a$ implies the existence of a conservative density of impulse–energy. The evolution of this density of impulse–energy is governed only by electro-weak and strong forces.

But only a few frames are inertial. For instance if $x' = MxM^\dagger$ where $M = M(x)$ is a variable element, then we are no longer in an inertial frame. In such a frame we must replace ∇ by D'. Therefore we have additional forces and not only electro-weak and strong forces. These additional forces are named inertial forces. We shall study them in two particular cases.

9.1.1 *Uniform movement of rotation*

We consider a frame at the surface of a sphere, with a fixed third axis that is parallel to the axis σ_3 of the rotation which is also an axis of the sphere. At the point P the $\vec{n}$ vector is normal to the sphere. The first axis is assumed to be orthogonal to the axis of rotation and in the plane $(\sigma_3, \vec{n})$. We suppose that $(\sigma_1, \sigma_2, \sigma_3)$ is an orthonormal direct basis. The movement of the frame at P is made of a movement of translation in the direction σ_2 and a movement of rotation with axis σ_3. We name R the distance of P to the axis of rotation and ω the angular velocity. The velocity of the movement of translation is $v = \omega R$. We let

$$M_1 = e^{\delta\sigma_2} \; ; \;\; M_2 = e^{\omega dx^0 i\sigma_3/2c} \; ; \;\; M = M_2 M_1 \; ; \;\; x' = MxM^\dagger, \qquad (9.40)$$

$$\delta = \frac{1}{2}\text{atanh}(\frac{\omega R}{c}). \qquad (9.41)$$

And we get

$$x'^0 = \cosh(2\delta)x^0 + \sinh(2\delta)x^2 \; ; \;\; x'^3 = x^3, \qquad (9.42)$$

$$x'^1 = x^1 + \frac{\omega}{c}[\sinh(2\delta)x^0 + \cosh(2\delta)x^2]dx^0, \qquad (9.43)$$

$$x'^2 = [\sinh(2\delta)x^0 + \cosh(2\delta)x^2] - \frac{\omega}{c}x^1 dx^0. \qquad (9.44)$$

This gives

$$\frac{d^2 x'^1}{dt^2} = \omega c \sinh(2\delta) \approx \omega^2 R, \qquad (9.45)$$

which is the centrifugal acceleration. The limit to the validity of this approximation is the same as always in classical physics: the velocity must be negligible in comparison to c.

9.1.2 *Uniformly accelerated movement of translation*

We consider here a constant acceleration $g = ac^2$ in the direction σ_1 and an M matrix:

$$v = gt = \frac{g}{c}x^0 \; ; \;\; M = e^{gdx^0\sigma_1/2c^2}. \qquad (9.46)$$

This gives the transformation:

$$x'^0 = x^0 + ax^1 dx^0 \; ; \;\; x'^1 = x^1 + ax^0 dx^0, \qquad (9.47)$$

$$x'^2 = x^2 \; ; \;\; x'^3 = x^3. \qquad (9.48)$$

And we get

$$\frac{dx'^1}{dx^0} = \frac{d}{dx^0}(x^1 + ax^0 dx^0), \qquad (9.49)$$

$$\frac{d^2 x'^1}{(dt')^2} \approx \frac{d^2 x'^1}{(dt)^2} = g + \frac{d^2 x^1}{(dt)^2}. \qquad (9.50)$$

The acceleration seen in the non-inertial frame is then the sum of the acceleration of the frame and the acceleration coming from any other forces.

9.2 Wave normalization

The invariance of the Lagrangian under all translations, as in the linear Dirac theory, induces the existence of a conservative impulse–energy tensor, the Tetrode's tensor:

$$T_\nu^\mu = \partial_\nu \overline{\psi} \frac{\partial \mathcal{L}}{\partial(\partial_\mu \overline{\psi})} + \frac{\partial \mathcal{L}}{\partial(\partial_\mu \psi)} \partial_\nu \psi - \delta_\nu^\mu \mathcal{L}. \tag{9.51}$$

Since the wave equation is homogeneous, the Lagrangian is null and we get:

$$T_\nu^\mu = \frac{i}{2}(-\overline{\psi}\gamma^\mu \partial_\nu \psi + \partial_\nu \overline{\psi}\gamma^\mu \psi). \tag{9.52}$$

For a stationary state with energy E we have:

$$\psi = e^{iEt/\hbar}\psi(x); \quad \overline{\psi} = e^{-iEt/\hbar}\overline{\psi}(x); \quad \partial_0 \psi = i\frac{E}{\hbar c}\psi; \quad \partial_0 \overline{\psi} = -i\frac{E}{\hbar c}\overline{\psi}. \tag{9.53}$$

So we get:

$$T_0^0 = \frac{i}{2}(-\overline{\psi}\gamma^0 (i\frac{E}{\hbar c})\psi - i\frac{E}{\hbar c}\overline{\psi}\gamma^0 \psi) = E\frac{J^0}{\hbar c}. \tag{9.54}$$

The condition normalizing the wave function must then be replaced by

$$\iiint \frac{J^0}{\hbar c}dv = 1, \tag{9.55}$$

that is equivalent, for a bound state, to

$$\iiint T_0^0 dv = E. \tag{9.56}$$

The left term of this equality is the total energy which is the sum of the local density of energy at each point of the wave, whilst the right term is the global energy of the electron. Therefore it is not because we must get a probability that the wave must be normalized. The physical wave is normalized because the inertial mass–energy which is moved by external fields, is equal to the gravitational mass–energy. The inertial mass–energy has then not an arbitrary, but a determined value. The necessity to normalize the wave may then be considered as a consequence of the principle of equality between inertial and gravitational mass, the principle that is the basis of general relativity. The presence of a probability density is then not a principle on which we must build any physical theory; it is only a consequence of the equality between gravitational and inertial mass–energy. We must also remark that the bad habit to suppose $\hbar = 1$ has hidden the correct dimension of the probability current: since J^0 has the numeric dimension 1 and $\hbar$ has the numeric dimension 4, $J^0/\hbar c$ has the correct numeric dimension -3 of a density of probability.

All this applies also to the Dirac equation as well as the homogeneous nonlinear equation which has the Dirac equation as linear approximation. This normalization applies [1] evidently to solutions in the case of the H atom that we study in appendix C.

9.3 Gravitation

The global energy E of the electron is the temporal component of a space-time vector, the energy–momentum vector. Since the integration has been made in a frame where this momentum is null, this vector reads $(E, 0, 0, 0,)$ but it is seen by other moving observers as (p_0, p_1, p_2, p_3). General relativity considers all particles of the universe as giving each observer such an energy–momentum space-time vector, and if there is no pressure the density of this fluid of particles (d_0, d_1, d_2, d_3) constitutes the material contribution to the symmetric tensor of energy $T_\mu^\nu = d_\mu d^\nu$. Einstein has linked this material tensor to the curvature of the space-time manifold:

$$\frac{1}{\chi}[R_\mu^\rho - \frac{1}{2}\delta_\mu^\rho(R - 2\Lambda)] = T_\mu^\rho, \tag{9.58}$$

where Λ is the cosmological constant and χ is the constant of gravitation. We have placed this constant on the left side, then Eq. (9.58) is invariant under Cl_3^*. The density of mass per unit volume μ_0 used in the Newtonian law of gravitation

$$\Delta U = -4\pi G \mu_0 \; ; \; \chi = 8\pi \frac{G}{c^4}, \tag{9.59}$$

gives, in the case of matter without pressure:

$$T_\mu^\rho = \mu_0 c^2 u_\mu u^\rho \; ; \; u^\mu = \frac{dx^\mu}{ds} \; ; \; ds^2 = g_{\mu\nu} dx^\mu dx^\nu. \tag{9.60}$$

The particle with reduced velocity u^μ may be an electron, an atom, a star or a galaxy in the cosmological case. We see no reason to change anything in General Relativity, as this theory has been successfully tested since a century ago, and not only in the case of low curvature.

Albert Einstein was searching a total unification between all domains of our physical universe. This work unifies only the conditions ruling all physical laws. Inertia and gravitation remain rather separate from other

---

1. The normalization of the wave of the positron is equivalent, in the case of a stationary state, to

$$\iiint T_0^0 dv = -E. \tag{9.57}$$

phenomena of our universe. A distinction appears between inertial and non-inertial frames. Inertial frames are those in which inertia seems absent, but also frames in which physical laws are simpler. In these frames the wave equation of fermions has its real scalar part as Lagrangian density. The invariance of this density under translations allows the existence of a conservative momentum–energy tensor. This tensor is not symmetric and is then not included in Eq. (9.58).

There are at least three ideas to account for, if we want to go further: the first one is the geometric difference between a charge, with numeric dimension 2, and a proper mass, with numeric dimension 3. The second idea is: in the quantum world the quantities having the correct relativistic transformations are tensorial densities, whilst the physical observables are obtained by integrating these tensorial densities on the physical space. This was remarked on very early by de Broglie [30]. Since a probability exists also for systems of quantum objects, since a frequency exists for the global wave of such a system, and since the equality between inertial and gravitational mass–energy is general, the reasoning of Sec. 9.2 may be generalized. The third idea is the fact that the physical space is oriented, like the physical time. The orientation of the space is ruled by an i that is present in the fundamental group of invariance of all physical laws, Cl_3^*. On the contrary this i is not present in the Christoffel symbols of differential geometry, because the kernel of the homomorphism $M \mapsto R$ is precisely the group generated by this i. None of these three things may be changed by any theoretical manipulation.

The tensor of Eq. (9.52) contains no term relative to the energy of the electromagnetic field, nor to other gauge fields. This is a direct consequence of the hypothesis of Einstein on the quanta of energy–momentum in electromagnetic waves [36]. Einstein established later that the fluctuations accounting for black body radiation is the sum of the fluctuations that we should get if light were only made of discreet particles and if it were only made of continuous fields. Nowadays these fluctuations are obtained by counting the number of photons in each mode of the quantized electromagnetic wave. It is possible to consider then that the momentum–energy of the electromagnetic field is entirely included in the symmetric tensorial density Eq. (9.60) coming from photons. This allows the symmetry of the gravitational law Eq. (9.58).

9.4   Unification

T. Socroun follows another way towards the unification [57]: in order to get the unification between gravitation and electromagnetism, he incorporates charges into potentials terms. This is equivalent, in fact, to the prescription made by Einstein that all laws of physics are covariant. With Cl_3^* as the fundamental group of invariance the difference between contravariant and covariant terms is a physical one: a contravariant vector in space-time transforms as x while a covariant vector transforms as ∇. These two transformations are not equivalent as soon as $\det(M) \neq 1$. It is easy to see that incorporation of the charges g_1, g_2, g_3 into the potentials can be made inside each of our preceding equations:

$$\nabla\widehat{\phi}\sigma_{21} + \mathbf{a}\widehat{\phi} + m\phi = 0 \ ; \quad \mathbf{a} = qA \ ; \quad \overline{\phi}\mathbf{a}\widehat{\phi} = \overline{\phi}'\mathbf{a}'\widehat{\phi}'. \qquad (9.61)$$

We let [24]

$$\underline{b} = \frac{g_1}{2}\underline{B} \ ; \quad \underline{w}^j = \frac{g_2}{2}\underline{W}^j \ ; \quad \underline{g}^k = \frac{g_3}{2}\underline{G}^k. \qquad (9.62)$$

Then the covariant derivative Eq. (7.52) is simply

$$\underline{D} = \underline{\partial} + \underline{b}\,\underline{P}_0 + \underline{w}^j\underline{P}_j + \underline{g}^k\,\mathrm{i}\Gamma_k. \qquad (9.63)$$

Why this has never been made? The reason is probably that theories of great unification consider the charges as slowly variable with the scale [2] of energy and hope that when these charges become equal the structure of the gauge group is enlarged. This is strictly impossible when charges are integrated into potential terms.

Now with Eq. (9.61) potential and mass term are similar, they have the same numeric dimension -1 of covariant terms. This is also the numeric dimension of an acceleration. With Eq. (9.29) we can consider

$$D = \nabla + \frac{1}{2}(p - f\sigma_1 - l\sigma_2 - a\sigma_3 + hi\sigma_1 + gi\sigma_2 + bi\sigma_3 - iq),$$

$$0 = D\widehat{\phi}\sigma_{21} + \mathbf{a}\widehat{\phi} + m\phi, \qquad (9.64)$$

as unifying the dynamics of the wave of the electron, with Lorentz forces and inertial ones. This is easily extensible to the wave of electron+neutrino and the wave of electron+neutrino+quarks , with:

$$\mathbf{D} = \begin{pmatrix} 0 & D \\ \widehat{D} & 0 \end{pmatrix} \ ; \quad \underline{D} = \begin{pmatrix} \mathbf{D} & 0 \\ 0 & \mathbf{D} \end{pmatrix} + \underline{b}\,\underline{P}_0 + \underline{w}^j\underline{P}_j + \underline{g}^k\,\mathrm{i}\Gamma_k. \qquad (9.65)$$

We must notice that this unification necessitates none constant. The constant of gravitation links gravitation to inertial acceleration.

---

2. Quantum physicists have built an axiomatic quantum theory, so as to copy the method of mathematicians. But they are not embarrassed to use constants of structures in the calculations of gauge groups that become variables if needed.

Chapter 10

Conclusion

Starting from the old flaws of relativistic quantum mechanics, we resume investigating the insights to the standard model that are allowed by our new approach with Clifford algebras. Physics using a principle of minimum is only a part of undulatory physics. Beyond the confrontation between theory and experiment, beyond future applications, the standard model appears both comforted and essential. Only novelties are the magnetic monopoles. Beyond the standard model emerges a new landscape of waves propagating in the physical space-time.

10.1   Old flaws

The discovery of the spin of the electron goes back to 1926 and was not predicted by the physical theory. Physicists have very naturally begun to get round the novelty by trying to reduce spinorial waves to tensors that were better known. The study was difficult, the field was cleared by the students of Louis de Broglie, mainly O. Costa de Beauregard [27] and T. Takabayasi [58]. He was able to give a set of tensorial equations equivalent to the Dirac equation. These tensorial equations however act on quantities which are quadratic on the wave. When we add the waves these tensors do not add. Therefore the spinorial wave itself is essential from the physical point of view, propagating and interfering. Only the solutions of the spinorial wave explain quanta, the true quantum numbers, the true number of bound states and the true energy levels. Let us go to the end of Takabayasi's attempt, let us replace completely spinors by a set of tensors and let us solve completely the tensorial equations in the case of the hydrogen atom. Should we get the true results, the true number of bound states, the true quantum numbers and the true energy levels? The answer is: no, because true representations of the rotation group $SO(3)$ use only integer numbers, not the half-integer numbers which are necessary to get the true results. These true results are obtained only by taking the representations of $SL(2, \mathbb{C})$, but then we are in Cl_3.

The second reason why scientists did not understand the novelty of the spinorial wave was the difficulty of the mathematical tools. Two different groups may be similar in the vicinity of their neutral element. $SL(2, \mathbb{C})$ and $\mathcal{L}_+^\uparrow$ or their subgroups $SU(2)$ and $SO(3)$ are globally different but locally identical. So the present study does not use infinitesimal operators, it is able to see the difference between a Lie group and its Lie algebra.

Physical waves imply the use of trigonometric functions, which in turn imply the complex exponential function that simplifies calculations. Going into a very unusual axiomatization, the quantum theory has been locked on the only use of complex numbers. This is equivalent to working only with plane geometry, with a unique i with square -1 that is the generator of all rotations of the Euclidean plane. It is in a sense a "2D software". The basic tool of the present study is a "3D software", the Clifford algebra of the 3-dimensional physical space. Next the building of Clifford algebras by recursion on the dimension allows us to use this basic tool in the algebra of space-time as in the algebra of the 6-dimensional space-time which is necessary and sufficient for all objects of the standard model. These algebras present all abilities of the linear spaces built on the complex field, because they are also linear spaces. But they also allow us to use products. The exponential function is then everywhere defined and allows us to study a large variety of wave phenomena. These algebras also allow us to use the inverse, when it exists.

10.2 Our work

Two kinds of particles, fermions and bosons, are used in the standard model. Each kind of fermion is a quantum object with a wave following the Dirac equation. This is the starting point of our work. Following de Broglie's initial idea of a physical wave linked to the movement of any particle, we have first made only a little change in the wave equation of the electron which concerns only its mass term. This modified wave equation is nonlinear, homogeneous and has the Dirac equation as a linear approximation.

First interesting result: the true sign of the mass–energy comes directly from the wave equation, and from the charge conjugation, which changes the sign of the derivative terms of the wave equation. This form of the charge conjugation was first gotten by the standard model itself, which uses, for the Dirac equation, the old bad frame of Dirac matrices.

The second result was very difficult, because the resolution of the Dirac equation is very accurate in the case of bound states of the H atom, and any change in this wave equation should imply total disaster. But for each bound state a solution of the linear equation exists such that the Yvon–Takabayasi β angle is everywhere defined and small. This result is very accurate and surprising. It means that physical bound states are the rare solutions of the homogeneous nonlinear equation (see Appendix C). We can therefore understand why there are privileged bound states and why an electron in a H atom is always in one of these states, never in a linear combination of states that is not a possible solution of the wave equation.

A second frame for the Dirac wave was introduced by D. Hestenes, the Clifford algebra of space-time, which is the second starting point of our work. A comparison between old and new frame is easy if we use the Dirac matrices as a matrix representation of the space-time vectors. We have reviewed in chapters 1 and 2 how the relativistic invariance is gotten for fermion waves. These waves appear very different: they are not vectors or tensors of the space-time, but a different kind of object, spinors.

The spinorial form of the fermion wave is included in the standard model, and it is one of its main features. What we have done here is only to fully account for all consequences of this fact. The form invariance of the Dirac equation necessitates the use of the $SL(2, \mathbb{C})$ group that is a subset of the Clifford algebra of the physical space, Cl_3. We have learned to read all the Dirac theory in this frame. This algebra is isomorphic to the matrix algebra $M_2(\mathbb{C})$ of 2×2 complex matrices. This algebra allows us to see its multiplicative group $GL(2, \mathbb{C}) = Cl_3^*$ as the true group of form invariance of the Dirac theory.

Then we have explained in a simple way how this form invariance, that is an enlargement of relativistic invariance, rules not only the Dirac wave equation, but all of electromagnetism (chapter 4). This is well hidden in the case of the electromagnetic field itself because only the $SL(2, \mathbb{C})$ part of the Cl_3^* group acts upon this boson field (and this is true of any other boson field). The electromagnetic field has properties resulting from its antisymmetric building from a pair of spinors. It is a pure bivector, sum $\vec{E} + i\vec{H}$ of a vector $\vec{E}$ and a pseudo-vector $i\vec{H}$, with neither scalar nor pseudo-scalar term. It rotates under a Lorentz rotation but it is insensitive to the ratio r of a dilation as well as to the chiral angle. This behavior is imposed by the form of its invariance under Cl_3^*.

More generally boson fields may be gotten by antisymmetric product of an even number of fermions. Their wave is then a physical wave, a function of space and time with value in the Clifford algebra of the usual 4-dimensional space-time, or of the complete 6-dimensional space-time. Such a construction was impossible in the old formalism of Dirac matrices, because the wave had values into a linear space which is not an algebra. The frame of Clifford algebras allows us to use the internal multiplication and the inverse to account for waves of systems of particles. This has interesting physical consequences: all waves are true functions of the space-time into well-defined sets. These sets of functions are the Hilbert spaces whose existence is supposed in the standard model. Clifford algebras have matrix representations, and their elements can always be considered as operators. The use of creation and annihilation operators is a consequence of the fact that products of an even number of elements belong to the same linear space.

The form invariance of electromagnetism uses a group which fully accounts for two main aspects of the modern physical results: the conservation of the orientation of time, and the conservation of the orientation of space. These two orientations are not a consequence of the invariance group. The form invariance is only compatible with the conservation of the orientation of physical space, which is the main experimental discovery of the second part of the twentieth century. The oriented space is fully compatible with a gauge group which acts differently on left and right waves. The conservation of an oriented time is compatible with laws of thermodynamics, with the emission or absorption of light, and with the propagation of retarded waves.

The extended form invariance allows a better understanding of old questions like: why there is a Planck constant? What is a charge, or a mass, and what is the difference between a charge and a mass? Charges appear in terms necessary to link a contravariant vector such as A to a covariant vector such as qA or to link a $\nabla \widehat{A}_j$ term to a $q(A_k \widehat{A}_l - A_l \widehat{A}_k)$ term of a Yang–Mills gauge group. The invariance under Cl_3^* necessitates $q = r^2 q'$. Mass appears in a term necessary to link a differential term to a constant term in the invariant wave equation. The invariance under Cl_3^* necessitates $m = rm'$. Therefore a charge is not a mass, a mass is not a charge: they have a different behavior under Cl_3^*. These new aspects of old concepts come from the supplementary strains added by a greater invariance group. They are fully compatible with classical and relativistic mechanics and with laws of electromagnetism.

Anything in the previous review is compatible with the experimental part of the standard model, and also with the CPT theorem which is now trivially satisfied. Nevertheless several bad habits must be avoided. For instance the Planck factor linking proper mass to frequency is variable if we consider the full invariance group. This is completely hidden if the Planck factor is changed into a constant number: we must put back the $\hbar$ term everywhere it is necessarily present. Tensors constructed from the Dirac waves with Dirac matrices do not have a behavior allowing the full invariance group. This implies we must use Clifford algebras, which are the only frames where the full invariance group acts. Another bad habit to abandon is the habit of going up or down an index of tensor, because covariant and contravariant vectors vary differently under the full invariance group.

If you have paid the price of accounting for the full invariance group you are able to get much in return. The first award is the possibility to read the electro-weak theory in a much simpler way, with a wave which is a function of space-time into a Clifford algebra: firstly of the usual space-time if you account only for the electron-neutrino case and secondly of a 6-dimensional space-time to account for all fermions and all anti-fermions of one generation. In this enlarged frame the gauge group is exactly the $U(1) \times SU(2) \times SU(3)$ group of the standard model, the lepton part of the complete wave sees only the $U(1) \times SU(2)$ part of the gauge group. Therefore electron and neutrino are automatically unable to see strong interactions and the right wave of the neutrino does not interact at all. A greater gauge group is not available, this accounts for the fact that no way exists to transform a quark into a lepton. Then this justifies the empirical construction, in the standard model, of conservative quantum numbers such as the baryonic number. Another award is the ability to understand the existence of both exactly three generations of fermions, completely similar and having nevertheless a separate behavior in the gauge invariance and the Lagrangian formalism, and four kinds of neutrinos.

The generalization to the complete wave of the geometric dilation linked to the electron wave is possible only if two additional dimensions are added [1] to the three usual dimensions of the physical space. The reward to this

---

1. We used in [17] two greater Clifford algebras, $Cl_{2,3}$ and $Cl_{3,4}$, which cannot allow the relation Eq. (7.73) between the reverse in space-time algebra and the reverse in the complete algebra. Relation Eq. (7.73) implies we should use only $Cl_{5,1}$, or $Cl_{1,5}$ but it happens to be the same algebra.

new strain on the theory is that this construction includes all parts of the standard model in a closed frame. Moreover the usual space-time is a well bounded part of the complete space-time, bounded by the fact that usual space-time is real in a complete space-time with a natural complex structure. A second reward is a unique dimension for the time in the complete space-time, this time is then our usual oriented time, oriented from past to future. Another reward is a much more general geometric transformation between the 6-dimensional space-time and the usual space-time, as soon as neutrinos and quarks are considered. The study of the wave of all fermions of the first generation introduces, at each point of the space-time, a geometric transformation from one manifold into the other one. With the complete wave the intrinsic manifold is 6-dimensional and the relative 4-dimensional space-time seems embedded into the 6-dimensional manifold.

Our nonlinear homogeneous wave equation for the electron is twice generalized, firstly as a wave equation with mass term for the electron+neutrino. This wave equation is both form invariant under Cl_3^* and gauge invariant under the gauge group of electro-weak interactions. This wave is then extended to a wave for electron+neutrino+quarks. The wave equation is both form invariant under Cl_3^* and gauge invariant under the gauge group of the standard model. The linear mass term of the Dirac equation was not able to give a wave equation compatible even with the electro-weak gauge. Now since we do have such an invariant wave equation with mass term, we are directly able to consider inertia and gravitation. Inertial frames are obtained with R transformations defined by any constant M element in Cl_3^* while other frames are obtained with variable M elements. The principle of equivalence between inertial and gravitational mass–energy is the origin of the normalization of the wave and of the density of probability. The proper mass–energy of the electron-particle is the integral over all space of the density of mass–energy of the electron-wave. This mechanism gives the symmetric tensor of impulse–energy of General Relativity.

10.3 Principle of minimum

Modern physics uses everywhere a principle of minimum. This was first seen by Fermat. He understood that light travels in such a way that the duration of the travel between two points is minimal. This was next seen in Hamiltonian mechanics, where the movement of any object is made in a way such that a quantity called action is minimal. These two principles of minimum were united by de Broglie and this led him to discover the wave linked to the movement of any material particle. So the Dirac wave equation of an electron or other fermions may also be gotten from such a principle of minimum. But why?

Clifford algebra and the form invariance group give a strange answer. The true wave equation of a unique fermion is the invariant form of the wave equation. And then the scalar part of this wave equation is exactly the Lagrangian density. This Lagrangian density is not truly minimal. In fact, it is exactly zero, because the wave equation is homogeneous. The second part of the answer comes from the fact that you can get the seven other equations by using the calculus of variations. This is probably not very correctly done, because an assumption is made that the infinitesimal variation of the wave is null on a boundary of the integration volume. It is easy to get this assumption in the case of a bound state, but nobody proved that it is always possible for a propagating wave. Since the true link between the Lagrangian density and the wave equation is not what we thought, an error there is not easily seen. The most important consequence is that the wave is more general than the principle of minimum: it is easy to get wave equations which cannot be obtained from a Lagrangian density (see Sec. 5.3). The Lagrangian density as real scalar part of the wave equation is our Ariadne's thread: The double link remains when we change the linear Dirac equation into our homogeneous nonlinear equation. Next it remains when we go from the wave equation of the electron alone into the equation of the electron+neutrino pair. It remains again when we go to the wave of electron+neutrino+quarks. The thread also allows us to find the true mass terms, for the case electron+neutrino and also for the case electron+neutrino+quarks.

Another part of this strange answer is the non-equality between the light speed as limit speed of any Dirac wave and the limit speed of other waves. We have gotten a limit speed different from the light speed in the frame of a wave equation coming not from a Lagrangian density.

And the Lagrangian domain of physics is actually not a very big domain. For instance in the Lagrangian formalism of quantum electrodynamics all Maxwell laws are thought to result from the Lagrange equations. But this is false, because the link between potentials and fields is postulated to simplify the second-order wave equations. The same remark was first made by de Broglie himself in his theory of the photon, where two Lagrangian densities are necessary to get all Maxwell laws. Then what is the true energy–momentum of the electromagnetic field? Usually physicists consider $\vec{E}^2 + \vec{H}^2$ as giving the density of energy. This is generalized in $F_a^{\mu\nu} F_{\mu\nu}^a$ in the case of a non-commutative gauge group. If the relation between potentials and fields comes also from a Lagrangian mechanism which is associated to another energy–momentum tensor, that quantum electrodynamics does not account for. If not, what proves that the energy–momentum tensor coming from the Lagrangian density is the true one? Is the Lagrangian formalism itself, with its reversible time, compatible with the transport by photons of energy–momentum?

10.4   Theory versus experiment

The world of the particles with magnetic charges is different and we are, in terms of the knowledge of this new world, probably no wiser than Colombus after his first journey. We know it exists and this is already something.

The two parts of this book, the theoretical part formed by nine chapters and three appendices, and the experimental part in one modest chapter, seem at the same time disconnected and disproportionate. We can expect that more experiments will discover many new properties of the magnetic world.

These two parts are however doubly linked. Firstly probabilities, essential in quantum theory, are very discreet here. In our powders, randomness does not seem to play a very big role. When titanium is transformed at some place, all that can be transformed is changed. Randomness may be the reason why results are variable, but getting copper instead of iron, chromium instead of manganese, this can also come from differences in temperature, pressure, duration of the discharge and so on. We know nothing until now about that.

The second link between our two parts is: space-time is different from what we thought it was. In the theoretical part we explained how a second space-time manifold appears, anisotropic, and how the spinor wave makes a bridge between these two manifolds. This bridge can be extended to the 6-dimensional space-time which allows us to describe electro-weak and strong interactions. The usual space-time is not the fundamental entity; it is only a part, well defined, of a complete space-time where complex numbers appear naturally. The geometric transformation linked to the wave contains a sum of direct and of inverse dilations. It is then very different from simple Lorentz rotations. Geometric transformations are only induced by elements of a group coming from the space algebra. They appear as secondary objects, the fundamental one being the wave.

In the experimental part also space seems very different from what we thought, since kernels of atoms that we think are separated by huge distances in comparison with their own sizes, seem able to put together their nucleons and to reallocate them as if the distance between them was insignificant.

10.5   Future applications

What will be the applications of magnetic monopoles is also a premature question and what we are able to imagine today will probably have little in common with the very practical applications which will come out of laboratories in the future.

Urutskoev who is a nuclear physicist thinks to new nuclear reactors, intrinsically safe, driven by very intense magnetic fields. Another exciting possibility is that magnetism may be linked to gravitation. There are some theoretical hints there, because G. Lochak [51] has explained that it is a magnetic photon which is linked to the graviton in the fusion theory of Louis de Broglie.

A third kind of possible practical consequence is about geology. The magnetic monopole of G. Lochak is a kind of excited neutrino. The Sun can produce magnetic monopoles which will arrive on our Earth, mainly at the magnetic poles. This was experimentally satisfied [1]. These monopoles are likely to induce the same transformations we see in the laboratory, notably producing hydrogen. This may change the process of fossilization and the creation of deposits.

Evidently if magnetic monopoles are produced in the heart of the Sun, they are able to produce effects on magnetic fields. The study of the magnetic monopoles is perhaps the best way to progress in the comprehension of our Sun.

We can also have a good idea of what will never allow the double space-time manifolds. The existence of a second space-time manifold changes nothing in the properties of the first manifold, the one in which we move and observe our universe. We must then not dream of things we know are forbidden by physical laws, such as overshooting the limit speed. Accelerating until this limit remains impossible because the mass approaches infinity when the speed approaches the limit even if this limit is not necessarily equal to the speed of light in the vacuum. Another restriction which has no chance of changing is the one linked to the time arrow. Any journey back into past will stay forbidden since the invariance group conserves the orientation of our time.

10.6　Improved standard model

The different parts of the work that we present here are strongly interacting and reinforce one another. The Cl_3^* group is easier to see from the invariant form Eq. (3.1) of the wave equation. The behavior of the mass term used in chapter 4 is a sufficient reason to prefer the nonlinear homogeneous wave equation to its linear approximation, the Dirac equation. We got also the electro-weak gauge group in a much simpler way. It was easy to extend this model from the lepton sector to the quark sector. We can explain not only why there are three completely similar generations (and four [2] neutrinos), but also why the different generations must be separately treated in the gauge theory. We can explain why the complete gauge group is the $U(1) \times SU(2) \times SU(3)$ gauge group found from experiments and why the $SU(3)$ gauge group does not act upon the lepton part of the complete wave, which has been postulated before. The link between the wave and the space-time geometry is reinforced by the fact that this link survives to the extension of the wave to all fermions of a generation, which necessitates the use of two supplementary dimensions of space. The generalization of the geometric transformation necessitates the use of the link between the wave of the particle and the wave of the anti-particle. From the physical point of view this changes the meaning of the charge conjugation. From the mathematical point of view this divides by two the number of parameters, then the dimension n of the space-time is reduced to $n - 1$. The criterion of simplicity then induces a preference for the algebra $Cl_{1,5}$. Then the manifold of space-time appears as embedded into a manifold with two supplementary dimensions of space.

A greater invariance group implies strong new strains. These new strains imply a better understanding of old concepts and indicate the only way to go further. For instance the relation $\phi' = M\phi$ is not new, it was used in equivalent form since the Pauli equation in the 1920s. But it indicates to us that the relation between the ϕ wave and the Weyl's spinors ξ and η is invariant, that the right and left parts of the wave are invariant, and that M and ϕ are similar. When you know that, it is evident that the wave of

---

2. The fourth neutrino is not able to interact by the electro-weak or strong force, it is then a part of **dark matter.**

a pair electron-neutrino must read $\Psi_l = \begin{pmatrix} \phi_e & \phi_n \\ \hat{\phi}_n & \hat{\phi}_e \end{pmatrix}$, and this equality gives then the form of the projectors P_μ. Next these projectors have naturally the $U(1) \times SU(2)$ structure of the electroweak gauge group. It is also easy to get the charges of quarks u and d simply by changing P_0 to P_0', changing only one coefficient from 1 into $-1/3$. All values of the charges of quarks are obtained by the choice of only one number, and the value of this number is necessary to simplify the operator of the chiral gauge. We recall that these values of the charges explain why the charge of the proton is exactly opposite to the charge of the electron, why the charge of the neutron is zero, and therefore why the electric charge of any unionized atom is zero.

Strains coming from the invariance group imply also that you have only one simple way to get a wave with all fermions of one generation, which is $\Psi = \begin{pmatrix} \Psi_l & \Psi_r \\ \Psi_g & \Psi_b \end{pmatrix}$. But after that if you want to have the same link as before between the wave and the geometry of space-time, it is necessary to dispose of Eq. (7.73). This in fact requires to use the link Eq. (2.90), well known in the standard model, existing between the wave of the particle and the wave of the antiparticle. This link restricts the value of the wave from $Cl_{2,3} = M_4(\mathbb{C})$ to its sub-algebra $Cl_{1,3}$ and from $Cl_{5,2} = M_8(\mathbb{C})$ to its sub-algebra $Cl_{5,1}$. It happens that this Clifford algebra is isomorphic to $Cl_{1,5}$ and this isomorphism is both the reason why the non-isomorphic sub-algebras $Cl_{1,3}$ and $Cl_{3,1}$ are equally used, and a reason to be more confident of the standard model and its precepts issued from a long history of experiments. Another reason to be confident both in the standard model and in the use of Clifford algebra is the link between the cancellation of right waves, except the electron wave, and the existence of a mathematical inverse, used to build waves of systems. This should allow us to build the wave of a proton or a neutron from their internal quarks.

Questions asked of the initial quantum theory are today nearly forgotten. Why there is a Planck constant and why there are complex numbers were two of them. Since nobody had a clear and simple answer, these questions were put "under the table". The existence of the Planck factor, which links proper mass to frequency of the wave, is directly linked to supplementary strains of the invariance group. Complex numbers are also simply explained, first by the isomorphism between Cl_3 and the algebra generated by Pauli matrices, next by the matrix representation of $Cl_{1,3}$, finally by the matrix representation of $Cl_{1,5}$. This algebra is isomorphic to a sub-algebra of the algebra $M_8(\mathbb{C})$ of 8×8 matrices on the complex field. The complete wave is then a function of the space-time with value into $Cl_{1,5}$, this justifies most of the mathematical apparatus of the standard model. From the physical point of view, this allows us to build the boson fields by antisymmetric products of fermions in even numbers.

New strains used here also explain why old attempts were not successful. We think to the numerous attempts made in the 1930s to unify mechanics and electromagnetism. Such a unification is limited by the fact that a charge is not a mass when you use the full invariance group. Another attempt, made to unify the different parts of the $U(1) \times SU(2) \times SU(3)$ gauge group as subgroups of $SU(5)$ or $SO(10)$ had no more success, predicting a possible disintegrating proton which was not experimentally found. The structure of the gauge group comes from the structure of the complete wave and does not change when you increase the energy. The structure of the wave is then fully compatible with protons without disintegration, and more generally with all known aspects of modern physics.

The standard model shall remain so more essential because the wave equations with a mass term are compatible both with the form invariance and with the gauge invariance. This makes the spontaneously broken symmetry useless.

10.7   Algorithmic and data structures

Even if the big accelerators of particle physics are today inseparable from the software necessary to get and treat an enormous quantity of data, the theory of the standard model is old and was made in a world without computers. Today all students learn algorithmic and data structures, but the consideration of data structures was not known when quantum field theory was growing. Then nothing was made to secure the formalism of bra and ket. The spirit of the second quantification is to consider only algorithms and calculations, and to refuse any consideration about things that are used in these calculations. It is a software with algorithmic but without data structures!

Our work proves that the objects used in the standard model are numerous and different. Then it shall be certainly necessary to think about them. The wave of the electron is a function of space-time in the Clifford algebra of space (8-dimensional), with numeric dimension $1/2$. It is not at all the same object as the wave of a photon, which is a composite object made of a space-time vector potential and a bivector field. The idea of a set of bra and ket working indifferently on waves of electrons and quarks as well as systems of them and photons and Z_0 and gluons as well as systems of them is a purely theoretical dream. It is in fact impossible to apply the same calculations to photons needing symmetrization and electrons needing anti-symmetrization. Even with the names ket and bra it is impossible to treat globally the different things that are fermions and bosons.

The correct use of the different data structures gives not only new strains. It also brings understanding. Why does the Dirac theory take place in Cl_3 when it should be in a relativistic frame? Why is the invariance group Cl_3^* and no more? Why are there three generations of fundamental fermions and no more? To these questions the response comes from data structures: the 3 is simply the dimension of the physical space, this means the number of real numbers necessary to describe a point. Next the 8 that is the number of parameters of the Dirac wave and the dimension of the invariance group is 2^3 and comes from a general link between dimension of the linear space and dimension of the algebra built on this linear space. The presence of complex numbers in physics also results from data structures, since it comes from the isomorphism between the Clifford algebra of space and the algebra of 2×2 complex matrices.

Space and time take place in Cl_3, but this is made in a very precise part of this algebra, the self-adjoint part. From this a link results between the reversion in the Clifford algebra of space-time and the reversion in the algebra of space. Since this reversion in Cl_3 is the adjoint matrix, this conjugation has had a great place in quantum theory, giving the hermitian and unitary theory. But the algebra of space, even if it is sufficient to study the electron alone, is not the only data structure to consider. Two homomorphisms exist from Cl_3^* into the group of dilations on space-time. This is why left and right waves exist. Since these homomorphisms respect the orientation of the space and the orientation of the time, all physical laws are made of this oriented space and of this oriented time.

The group of invariance induces many consequences. The strains com ing from the different variances, covariance, contra-variance, invariance, half-variance of the wave, are summarized in the numeric dimension. This concept, since it comes from the invariance itself, rules all parts of the phys-ical reality. The concept of numeric dimension is not arbitrary: General Relativity is based on physical lengths whilst differential Geometry is based on real numbers. Necessarily something must compensate this difference. Our definition of the numeric dimension simply gives the dimension 1 to any length in space (or time). The product-measure gives then the dimension 2 for surfaces, the dimension 3 for volumes in space and the dimension 4 for volumes in space-time. The derivation uses a quotient, then a numeric dimension -1 is added for each derivation. It results that an acceleration has dimension -1 since it is the derivative of a velocity, the quotient of a length of space by a length of time, with numeric dimension 0. And so on... Moreover the integration over the space links the Lagrangian density of the quantum wave to the Lagrangian of the particle in relativistic mechanics. Then the proper mass of this Lagrangian has necessarily the numeric dimension 3 that has been obtained by supposing that gauge fields have numeric dimension 0. We could have obtained this dimensionality 0 from other considerations. For instance all gauge fields are boson fields. This means that the waves of 2, 3,... n photons are similar and may be added. Since the general rule to get the wave of systems of particles is the multiplication of the waves, each wave has necessarily the numeric dimension 0.

The Clifford algebra of space-time is built on a 4-dimensional linear space. It is then 16-dimensional. This $16 = 2^4$ is also a 4×4 that has given for instance the Majorana real matrices or the 4×4 complex matrices of the Dirac equation. Unhappily for them it is 2^4 that is important and general, and 4×4 is only an accidental coincidence. All laws of electromagnetism take place naturally in space-time algebra and this may be extended to the quantum wave of electron+neutrino, and then to electro-weak interactions. It happens that the quantum wave is not only reserved to a lone particle, it concerns also systems of particles. Then the wave must be invertible, and is is much easier to get such an invertible wave for an electron+neutrino pair if a neutrino has only one of the two possible right and left waves. And this is exactly what happens since the neutrino has only a left wave.

The double link between Lagrangian density and wave equation makes that most of the information present in one equation, the scalar one, gives the other equations. This also may be considered as a mathematical accident, because if a similar wave equation with operator ∂ is used with another structure of space-time, the double link is not automatic. We must be not too confident from the fact that the double link exists for the electron alone, for electron+neutrino, and for electron+neutrino+quarks. First it is only one wave equation with particular cases. Next the dimension and signature of $Cl_{1,5}$ is not any dimension and not any signature. Two supplementary dimensions, not only one, are necessary for embedding all solutions of the equations of General Relativity. It is the same in the quantum domain, we could not have obtained the $U(1) \times SU(2) \times SU(3)$ gauge group in a

$Cl_{p,q}$ with $p + q < 6$. Next more than two supplementary dimensions are difficult to manage and still obtain correct physical results, this comes from the necessary link between reversions in the different algebras.

The form of the wave equation itself is not just anything: the differential term is linear and linear applications are the natural first approximation of any more complicated equation with derivatives. Next the invariance under Cl_3^* adds so many new strains that the partial differential operator is necessary: a $\nabla = \sigma^\mu \partial_\mu$, or a $\partial = \gamma^\mu \partial_\mu$, or a $\underline{\partial} = L^\mu \partial_\mu$, respectively in Cl_3, $Cl_{1,3}$ or $Cl_{1,5}$. The form of the gauge invariance itself implies that linear transformations must also be used there. In space-time algebra, 16-dimensional, the linear space of all linear transformations of $Cl_{1,3}$ into $Cl_{1,3}$ is 256-dimensional. It happens that the 256 transformations $\Psi \mapsto A\Psi B$ where A and B are generators of $Cl_{1,3}$ form a basis of this linear space. Then any linear transformation is necessarily a linear combination of these transformations. This gives the necessary form of the gauge derivative.

The difference existing between right and left multiplication has an exception since the i of Cl_3 commutes with any element of this algebra. This has a great implication in the electro-weak theory: it is just this part of the gauge group that separates leptons from quarks, just as it is this part of the inertial potentials that is present in the quantum world but is absent in differential geometry. Consequently it is impossible to reduce the quantum world to differential geometry.

10.8 Beyond the standard model, back to physical reality

This work is more similar to Einstein's views than to those of the Copenhagen school. Waves obeying partial differential equations may be considered as deterministic. A probability density exists, but it is only a derived consequence of the equality between inertial and gravitational mass–energy. Moreover this density of probability concerns both electron and neutrino in the lepton case. In the quark case we get a second density of probability, but is a also a sum of 6 currents, for two quarks in three color states each. A lot of work has been done on Bell's inequalities and their invalidation. Bell's inequalities, as predicted, are violated. This proves only that at least one of the numerous conditions giving these inequalities is false. We have the right to think that the first of these false conditions was believing that the wave of two electrons, and a fortiori two photons, must obey a Schrödinger equation for a function with value in the complex field, with a probability equal to the square of the modulus of this complex value. The wave equations that we have found do not reduce to the Schrödinger equation. To get the physical wave equation of two electrons or two photons is a very difficult problem that must still be solved.

The standard model present many aspects that have induced to search beyond. There are too many suppositions, too many parameters. This study has reduced the number of parameters, for instance we have only one parameter giving all charges of quarks and only 6 mass parameters for all leptons and quarks of the three generations, and only 3 wave equations for all that. Then the need to search "beyond the standard model" is rather reduced. We have explained why the attempts of "great unification" with a gauge group simpler than the group of the standard model could not have succeeded. Theories of chords with many supplementary dimensions will also have problems with the reversion. The super-symmetry transforming fermions into bosons is also completely out of the frame of this work. This implies that we await no new particle or system of particles. The only new objects that remain to study are then the magnetic monopoles. As this study is only beginning, inventive researchers may obtain interesting results without needing much money.

Theories of chords, branes, supersymmetry and other attempts made to include gravitation in the frame of quantum physics were unlucky: they were based on old concepts coming from the beginning of quantum mechanics as a non-relativistic theory, while electromagnetism, weak interactions, strong interactions, gravitation are all based on relativistic waves and relativistic invariance. All these attempts are also based on a unique Lagrangian that is thought of as the origin of all partial differential equations. This is also a false track because Lagrangians are consequences of wave equations of fermions, and only for one fermion and only in an inertial frame. A third false track is unitarity; the true one is reversion, and it is the same only in Cl_3. The use of a greater invariance group is enough to gather gravitation, electromagnetism, weak interactions and strong interactions in the same frame.

In the domain of particle physics, we have still many things to understand and parameters to measure precisely. The response that we have made about the three generations is only a theoretical response. Physical quarks in the protons and the neutrons seem much more complicated and not exactly in one unique generation. This may come from the fact that the three terms at the right of the differential term in Eq. (7.111) are not orthogonal.

What about the particle-wave dualism? The Dirac equation and our new wave equations are partial derivative equations of the first order and describe the evolution of waves. The only part of this work where the electron appears as a particle is when we use the equality between inertial and gravitational mass–energy. After integration to the physical space of the density of energy of the wave, the non-symmetric energy–momentum tensor of the wave gives the momentum–energy space-time vector of the electron-particle. We know no other reason for the coexistence in the electron wave of an electron-particle. But we also have no reason to pretend that the energy of the electron cannot be concentrated in a little portion of the wave.

We have studied only the most simple part of the standard model, the fermion wave. We know that it will be necessary to understand also the boson part of the physical world. It is a much more complicated problem: a boson like a photon, Z^0 or gluon is a composite object with a potential term that is a space-time vector and that has the numeric dimension 1, and a bivector field with numeric dimension 0. Therefore a lot of work will be necessary to fully understand what bosons are and how they interact with the fermion and boson parts of our physical Universe. We have more questions without answer than questions with them. For instance we saw two ways to get a wave with numeric dimension 0 from waves with numeric dimension $1/2$. One of them was described in chapter 4: $m(\phi_2\overline{\phi_1} - \phi_1\overline{\phi_2})$; the other was described in chapter 5: $\phi_{12} = \phi_1\sigma_1\phi_2^{-1} - \phi_2\sigma_1\phi_1^{-1}$. These two ways use the Pauli principle of anti-symmetrization. Are these waves similar, can we add them? Shall we understand why we must anti-symmetrize spinor waves?

The quest continues.

Appendix A

Calculations in Clifford algebras

A.1 Invariant equation and Lagrangian

Let M be an invertible matrix element of Cl_3^*, with determinant $re^{i\theta}$. Let R and $\overline{R}$ be Lorentz dilations such as :

$$R : x \mapsto x' = R(x) = MxM^\dagger \; ; \; \overline{R} : x \mapsto x' = \overline{R}(x) = \overline{M}x\widehat{M}. \quad (A.1)$$

Let P be the matrix such as:

$$M = \sqrt{r}e^{i\frac{\theta}{2}}P, \quad (A.2)$$

and let L and $\overline{L}$ be dilations such as :

$$L : x \mapsto x' = L(x) = PxP^\dagger \; ; \; \overline{L} : x \mapsto x' = \overline{L}(x) = \overline{P}x\widehat{P}. \quad (A.3)$$

We have:

$$re^{i\theta} = \det(M) = M\overline{M} = \sqrt{r}e^{i\frac{\theta}{2}}P\sqrt{r}e^{i\frac{\theta}{2}}\overline{P} = re^{i\theta}P\overline{P}. \quad (A.4)$$

We get then

$$P\overline{P} = 1 \; ; \; \overline{P} = P^{-1} \; ; \; \overline{L} = L^{-1}. \quad (A.5)$$

P is then an element of $SL(2, \mathbb{C})$ and L is a Lorentz rotation. We know, for such a rotation, that:

$$g = \begin{pmatrix} 1 & 0 & 0 & 0 \\ 0 & -1 & 0 & 0 \\ 0 & 0 & -1 & 0 \\ 0 & 0 & 0 & -1 \end{pmatrix} \quad (A.6)$$

where g is the signature-matrix. Denoting (L) the matrix of L in an orthonormal basis and M^t the transposed [1] matrix of M:

$$(L)^{-1} = g(L)^t g \; ; \; (\overline{L})g = g(L)^t. \quad (A.7)$$

---

1. The transposition exchanges lines and columns of matrices: if $M = \begin{pmatrix} a & b \\ c & d \end{pmatrix}$ then $M^t = \begin{pmatrix} a & c \\ b & d \end{pmatrix}$. We have, for any matrices A and B, $(AB)^t = B^t A^t$ and $\det(A^t) = \det(A)$.

But we have also:

$$R(x) = MxM^\dagger = \sqrt{r}e^{i\frac{\theta}{2}}Px\sqrt{r}e^{-i\frac{\theta}{2}}P^\dagger = rPxP^\dagger = rL(x), \qquad (A.8)$$

therefore

$$R = rL \ ; \quad (R) = r(L). \qquad (A.9)$$

We have also:

$$\overline{R}(x) = \overline{M}x\widehat{M} = \sqrt{r}e^{i\frac{\theta}{2}}\overline{P}x\sqrt{r}e^{-i\frac{\theta}{2}}\widehat{P} = r\overline{P}x\widehat{P} = r\overline{L}(x), \qquad (A.10)$$

$$\overline{R} = r\overline{L} \ ; \quad (\overline{R}) = r(\overline{L}). \qquad (A.11)$$

Multiplying Eq. (A.7) by r we get :

$$(\overline{R})g = g(R)^t \ ; \quad (\overline{R}) = g(R)^t g, \qquad (A.12)$$

which gives for $j = 1,\ 2,\ 3$ and $k = 1,\ 2,\ 3$:

$$\overline{R}_0^0 = R_0^0 \ ; \quad \overline{R}_0^j = -R_j^0 \ ; \quad \overline{R}_j^0 = -R_0^j \ ; \quad \overline{R}_j^k = R_k^j. \qquad (A.13)$$

Consequently lines as well as columns of the matrix R_μ^ν are orthogonal, because we have, for R and $\overline{R}$:

$$R_\mu = M\sigma_\mu M^\dagger = R_\mu^\nu \sigma_\nu \ ; \quad \overline{R}_\mu = \overline{M}\sigma_\mu \widehat{M} = \overline{R}_\mu^\nu \sigma_\nu, \qquad (A.14)$$

$$R_\mu \cdot R_\nu = \overline{R}_\mu \cdot \overline{R}_\nu = \delta_{\mu\nu}\rho^2, \qquad (A.15)$$

where $\delta_{00} = 1$, $\delta_{11} = \delta_{22} = \delta_{33} = -1$, $\delta_{\mu\nu} = 0$ if $\mu \neq \nu$. We have

$$\overline{\phi}A\widehat{\phi} = A^\mu \overline{\phi}\sigma_\mu \widehat{\phi} = A_0 \overline{D}_0 - \sum_{j=1}^{j=3} A_j \overline{D}_j = A_0(\overline{D}_0^\mu \sigma_\mu) - \sum_{j=1}^{j=3} A_j(\overline{D}_j^\mu \sigma_\mu). \qquad (A.16)$$

But the link between the D_μ and $\overline{D}_\mu$ is the same as between R_μ and $\overline{R}_\mu$ and we get with Eq. (A.13) for $j = 1,\ 2,\ 3$ and $k = 1,\ 2,\ 3$:

$$\overline{D}_0^0 = D_0^0 \ ; \quad \overline{D}_0^j = -D_j^0 \ ; \quad \overline{D}_j^0 = -D_0^j \ ; \quad \overline{D}_j^k = D_k^j, \qquad (A.17)$$

which gives

$$\overline{\phi}A\widehat{\phi} = A_0\left(\overline{D}_0^0 + \sum_{j=1}^{j=3}\overline{D}_0^j \sigma_j\right) - \sum_{j=1}^{j=3} A_j\left(\overline{D}_j^0 + \sum_{k=1}^{k=3}\overline{D}_j^k \sigma_k\right) \qquad (A.18)$$

$$= A_0\left(D_0^0 - \sum_{j=1}^{j=3} D_j^0 \sigma_j\right) - \sum_{j=1}^{j=3} A_j\left(-D_0^j + \sum_{k=1}^{k=3} D_k^j \sigma_k\right) = A_\nu D_\mu^\nu \sigma^\mu.$$

The scalar part is then

$$\langle \overline{\phi}A\widehat{\phi}\rangle_0 = D_0^\nu A_\nu = A_\mu J^\mu. \qquad (A.19)$$

The corresponding term with the Dirac matrices is

$$\frac{1}{2}[(\overline{\psi}\gamma^\mu qA_\mu)\psi) + (\overline{\psi}\gamma^\mu qA_\mu\psi)^\dagger]$$

$$= \frac{q}{2}A_\mu[\overline{\psi}\gamma^\mu\psi + (\overline{\psi}\gamma^\mu\psi)] = qA_\mu\overline{\psi}\gamma^\mu\psi = qA_\mu J^\mu. \qquad (A.20)$$

We get next

$$\frac{1}{2}[(\overline{\psi}\gamma^\mu(-i)\partial_\mu\psi) + (\overline{\psi}\gamma^\mu(-i)\partial_\mu\psi)^\dagger] = \frac{i}{2}(-\overline{\psi}\gamma^\mu\partial_\mu\psi + \partial_\mu\overline{\psi}\gamma^\mu\psi)$$

$$= \frac{i}{2}[-\xi^\dagger\partial_0\xi - \eta^\dagger\partial_0\eta + (\partial_0\xi^\dagger)\xi + (\partial_0\eta^\dagger)\eta] \qquad (A.21)$$

$$+ \frac{i}{2}\sum_{j=1}^{j=3}[-\xi^\dagger\sigma_j\partial_j\xi + \eta^\dagger\sigma_j\partial_j\eta + (\partial_j\xi^\dagger)\sigma_j\xi - (\partial_j\eta^\dagger)\sigma_j\eta]$$

which gives

$$- i[(\overline{\psi}\gamma^\mu(-i)\partial_\mu\psi) + (\overline{\psi}\gamma^\mu(-i)\partial_\mu\psi)^\dagger] = \qquad (A.22)$$

$$\xi_1\partial_0\xi_1^* + \xi_2\partial_0\xi_2^* + \eta_1\partial_0\eta_1^* + \eta_2\partial_0\eta_2^* - \xi_1^*\partial_0\xi_1 - \xi_2^*\partial_0\xi_2 - \eta_1^*\partial_0\eta_1 - \eta_2^*\partial_0\eta_2$$

$$\xi_1\partial_1\xi_2^* + \xi_2\partial_1\xi_1^* - \eta_1\partial_1\eta_2^* - \eta_2\partial_1\eta_1^* - \xi_1^*\partial_1\xi_2 - \xi_2^*\partial_1\xi_1 + \eta_1^*\partial_1\eta_2 + \eta_2^*\partial_1\eta_1$$

$$- i(-\xi_1\partial_2\xi_2^* + \xi_2\partial_2\xi_1^* + \eta_1\partial_2\eta_2^* - \eta_2\partial_2\eta_1^* - \xi_1^*\partial_2\xi_2 + \xi_2^*\partial_2\xi_1 + \eta_1^*\partial_2\eta_2 - \eta_2^*\partial_2\eta_1)$$

$$\xi_1\partial_3\xi_1^* - \xi_2\partial_3\xi_2^* - \eta_1\partial_3\eta_1^* + \eta_2\partial_3\eta_2^* - \xi_1^*\partial_3\xi_1 + \xi_2^*\partial_3\xi_2 + \eta_1^*\partial_3\eta_1 - \eta_2^*\partial_3\eta_2.$$

In the Pauli algebra we have

$$\overline{\phi}(\nabla\widehat{\phi})\sigma_{21} = 2i\begin{pmatrix} \eta_1^* & \eta_2^* \\ -\xi_2 & \xi_1 \end{pmatrix}\begin{pmatrix} \partial_0 - \partial_3 & -\partial_1 + i\partial_2 \\ -\partial_1 - i\partial_2 & \partial_0 + \partial_3 \end{pmatrix}\begin{pmatrix} -\eta_1 & -\xi_2^* \\ -\eta_2 & \xi_1^* \end{pmatrix}, \qquad (A.23)$$

and with Eq. (2.74) we get

$$\overline{\phi}(\nabla\widehat{\phi})\sigma_{21} = \begin{pmatrix} w^3 + w^0 - iv^3 - iv^0 & v^2 + iv^1 + iw^2 - w^1 \\ v^2 - iv^1 + iw^2 + w^1 & w^3 - w^0 - iv^3 + iv^0 \end{pmatrix} = \qquad (A.24)$$

$$2i\begin{pmatrix} \begin{matrix}\eta_1^*(-\partial_0\eta_1+\partial_1\eta_2-i\partial_2\eta_2+\partial_3\eta_1) \\ +\eta_2^*(-\partial_0\eta_2+\partial_1\eta_1+i\partial_2\eta_1-\partial_3\eta_2)\end{matrix} & \begin{matrix}\eta_1^*(-\partial_0\xi_2^*-\partial_1\xi_1^*+i\partial_2\xi_1^*+\partial_3\xi_2^*) \\ +\eta_2^*(\partial_0\xi_1^*+\partial_1\xi_2^*+i\partial_2\xi_2^*+\partial_3\xi_1^*)\end{matrix} \\ \\ \begin{matrix}-\xi_2(-\partial_0\eta_1+\partial_1\eta_2-i\partial_2\eta_2+\partial_3\eta_1) \\ +\xi_1(-\partial_0\eta_2+\partial_1\eta_1+i\partial_2\eta_1-\partial_3\eta_2)\end{matrix} & \begin{matrix}-\xi_2(-\partial_0\xi_2^*-\partial_1\xi_1^*+i\partial_2\xi_1^*+\partial_3\xi_2^*) \\ +\xi_1(\partial_0\xi_1^*+\partial_1\xi_2^*+i\partial_2\xi_2^*+\partial_3\xi_1^*)\end{matrix} \end{pmatrix}.$$

This gives

$$w^3 + w^0 - iv^3 - iv^0 = \quad 2i(-\eta_1^*\partial_0\eta_1 - \eta_2^*\partial_0\eta_2 + \eta_1^*\partial_1\eta_2 + \eta_2^*\partial_1\eta_1$$

$$-i\eta_1^*\partial_2\eta_2 + i\eta_2^*\partial_2\eta_1 + \eta_1^*\partial_3\eta_1 - \eta_2^*\partial_3\eta_2), \quad (A.25)$$

$$w^3 - w^0 - iv^3 + iv^0 = \quad 2i(\xi_2\partial_0\xi_2^* + \xi_1\partial_0\xi_1^* + \xi_2\partial_1\xi_1^* + \xi_1\partial_1\xi_2^*$$

$$-i\xi_2\partial_2\xi_1^* + i\xi_1\partial_2\xi_2^* - \xi_2\partial_3\xi_2^* + \xi_1\partial_3\xi_1^*), \quad (A.26)$$

$$v^2 - iv^1 + iw^2 + w^1 = \quad 2i(\xi_2\partial_0\eta_1 - \xi_1\partial_0\eta_2 - \xi_2\partial_1\eta_2 + \xi_1\partial_1\eta_1$$
$$+i\xi_2\partial_2\eta_2 + i\xi_1\partial_2\eta_1 - \xi_2\partial_3\eta_1 - \xi_1^*\partial_3\eta_2), \quad (A.27)$$

$$v^2 + iv^1 + iw^2 - w^1 = \quad 2i(-\eta_1^*\partial_0\xi_2^* - \eta_2^*\partial_0\xi_1^* - \eta_1^*\partial_1\xi_1^* + \eta_2^*\partial_1\xi_2^*$$
$$+i\eta_1^*\partial_2\xi_1^* + i\eta_2^*\partial_2\xi_2^* + \eta_1^*\partial_3\xi_2^* + \eta_2^*\partial_3\xi_1^*). \quad (A.28)$$

Adding and subtracting Eq. (A.25) and Eq. (A.26) we get

$$w^3 - iv^3 = -i\eta_1^*\partial_0\eta_1 - i\eta_2^*\partial_0\eta_2 + i\xi_2\partial_0\xi_2^* + i\xi_1\partial_0\xi_1^*$$
$$+ i\eta_1^*\partial_1\eta_2 + i\eta_2^*\partial_1\eta_1 + i\xi_2\partial_1\xi_1^* + i\xi_1\partial_1\xi_2^* \qquad (A.29)$$
$$+ \eta_1^*\partial_2\eta_2 - \eta_2^*\partial_2\eta_1 + \xi_2\partial_2\xi_1^* - \xi_1\partial_2\xi_2^*$$
$$+ i\eta_1^*\partial_3\eta_1 - i\eta_2^*\partial_3\eta_2 - i\xi_2\partial_3\xi_2^* + i\xi_1\partial_3\xi_1^*$$

$$w^0 - iv^0 = -i\eta_1^*\partial_0\eta_1 - i\eta_2^*\partial_0\eta_2 - i\xi_2\partial_0\xi_2^* - i\xi_1\partial_0\xi_1^*$$
$$+ i\eta_1^*\partial_1\eta_2 + i\eta_2^*\partial_1\eta_1 - i\xi_2\partial_1\xi_1^* - i\xi_1\partial_1\xi_2^* \qquad (A.30)$$
$$+ \eta_1^*\partial_2\eta_2 - \eta_2^*\partial_2\eta_1 - \xi_2\partial_2\xi_1^* + \xi_1\partial_2\xi_2^*$$
$$+ i\eta_1^*\partial_3\eta_1 - i\eta_2^*\partial_3\eta_2 + i\xi_2\partial_3\xi_2^* - i\xi_1\partial_3\xi_1^*.$$

Separating the real and the imaginary part of Eq. (A.29) we get

$$\frac{2}{i}w^3 = \quad \xi_1\partial_0\xi_1^* + \xi_2\partial_0\xi_2^* + \eta_1\partial_0\eta_1^* + \eta_2\partial_0\eta_2^* - \xi_1^*\partial_0\xi_1 - \xi_2^*\partial_0\xi_2 - \eta_1^*\partial_0\eta_1$$
$$- \eta_2^*\partial_0\eta_2 + \xi_1\partial_1\xi_2^* + \xi_2\partial_1\xi_1^* - \eta_1\partial_1\eta_2^* - \eta_2\partial_1\eta_1^* - \xi_1^*\partial_1\xi_2 - \xi_2^*\partial_1\xi_1$$
$$+ \eta_1^*\partial_1\eta_2 + \eta_2^*\partial_1\eta_1 - i(-\xi_1\partial_2\xi_2^* + \xi_2\partial_2\xi_1^* + \eta_1\partial_2\eta_2^* - \eta_2\partial_2\eta_1^*$$
$$- \xi_1^*\partial_2\xi_2 + \xi_2^*\partial_2\xi_1 + \eta_1^*\partial_2\eta_2 - \eta_2^*\partial_2\eta_1) + \xi_1\partial_3\xi_1^* - \xi_2\partial_3\xi_2^*$$
$$- \eta_1\partial_3\eta_1^* + \eta_2\partial_3\eta_2^* - \xi_1^*\partial_3\xi_1 + \xi_2^*\partial_3\xi_2 + \eta_1^*\partial_3\eta_1 - \eta_2^*\partial_3\eta_2.$$
$$(A.31)$$

This gives with Eq. (A.22)

$$\frac{1}{2}[(\overline{\psi}\gamma^\mu(-i)\partial_\mu\psi) + (\overline{\psi}\gamma^\mu(-i)\partial_\mu\psi)^\dagger] = w^3, \qquad (A.32)$$

and with Eq. (A.20) we get Eq. (2.83). The Tetrode's impulse–energy tensor coming from the invariance of the Lagrangian density under translations satisfies

$$T_\lambda^\mu = \frac{1}{2}[(\overline{\psi}\gamma^\mu(-i\partial_\lambda + qA_\lambda)\psi + ((\overline{\psi}\gamma^\mu(-i\partial_\lambda + qA_\lambda)\psi)^\dagger]. \qquad (A.33)$$

We get then from Eq. (A.32)

$$w^3 = T_\mu^\mu - q\overline{\psi}\gamma^\mu A_\mu\psi$$
$$w^3 = T_\mu^\mu - V^0$$
$$w^3 + V^0 = \text{tr}(T). \qquad (A.34)$$

Now the imaginary part of Eq. (A.29) gives

$$-2v^3 = \partial_0(\xi_1\xi_1^* + \xi_2\xi_2^* - \eta_1\eta_1^* - \eta_2\eta_2^*) \tag{A.35}$$
$$+ \partial_1(\xi_1\xi_2^* + \xi_2\xi_1^* + \eta_1\eta_2^* + \eta_2\eta_1^*)$$
$$+ \partial_2 i(\xi_1\xi_2^* - \xi_2\xi_1^* + \eta_1\eta_2^* - \eta_2\eta_1^*)$$
$$+ \partial_3(\xi_1\xi_1^* - \xi_2\xi_2^* + \eta_1\eta_1^* - \eta_2\eta_2^*)$$
$$= \partial_\mu D_3^\mu = \nabla \cdot D_3. \tag{A.36}$$

The imaginary part of Eq. (A.30) gives

$$-2v^0 = \partial_0(\xi_1\xi_1^* + \xi_2\xi_2^* + \eta_1\eta_1^* + \eta_2\eta_2^*) \tag{A.37}$$
$$+ \partial_1(\xi_1\xi_2^* + \xi_2\xi_1^* - \eta_1\eta_2^* - \eta_2\eta_1^*)$$
$$+ \partial_2 i(\xi_1\xi_2^* - \xi_2\xi_1^* - \eta_1\eta_2^* + \eta_2\eta_1^*)$$
$$+ \partial_3(\xi_1\xi_1^* - \xi_2\xi_2^* - \eta_1\eta_1^* + \eta_2\eta_2^*)$$
$$= \partial_\mu D_0^\mu = \nabla \cdot D_0, \tag{A.38}$$

and we get the conservation of the current of probability. From Eq. (A.18) we get

$$qA_\nu D_\mu^\nu \sigma^\mu = \bar{\phi} qA\widehat{\phi} = V = V^\mu \sigma_\mu,$$
$$= V^0 - V^1\sigma^1 - V^2\sigma^2 - V^3\sigma^3 \tag{A.39}$$
$$V^j = -qA_\nu D_j^\nu = -qA \cdot D_j \; ; \; j = 1, 2, 3. \tag{A.40}$$

The real part of Eq. (A.30) gives with Eq. (2.78)

$$\frac{2}{i}w^0 = 2iV^3 = \frac{2}{i}qA \cdot D_3 = \tag{A.41}$$

$$- \xi_1\partial_0\xi_1^* - \xi_2\partial_0\xi_2^* + \eta_1\partial_0\eta_1^* + \eta_2\partial_0\eta_2^* + \xi_1^*\partial_0\xi_1 + \xi_2^*\partial_0\xi_2 - \eta_1^*\partial_0\eta_1 - \eta_2^*\partial_0\eta_2$$
$$- \xi_1\partial_1\xi_2^* - \xi_2\partial_1\xi_1^* - \eta_1\partial_1\eta_2^* - \eta_2\partial_1\eta_1^* + \xi_1^*\partial_1\xi_2 + \xi_2^*\partial_1\xi_1 + \eta_1^*\partial_1\eta_2 + \eta_2^*\partial_1\eta_1$$
$$- i(\xi_1\partial_2\xi_2^* - \xi_2\partial_2\xi_1^* + \eta_1\partial_2\eta_2^* - \eta_2\partial_2\eta_1^* + \xi_1^*\partial_2\xi_2 - \xi_2^*\partial_2\xi_1 + \eta_1^*\partial_2\eta_2 - \eta_2^*\partial_2\eta_1)$$
$$- \xi_1\partial_3\xi_1^* + \xi_2\partial_3\xi_2^* - \eta_1\partial_3\eta_1^* + \eta_2\partial_3\eta_2^* + \xi_1^*\partial_3\xi_1 - \xi_2^*\partial_3\xi_2 + \eta_1^*\partial_3\eta_1 - \eta_2^*\partial_3\eta_2$$

Now adding and subtracting Eq. (A.27) and Eq. (A.28) we get

$$v^2 + iw^2 = i\xi_2\partial_0\eta_1 - i\xi_1\partial_0\eta_2 - i\eta_1^*\partial_0\xi_2^* + i\eta_2^*\partial_0\xi_1^*$$
$$- i\xi_2\partial_1\eta_2 + i\xi_1\partial_1\eta_1 - i\eta_1^*\partial_1\xi_1^* + i\eta_2^*\partial_1\xi_2^* \tag{A.42}$$
$$- \xi_2\partial_2\eta_2 - \xi_1\partial_2\eta_1 - \eta_1^*\partial_2\xi_1^* - \eta_2^*\partial_2\xi_2^*$$
$$- i\xi_2\partial_3\eta_1 - i\xi_1\partial_3\eta_2 + i\eta_1^*\partial_3\xi_2^* + i\eta_2^*\partial_3\xi_1^*,$$

$$w^1 - iv^1 = i\xi_2\partial_0\eta_1 - i\xi_1\partial_0\eta_2 + i\eta_1^*\partial_0\xi_2^* - i\eta_2^*\partial_0\xi_1^*$$
$$- i\xi_2\partial_1\eta_2 + i\xi_1\partial_1\eta_1 + i\eta_1^*\partial_1\xi_1^* - i\eta_2^*\partial_1\xi_2^* \tag{A.43}$$
$$- \xi_2\partial_2\eta_2 - \xi_1\partial_2\eta_1 + \eta_1^*\partial_2\xi_1^* + \eta_2^*\partial_2\xi_2^*$$
$$- i\xi_2\partial_3\eta_1 - i\xi_1\partial_3\eta_2 - i\eta_1^*\partial_3\xi_2^* - i\eta_2^*\partial_3\xi_1^*.$$

The real part of Eq. (A.42) gives

$$2v^2 = \partial_0 i(-\xi_1\eta_2 + \xi_2\eta_1 + \xi_1^*\eta_2^* - \xi_2^*\eta_1^*) \tag{A.44}$$
$$+ \partial_1 i(\xi_1\eta_1 - \xi_2\eta_2 - \xi_1^*\eta_1^* + \xi_2^*\eta_2^*)$$
$$+ \partial_2(-\xi_1\eta_1 - \xi_2\eta_2 - \xi_1^*\eta_1^* - \xi_2^*\eta_2^*)$$
$$+ \partial_3 i(-\xi_1\eta_2 - \xi_2\eta_1 + \xi_1^*\eta_2^* + \xi_2^*\eta_1^*)$$
$$= -\partial_\mu D_2^\mu = -\nabla \cdot D_2 \tag{A.45}$$

which gives, with Eq. (2.76)

$$\nabla \cdot D_2 = -2v^2 = 2V^1, \tag{A.46}$$

and we get with Eq. (A.40)

$$\nabla \cdot D_2 + 2qA \cdot D_1 = 0, \tag{A.47}$$

which is 2.87. The imaginary part of Eq. (A.42) gives with Eq. (2.80)

$$2w^2 = 0 = -\xi_1\partial_0\eta_2 + \xi_2\partial_0\eta_1 - \eta_1\partial_0\xi_2 - \eta_2\partial_0\xi_1 - \xi_1^*\partial_0\eta_2^* + \xi_2^*\partial_0\eta_1^* - \eta_1^*\partial_0\xi_2^*$$
$$+ \eta_2^*\partial_0\xi_1^* + \xi_1\partial_1\eta_1 - \xi_2\partial_1\eta_2 - \eta_1\partial_1\xi_1 + \eta_2\partial_1\xi_2 + \xi_1^*\partial_1\eta_1^* - \xi_2^*\partial_1\eta_2^*$$
$$- \eta_1^*\partial_1\xi_1^* + \eta_2^*\partial_1\xi_2^* + i(\xi_1\partial_2\eta_1 + \xi_2\partial_2\eta_2 - \eta_1\partial_2\xi_1 - \eta_2\partial_2\xi_2 - \xi_1^*\partial_2\eta_1^*$$
$$- \xi_2^*\partial_2\eta_2^* + \eta_1^*\partial_2\xi_1^* + \eta_2^*\partial_2\xi_2^*) - \xi_1\partial_3\eta_2 - \xi_2\partial_3\eta_1 + \eta_1\partial_3\xi_2$$
$$- \eta_2\partial_3\xi_1 - \xi_1^*\partial_3\eta_2^* - \xi_2^*\partial_3\eta_1^* + \eta_1^*\partial_3\xi_2^* + \eta_2^*\partial_3\xi_1^*. \tag{A.48}$$

The real part of Eq. (A.43) gives with Eq. (2.81)

$$2w^1 = 0 = i(-\xi_1\partial_0\eta_2 + \xi_2\partial_0\eta_1 - \eta_1\partial_0\xi_2 + \eta_2\partial_0\xi_1 + \xi_1^*\partial_0\eta_2^* - \xi_2^*\partial_0\eta_1^* + \eta_1^*\partial_0\xi_2^*$$
$$- \eta_2^*\partial_0\xi_1^*) + i(\xi_1\partial_1\eta_1 - \xi_2\partial_1\eta_2 - \eta_1\partial_1\xi_1 + \eta_2\partial_1\xi_2 - \xi_1^*\partial_1\eta_1^* + \xi_2^*\partial_1\eta_2^*$$
$$+ \eta_1^*\partial_1\xi_1^* - \eta_2^*\partial_1\xi_2^*) - \xi_1\partial_2\eta_1 - \xi_2\partial_2\eta_2 + \eta_1\partial_2\xi_1 + \eta_2\partial_2\xi_2 - \xi_1^*\partial_2\eta_1^*$$
$$- \xi_2^*\partial_2\eta_2^* + \eta_1^*\partial_2\xi_1^* + \eta_2^*\partial_2\xi_2^* + i(-\xi_1\partial_3\eta_2 - \xi_2\partial_3\eta_1 + \eta_1\partial_3\xi_2 + \eta_2\partial_3\xi_1$$
$$+ \xi_1^*\partial_3\eta_2^* + \xi_2^*\partial_3\eta_1^* - \eta_1^*\partial_3\xi_2^* - \eta_2^*\partial_3\xi_1^*), \tag{A.49}$$

The imaginary part of Eq. (A.43) gives

$$-2v^1 = \partial_0(-\xi_1\eta_2 + \xi_2\eta_1 - \xi_1^*\eta_2^* + \xi_2^*\eta_1^*) \tag{A.50}$$
$$+ \partial_1(\xi_1\eta_1 - \xi_2\eta_2 + \xi_1^*\eta_1^* - \xi_2^*\eta_2^*)$$
$$+ \partial_2 i(\xi_1\eta_1 + \xi_2\eta_2 - \xi_1^*\eta_1^* - \xi_2^*\eta_2^*)$$
$$+ \partial_3(-\xi_1\eta_2 - \xi_2\eta_1 - \xi_1^*\eta_2^* - \xi_2^*\eta_1^*)$$
$$= \partial_\mu D_1^\mu = \nabla \cdot D_1, \tag{A.51}$$

which gives, with Eq. (2.77)

$$\nabla \cdot D_1 = -2v^1 = -2V^2, \tag{A.52}$$

and we get with Eq. (A.40)

$$\nabla \cdot D_1 - 2qA \cdot D_2 = 0, \tag{A.53}$$

which is Eq. (2.86).

A.2   Calculation of the reverse in $Cl_{1,5}$

Here, as in Sec. 1.5, indices $\mu, \nu, \rho \ldots$ have value $0, 1, 2, 3$ and indices a, b, c, d, e have value $0, 1, 2, 3, 4, 5$. We get [2]

$$L_{\mu\nu} = L_\mu L_\nu = \begin{pmatrix} 0 & \gamma_\mu \\ \gamma_\mu & 0 \end{pmatrix} \begin{pmatrix} 0 & \gamma_\nu \\ \gamma_\nu & 0 \end{pmatrix} = \begin{pmatrix} \gamma_{\mu\nu} & 0 \\ 0 & \gamma_{\mu\nu} \end{pmatrix}, \tag{A.54}$$

$$L_{\mu\nu\rho} = L_{\mu\nu} L_\rho = \begin{pmatrix} \gamma_{\mu\nu} & 0 \\ 0 & \gamma_{\mu\nu} \end{pmatrix} \begin{pmatrix} 0 & \gamma_\rho \\ \gamma_\rho & 0 \end{pmatrix} = \begin{pmatrix} 0 & \gamma_{\mu\nu\rho} \\ \gamma_{\mu\nu\rho} & 0 \end{pmatrix}, \tag{A.55}$$

$$L_{0123} = L_{01} L_{23} = \begin{pmatrix} \gamma_{0123} & 0 \\ 0 & \gamma_{0123} \end{pmatrix} = \begin{pmatrix} \mathbf{i} & 0 \\ 0 & \mathbf{i} \end{pmatrix}. \tag{A.56}$$

We get also

$$L_{45} = L_4 L_5 = \begin{pmatrix} 0 & -I_4 \\ I_4 & 0 \end{pmatrix} \begin{pmatrix} 0 & \mathbf{i} \\ \mathbf{i} & 0 \end{pmatrix} = \begin{pmatrix} -\mathbf{i} & 0 \\ 0 & \mathbf{i} \end{pmatrix} = -L_{54}, \tag{A.57}$$

$$L_{012345} = \begin{pmatrix} \mathbf{i} & 0 \\ 0 & \mathbf{i} \end{pmatrix} \begin{pmatrix} -\mathbf{i} & 0 \\ 0 & \mathbf{i} \end{pmatrix} = \begin{pmatrix} I_4 & 0 \\ 0 & -I_4 \end{pmatrix}, \tag{A.58}$$

$$L_{01235} = L_{0123} L_5 = \begin{pmatrix} \mathbf{i} & 0 \\ 0 & \mathbf{i} \end{pmatrix} \begin{pmatrix} 0 & \mathbf{i} \\ \mathbf{i} & 0 \end{pmatrix} = \begin{pmatrix} 0 & -I_4 \\ -I_4 & 0 \end{pmatrix}. \tag{A.59}$$

Similarly we get [3]

$$L_{\mu4} = \begin{pmatrix} \gamma_\mu & 0 \\ 0 & -\gamma_\mu \end{pmatrix} \ ; \ L_{\mu5} = \begin{pmatrix} \gamma_\mu \mathbf{i} & 0 \\ 0 & \gamma_\mu \mathbf{i} \end{pmatrix}, \tag{A.60}$$

$$L_{\mu\nu4} = \begin{pmatrix} 0 & -\gamma_{\mu\nu} \\ \gamma_{\mu\nu} & 0 \end{pmatrix} \ ; \ L_{\mu\nu5} = \begin{pmatrix} 0 & \gamma_{\mu\nu}\mathbf{i} \\ \gamma_{\mu\nu}\mathbf{i} & 0 \end{pmatrix}, \tag{A.61}$$

$$L_{\mu\nu\rho4} = \begin{pmatrix} \gamma_{\mu\nu\rho} & 0 \\ 0 & -\gamma_{\mu\nu\rho} \end{pmatrix} \ ; \ L_{\mu\nu\rho5} = \begin{pmatrix} \gamma_{\mu\nu\rho}\mathbf{i} & 0 \\ 0 & \gamma_{\mu\nu\rho}\mathbf{i} \end{pmatrix}, \tag{A.62}$$

$$L_{\mu45} = \begin{pmatrix} 0 & \gamma_\mu \mathbf{i} \\ -\gamma_\mu \mathbf{i} & 0 \end{pmatrix} \ ; \ L_{\mu\nu45} = \begin{pmatrix} -\gamma_{\mu\nu}\mathbf{i} & 0 \\ 0 & \gamma_{\mu\nu}\mathbf{i} \end{pmatrix}, \tag{A.63}$$

$$L_{\mu\nu\rho45} = \begin{pmatrix} 0 & \gamma_{\mu\nu\rho}\mathbf{i} \\ -\gamma_{\mu\nu\rho}\mathbf{i} & 0 \end{pmatrix} \ ; \ L_{01234} = \begin{pmatrix} 0 & -\mathbf{i} \\ \mathbf{i} & 0 \end{pmatrix}. \tag{A.64}$$

Scalar and pseudo-scalar terms read

$$\alpha I_8 + \omega L_{012345} = \begin{pmatrix} (\alpha + \omega)I_4 & 0 \\ 0 & (\alpha - \omega)I_4 \end{pmatrix}, \tag{A.65}$$

$$\alpha I_8 - \omega L_{012345} = \begin{pmatrix} (\alpha - \omega)I_4 & 0 \\ 0 & (\alpha + \omega)I_4 \end{pmatrix}. \tag{A.66}$$

---

2. I_2, I_4, I_8 are unit matrices. The identification process allowing us to include $\mathbb{R}$ in each real Clifford algebra allows us to read a instead of aI_n for any complex number a.

3. $\mathbf{i}$ anti-commutes with any odd element in space-time algebra and commutes with any even element.

For the calculation of the 1-vector term

$$N^a L_a = N^4 L_4 + N^5 L_5 + N^\mu L_\mu$$

we let

$$\beta = N^4 \; ; \quad \delta = N^5 \; ; \quad \mathbf{a} = N^\mu \gamma_\mu. \tag{A.67}$$

This gives

$$N^a L_a = \begin{pmatrix} 0 & -\beta I_4 + \delta \mathbf{i} + \mathbf{a} \\ \beta I_4 + \delta \mathbf{i} + \mathbf{a} & 0 \end{pmatrix}. \tag{A.68}$$

For the calculation of the 2-vector term

$$N^{ab} L_{ab} = N^{45} L_{45} + N^{\mu 4} L_{\mu 4} + N^{\mu 5} L_{\mu 5} + N^{\mu\nu} L_{\mu\nu},$$

we let

$$\epsilon = N^{45} \; ; \quad \mathbf{b} = N^{\mu 4} \gamma_\mu \; ; \quad \mathbf{c} = N^{\mu 5} \gamma_\mu \; ; \quad \mathbf{A} = N^{\mu\nu} \gamma_{\mu\nu}. \tag{A.69}$$

This gives with Eq. (A.54) and Eq. (A.60)

$$N^{ab} L_{ab} = \begin{pmatrix} -\epsilon \mathbf{i} + \mathbf{b} - \mathbf{ic} + \mathbf{A} & 0 \\ 0 & \epsilon \mathbf{i} - \mathbf{b} - \mathbf{ic} + \mathbf{A} \end{pmatrix}. \tag{A.70}$$

For the calculation of the 3-vector term

$$N^{abc} L_{abc} = N^{\mu 45} L_{\mu 45} + N^{\mu\nu 4} L_{\mu\nu 4} + N^{\mu\nu 5} L_{\mu\nu 5} + N^{\mu\nu\rho} L_{\mu\nu\rho},$$

we let

$$\mathbf{d} = N^{\mu 45} \gamma_\mu \; ; \quad \mathbf{B} = N^{\mu\nu 4} \gamma_{\mu\nu} \; ; \quad \mathbf{C} = N^{\mu\nu 5} \gamma_{\mu\nu} \; ; \quad \mathbf{ie} = N^{\mu\nu\rho} \gamma_{\mu\nu\rho}. \tag{A.71}$$

This gives with Eq. (A.55) and Eq. (A.61)

$$N^{abc} L_{abc} = \begin{pmatrix} 0 & \mathbf{di} - \mathbf{B} + \mathbf{iC} + \mathbf{ie} \\ \mathbf{id} + \mathbf{B} + \mathbf{iC} + \mathbf{ie} & 0 \end{pmatrix}. \tag{A.72}$$

For the calculation of the 4-vector term

$$N^{abcd} L_{abcd} = N^{\mu\nu 45} L_{\mu\nu 45} + N^{\mu\nu\rho 4} L_{\mu\nu\rho 4} + N^{\mu\nu\rho 5} L_{\mu\nu\rho 5} + N^{0123} L_{0123},$$

we let

$$\mathbf{D} = N^{\mu\nu 45} \gamma_{\mu\nu} \; ; \quad \mathbf{if} = N^{\mu\nu\rho 4} \gamma_{\mu\nu\rho} \; ; \quad \mathbf{ig} = N^{\mu\nu\rho 5} \gamma_{\mu\nu\rho} \; ; \quad \zeta = N^{0123}. \tag{A.73}$$

This gives with Eq. (A.56) and Eq. (A.62)

$$N^{abcd} L_{abcd} = \begin{pmatrix} -\mathbf{iD} + \mathbf{if} + \mathbf{g} + \zeta \mathbf{i} & 0 \\ 0 & \mathbf{iD} - \mathbf{if} + \mathbf{g} + \zeta \mathbf{i} \end{pmatrix}. \tag{A.74}$$

For the calculation of the pseudo-vector term

$$N^{abcde} L_{abcde} = N^{\mu\nu\rho 45} L_{\mu\nu\rho 45} + N^{01234} L_{01234} + N^{01235} L_{01235},$$

we let

$$\mathbf{ih} = N^{\mu\nu\rho45}\gamma_{\mu\nu\rho} \; ; \quad \eta = N^{01234} \; ; \quad \theta = N^{01235}. \tag{A.75}$$

This gives with Eq. (A.59) and Eq. (A.64)

$$N^{abcde}L_{abcde} = \begin{pmatrix} 0 & \mathbf{h} - \eta\mathbf{i} - \theta I_4 \\ -\mathbf{h} + \eta\mathbf{i} - \theta I_4 \end{pmatrix}. \tag{A.76}$$

We then get

$$\Psi = \begin{pmatrix} \Psi_l & \Psi_r \\ \Psi_g & \Psi_b \end{pmatrix} \tag{A.77}$$

$$= \begin{pmatrix} (\alpha + \omega)I_4 + (\mathbf{b} + \mathbf{g}) + (\mathbf{A} - \mathbf{iD}) & -(\beta + \theta)I_4 + (\mathbf{a} + \mathbf{h}) + (-\mathbf{B} + \mathbf{iC}) \\ \quad + \mathbf{i}(-\mathbf{c} + \mathbf{f}) + (\zeta - \epsilon)\mathbf{i} & \quad + \mathbf{i}(-\mathbf{d} + \mathbf{e}) + (\delta - \eta)\mathbf{i} \\ \\ (\beta - \theta)I_4 + (\mathbf{a} - \mathbf{h}) + (\mathbf{B} + \mathbf{iC}) & (\alpha - \omega)I_4 + (-\mathbf{b} + \mathbf{g}) + (\mathbf{A} + \mathbf{iD}) \\ \quad + \mathbf{i}(\mathbf{d} + \mathbf{e}) + (\delta + \eta)\mathbf{i} & \quad + \mathbf{i}(-\mathbf{c} - \mathbf{f}) + (\zeta + \epsilon)\mathbf{i} \end{pmatrix}.$$

This implies

$$\Psi_l = (\alpha + \omega) + (\mathbf{b} + \mathbf{g}) + (\mathbf{A} - \mathbf{iD}) + \mathbf{i}(-\mathbf{c} + \mathbf{f}) + (\zeta - \epsilon)\mathbf{i}, \tag{A.78}$$
$$\Psi_r = -(\beta + \theta) + (\mathbf{a} + \mathbf{h}) + (-\mathbf{B} + \mathbf{iC}) + \mathbf{i}(-\mathbf{d} + \mathbf{e}) + (\delta - \eta)\mathbf{i}, \tag{A.79}$$
$$\Psi_g = (\beta - \theta) + (\mathbf{a} - \mathbf{h}) + (\mathbf{B} + \mathbf{iC}) + \mathbf{i}(\mathbf{d} + \mathbf{e}) + (\delta + \eta)\mathbf{i}, \tag{A.80}$$
$$\Psi_b = (\alpha - \omega) + (-\mathbf{b} + \mathbf{g}) + (\mathbf{A} + \mathbf{iD}) + \mathbf{i}(-\mathbf{c} - \mathbf{f}) + (\zeta + \epsilon)\mathbf{i}. \tag{A.81}$$

In $Cl_{1,3}$ the reverse of

$$A = \langle A \rangle_0 + \langle A \rangle_1 + \langle A \rangle_2 + \langle A \rangle_3 + \langle A \rangle_4$$

is

$$\widetilde{A} = \langle A \rangle_0 + \langle A \rangle_1 - \langle A \rangle_2 - \langle A \rangle_3 + \langle A \rangle_4$$

. We must change the sign of bivectors $\mathbf{A}$, $\mathbf{B}$, $\mathbf{iC}$, $\mathbf{iD}$, and trivectors $\mathbf{ic}$, $\mathbf{id}$, $\mathbf{ic}$, $\mathbf{if}$ and we then get

$$\widetilde{\Psi}_l = (\alpha - \omega)I_4 + (-\mathbf{b} + \mathbf{g}) + (\mathbf{A} + \mathbf{iD}) - \mathbf{i}(\mathbf{c} + \mathbf{f}) + (\zeta + \epsilon)\mathbf{i}, \tag{A.82}$$
$$\widetilde{\Psi}_r = (\beta + \theta)I_4 - (\mathbf{a} + \mathbf{h}) + (\mathbf{B} - \mathbf{iC}) + \mathbf{i}(\mathbf{d} - \mathbf{e}) + (-\delta + \eta)\mathbf{i}, \tag{A.83}$$
$$\widetilde{\Psi}_g = (\beta - \theta)I_4 + (\mathbf{a} - \mathbf{h}) + (\mathbf{B} + \mathbf{iC}) + \mathbf{i}(\mathbf{d} + \mathbf{e}) + (\delta + \eta)\mathbf{i}, \tag{A.84}$$
$$\widetilde{\Psi}_b = (\alpha + \omega)I_4 + (\mathbf{b} + \mathbf{g}) + (\mathbf{A} - \mathbf{iD}) + \mathbf{i}(-\mathbf{c} + \mathbf{f}) + (\zeta - \epsilon)\mathbf{i}. \tag{A.85}$$

The reverse, in $Cl_{1,5}$ now, of

$$A = A_0 + A_1 + A_2 + A_3 + A_4 + A_5 + A_6$$

is

$$\widetilde{A} = A_0 + A_1 - A_2 - A_3 + A_4 + A_5 - A_6$$

Only terms which change [4] sign, with Eq. (A.65), Eq. (A.70) and Eq. (A.72), are scalars ϵ and ω, vectors $\mathbf{b}$, $\mathbf{c}$, $\mathbf{d}$, $\mathbf{e}$ and bivectors $\mathbf{A}$, $\mathbf{B}$, $\mathbf{C}$. We then get from Eq. (A.78)

$$\widetilde{\Psi} = \begin{pmatrix} (\alpha - \omega)I_4 + (-\mathbf{b} + \mathbf{g}) + (-\mathbf{A} - i\mathbf{D}) & -(\beta + \theta)I_4 + (\mathbf{a} + \mathbf{h}) + (\mathbf{B} - i\mathbf{C}) \\ \quad + i(\mathbf{c} + \mathbf{f}) + (\zeta + \epsilon)i & \quad + i(\mathbf{d} - \mathbf{e}) + (\delta - \eta)i \\ \\ (\beta - \theta)I_4 + (\mathbf{a} - \mathbf{h}) - (\mathbf{B} + i\mathbf{C}) & (\alpha + \omega)I_4 + (\mathbf{b} + \mathbf{g}) + (-\mathbf{A} + i\mathbf{D}) \\ \quad - i(\mathbf{d} + \mathbf{e}) + (\delta + \eta)i & \quad + i(\mathbf{c} - \mathbf{f}) + (\zeta - \epsilon)i \end{pmatrix}$$

$$= \begin{pmatrix} \widetilde{\Psi}_b & \widetilde{\Psi}_r \\ \widetilde{\Psi}_g & \widetilde{\Psi}_l \end{pmatrix}. \tag{A.86}$$

And we have proved Eq. (7.73).

---

4. These changes of sign are not the same in $Cl_{1,5}$ as in $Cl_{1,3}$. Differences are corrected by the fact that the reversion in $Cl_{1,5}$ also exchanges the place of the Ψ_l and Ψ_b terms.

Appendix B

Electron+neutrino+quarks

B.1 Gauge generated by i

The operators P_0 in Eq. (6.13) and Eq. (8.25) and P_0' in Eq. (7.13) have the form Eq. (8.25). They satisfy, with constant numbers a and b:

$$P_0(\Psi) = a\Psi\gamma_{21} + bP_-(\Psi)\mathbf{i}.$$

Applied to

$$\Psi = \begin{pmatrix} \phi_e & \phi_n \\ \hat{\phi}_n & \hat{\phi}_e \end{pmatrix} = \sqrt{2}\begin{pmatrix} \xi_{1e} & -\eta_{2e}^* & 0 & -\eta_{2n}^* \\ \xi_{2e} & \eta_{1e}^* & 0 & \eta_{1n}^* \\ \eta_{1n} & 0 & \eta_{1e} & -\xi_{2e}^* \\ \eta_{2n} & 0 & \eta_{2e} & \xi_{1e}^* \end{pmatrix}, \tag{B.1}$$

this gives

$$P_0(\Psi) = ia\begin{pmatrix} \phi_e\sigma_3 & \phi_n\sigma_3 \\ \hat{\phi}_n\sigma_3 & \hat{\phi}_e\sigma_3 \end{pmatrix} + ib\begin{pmatrix} \phi_{eR} & 0 \\ 0 & -\hat{\phi}_{eR} \end{pmatrix}, \tag{B.2}$$

$$P_0(\Psi) = i\sqrt{2}\begin{pmatrix} (a+b)\xi_{1e} & (-a)(-\eta_{2e}^*) & 0 & (-a)(-\eta_{2n}^*) \\ (a+b)\xi_{2e} & (-a)\eta_{1e}^* & 0 & (-a)\eta_{1n}^* \\ a\eta_{1n} & 0 & a\eta_{1e} & -(a+b)(-\xi_{2e}^*) \\ a\eta_{2n} & 0 & a\eta_{2e} & -(a+b)\xi_{1e}^* \end{pmatrix}, \tag{B.3}$$

$$P_0(\Psi) = \frac{b\mathbf{i}}{2}\Psi + \Psi(a + \frac{b}{2})\gamma_{21}. \tag{B.4}$$

Since by $\exp(a^0 P_0)$

$$\xi_e \mapsto e^{ia^0(a+b)}\xi_e; \quad \eta_e \mapsto e^{ia^0 a}\eta_e; \quad \eta_n \mapsto e^{ia^0 a}\eta_n, \tag{B.5}$$

we get

$$[\exp(a^0 P_0)](\Psi) = e^{a^0\frac{b}{2}\mathbf{i}}\Psi e^{a^0(a+\frac{b}{2})\gamma_{21}}. \tag{B.6}$$

171

We then get

$$\partial_\mu \Big[[\exp(a^0 P_0)](\Psi) \Big]$$

$$= \partial_\mu a^0 \frac{b}{2} i e^{a^0 \frac{b}{2} i} \Psi e^{a^0(a+\frac{b}{2})\gamma_{21}} + e^{a^0 \frac{b}{2} i} \partial_\mu \Psi e^{a^0(a+\frac{b}{2})\gamma_{21}}$$

$$+ e^{a^0 \frac{b}{2} i} \Psi \partial_\mu a^0 (a+\frac{b}{2}) \gamma_{21} e^{a^0(a+\frac{b}{2})\gamma_{21}}$$

$$= \partial_\mu a^0 e^{a^0 \frac{b}{2} i} [\frac{b i}{2} \Psi + \Psi(a+\frac{b}{2}) \gamma_{21}] e^{a^0(a+\frac{b}{2})\gamma_{21}} + e^{a^0 \frac{b}{2} i} \partial_\mu \Psi e^{a^0(a+\frac{b}{2})\gamma_{21}}$$

$$= \partial_\mu a^0 e^{a^0 \frac{b}{2} i} P_0(\Psi) e^{a^0(a+\frac{b}{2})\gamma_{21}} + e^{a^0 \frac{b}{2} i} \partial_\mu \Psi e^{a^0(a+\frac{b}{2})\gamma_{21}}$$

$$= e^{a^0 \frac{b}{2} i} [\partial_\mu a^0 P_0(\Psi) + \partial_\mu \Psi] e^{a^0(a+\frac{b}{2})\gamma_{21}}. \tag{B.7}$$

The gauge transformation defined as

$$B'_\mu = B_\mu - \frac{2}{g_1} \partial_\mu a^0, \tag{B.8}$$

$$\Psi' = [\exp(a^0 P_0)](\Psi) = e^{a^0 \frac{b}{2} i} \Psi e^{a^0(a+\frac{b}{2})\gamma_{21}}, \tag{B.9}$$

gives:

$$D_\mu \Psi = \partial_\mu \Psi + \frac{g_1}{2} B_\mu P_0(\Psi), \tag{B.10}$$

$$D'_\mu \Psi' = \partial_\mu \Psi' + (\frac{g_1}{2} B_\mu - \partial_\mu a^0) P_0(\Psi') \tag{B.11}$$

$$= \partial_\mu \Big[[\exp(a^0 P_0)](\Psi) \Big] + (\frac{g_1}{2} B_\mu - \partial_\mu a^0) [\frac{b}{2} i \Psi' + \Psi'(a+\frac{b}{2}) \gamma_{21}] \tag{B.12}$$

$$= e^{a^0 \frac{b}{2} i} [\partial_\mu a^0 P_0(\Psi) + \partial_\mu \Psi + (\frac{g_1}{2} B_\mu - \partial_\mu a^0) P_0(\Psi)] e^{a^0(a+\frac{b}{2})\gamma_{21}}$$

$$= e^{a^0 \frac{b}{2} i} (D_\mu \Psi) e^{a^0(a+\frac{b}{2})\gamma_{21}}. \tag{B.13}$$

We deduce:

$$\mathbf{D}'\Psi' = \gamma^\mu D'_\mu \Psi' = \gamma^\mu e^{a^0 \frac{b}{2} i} (D_\mu \Psi) e^{a^0(a+\frac{b}{2})\gamma_{21}}$$

$$= e^{-a^0 \frac{b}{2} i} (\mathbf{D}\Psi) e^{a^0(a+\frac{b}{2})\gamma_{21}}, \tag{B.14}$$

because $\mathbf{i}$ anti-commutes with each γ^μ. Next we have

$$\widetilde{\Psi}' = e^{a^0(a+\frac{b}{2})\widetilde{\gamma}_{21}} \widetilde{\Psi} e^{a^0 \frac{b}{2} \widetilde{i}}$$

$$= e^{-a^0(a+\frac{b}{2})\gamma_{21}} \widetilde{\Psi} e^{a^0 \frac{b}{2} i}, \tag{B.15}$$

$$\widetilde{\Psi}' \mathbf{D}'\Psi' = e^{-a^0(a+\frac{b}{2})\gamma_{21}} \widetilde{\Psi} e^{a^0 \frac{b}{2} i} e^{-a^0 \frac{b}{2} i} (\mathbf{D}\Psi) e^{a^0(a+\frac{b}{2})\gamma_{21}}$$

$$= e^{-a^0(a+\frac{b}{2})\gamma_{21}} \widetilde{\Psi} (\mathbf{D}\Psi) e^{a^0(a+\frac{b}{2})\gamma_{21}}. \tag{B.16}$$

In the magnetic monopole case the lepton wave reads:

$$\Psi = \begin{pmatrix} \phi_L & \phi_n \\ \hat{\phi}_n & \hat{\phi}_L \end{pmatrix} = \sqrt{2} \begin{pmatrix} 0 & -\eta_{2L}^* & \xi_{1n} & -\eta_{2n}^* \\ 0 & \eta_{1L}^* & \xi_{2n} & \eta_{1n}^* \\ \eta_{1n} & -\xi_{2n}^* & \eta_{1L} & 0 \\ \eta_{2n} & \xi_{1n}^* & \eta_{2L} & 0 \end{pmatrix}, \tag{B.17}$$

which gives

$$P_0(\Psi) = a\Psi\gamma_{21} + bP_-(\Psi)\mathbf{i}$$

$$= i\sqrt{2} \begin{pmatrix} 0 & -a(-\eta_{2L}^*) & (a-b)\xi_{1n} & -a(-\eta_{2n}^*) \\ 0 & -a\eta_{1L}^* & (a-b)\xi_{2n} & -a\eta_{1n}^* \\ a\eta_{1n} & (-a+b)(-\xi_{2n}^*) & a\eta_{1L} & 0 \\ a\eta_{2n} & (-a+b)\xi_{1n}^* & a\eta_{2L} & 0 \end{pmatrix} \tag{B.18}$$

$$= -\frac{b}{2}\mathbf{i}\Psi + \Psi(a - \frac{b}{2})\gamma_{21}. \tag{B.19}$$

The following calculation is then the same, but with the change of b into $-b$.

B.2 Tensorial densities for electron+neutrino

In the case of the electron+neutrino pair, like in the case of the magnetic monopole, each spinor has four real parameters, which gives $4 \times 5/2 = 10$ components of tensors: one space-time vector (4 components) and a space-time bivector (6 components). With the right ϕ_R spinor of the electron (or [5] of the magnetic monopole)

$$\phi_R = \sqrt{2} \begin{pmatrix} \xi_1 & 0 \\ \xi_2 & 0 \end{pmatrix}, \tag{B.20}$$

we get the space-time vector D_R and [6] satisfying

$$D_R = \phi_R \phi_R^\dagger \; ; \quad S_R = \phi_R \sigma_1 \bar{\phi}_R. \tag{B.21}$$

the bivector $S_R D_R$ is a space-time vector, because it satisfies $D_R^\dagger = D_R$. Similarly, with the left spinor ϕ_L

$$\phi_L = \sqrt{2} \begin{pmatrix} 0 & -\eta_2^* \\ 0 & \eta_1^* \end{pmatrix} \; ; \quad \phi_e = \phi_R + \phi_L, \tag{B.22}$$

---

5. Since we can study both the case of the electron and the case of the magnetic monopole, we shall note without e index the components of the electron wave and we shall note here ζ_j what was noticed η_{jn} in the case of the left wave of the electronic neutrino, and η_{jL} in the case of the supplementary left wave of the magnetic monopole.

6. A detailed calculation of components is in B of [21].

we get the vector D_L and the bivector S_L satisfying

$$D_L = \phi_L \phi_L^\dagger \; ; \quad S_L = \phi_L \sigma_1 \overline{\phi}_L. \tag{B.23}$$

Here and for the Lochak's magnetic monopole [46] the D_R, D_L currents are the fundamental ones of the Dirac theory. The usual $J = D_0$ and $K = D_3$ currents are simply sum and difference of these chiral currents:

$$D_0 = D_R + D_L \; ; \quad D_3 = D_R - D_L. \tag{B.24}$$

With the $\phi_n = \phi_{nL}$ spinor of the electronic neutrino, that is here

$$\phi_n = \sqrt{2} \begin{pmatrix} 0 & -\zeta_2^* \\ 0 & \zeta_1^* \end{pmatrix}, \tag{B.25}$$

we get the space-time vector D_n and the bivector S_n such as

$$D_n = \phi_n \phi_n^\dagger \; ; \quad S_n = \phi_n \sigma_1 \overline{\phi}_n. \tag{B.26}$$

Next with two of the three spinors we can get 16 densities. We begin with ϕ_R and ϕ_L. We let

$$P = 2\phi_R \overline{\phi}_L = \mathbf{a} + S_{RL},$$
$$\overline{P} = 2\phi_L \overline{\phi}_R = \mathbf{a} - S_{RL}, \tag{B.27}$$
$$I = D_{RL} + i d_{RL} = 2\phi_R \sigma_1 \phi_L^\dagger,$$
$$I^\dagger = D_{RL} - i d_{RL} = 2\phi_L \sigma_1 \phi_R^\dagger. \tag{B.28}$$

$\mathbf{a}$ and S_{RL} are well known in the Dirac theory:

$$\mathbf{a} = \det(\phi_e) = \Omega_1 + i\Omega_2 = 2(\xi_1 \eta_1^* + \xi_2 \eta_2^*), \tag{B.29}$$

where Ω_1, Ω_2 are the relativistic invariants of Eq. (2.33) and Eq. (2.34). The $S_3 = S_{RL}$ bivector is the one in Eq. (2.44) which with Ω_1, Ω_2, D_0 and D_3, gives the 16 densities that were considered as the only possible densities without derivatives from the complex formalism. These densities are the invariant ones under the electric gauge. $D_{RL} = D_1$ and $d_{RL} = D_2$ are the space-time vectors defined in Eq. (2.37). Under the R dilation defined in Eq. (1.42) $\mathbf{a}$ is changed into $\mathbf{a}'$ such as

$$\mathbf{a}' = M\phi_e \overline{\phi}_e \overline{M} = M\mathbf{a}\overline{M} = \mathbf{a}M\overline{M} = re^{i\theta}\mathbf{a}. \tag{B.30}$$

Therefore

$$\mathbf{a}'\mathbf{a}'^* = re^{i\theta}\mathbf{a} r e^{-i\theta}\mathbf{a}^* = r^2 \mathbf{a}\mathbf{a}^*. \tag{B.31}$$

Next with ϕ_L and ϕ_n we let

$$P = 2\widehat{\phi}_n \sigma_1 \phi_L^\dagger = \mathbf{b} + S_{Ln},$$

$$\overline{P} = 2\widehat{\phi}_L \overline{\sigma}_1 \phi_n^\dagger = \mathbf{b} - S_{Ln}, \qquad (\text{B.32})$$

$$I = D_{Ln} + i d_{Ln} = 2\phi_n \phi_L^\dagger,$$

$$I^\dagger = D_{Ln} - i d_{Ln} = 2\phi_L \phi_n^\dagger. \qquad (\text{B.33})$$

D_{Ln} and d_{Ln} are contravariant space-time vectors, while S_{Ln} is a bivector. We shall need

$$\mathbf{b} = \widehat{\phi}_n \sigma_1 \phi_L^\dagger + \widehat{\phi}_L \overline{\sigma}_1 \phi_n^\dagger = 2(\eta_1 \zeta_2 - \eta_2 \zeta_1), \qquad (\text{B.34})$$

$$\mathbf{b}' = \widehat{M} \mathbf{b} M^\dagger = \widehat{M} M^\dagger \mathbf{b} = re^{-i\theta} \mathbf{b},$$

$$\mathbf{b}' \mathbf{b}'^* = r^2 \mathbf{b} \mathbf{b}^*. \qquad (\text{B.35})$$

Finally with ϕ_R and ϕ_n we let

$$P = 2\phi_R \overline{\phi}_n = \mathbf{c} + S_{Rn}$$

$$\overline{P} = 2\phi_n \overline{\phi}_R = \mathbf{c} - S_{Rn} \qquad (\text{B.36})$$

$$I = D_{Rn} + i d_{Rn} = 2\phi_R \sigma_1 \phi_n^\dagger$$

$$I^\dagger = D_{Rn} - i d_{Rn} = 2\phi_n \sigma_1 \phi_R^\dagger \qquad (\text{B.37})$$

D_{Rn} and d_{Rn} are also contravariant vectors, while S_{Rn} is a bivector. We shall use

$$\mathbf{c} = \phi_R \overline{\phi}_n + \phi_n \overline{\phi}_R = 2(\xi_1 \zeta_1^* + \xi_2 \zeta_2^*), \qquad (\text{B.38})$$

$$\mathbf{c}' = M \mathbf{c} \overline{M} = M \overline{M} \mathbf{c} = re^{i\theta} \mathbf{c} ; \quad \mathbf{c}' \mathbf{c}'^* = r^2 \mathbf{c} \mathbf{c}^*. \qquad (\text{B.39})$$

We have established in Eq. (3.42) that the main invariant of the wave of the electron is $m\rho$. Since we now have not only one but three similar terms, the natural generalization that is necessary and also sufficient to get the gauge invariance is:

$$\rho = \sqrt{\mathbf{a}\mathbf{a}^* + \mathbf{b}\mathbf{b}^* + \mathbf{c}\mathbf{c}^*}, \qquad (\text{B.40})$$

$$m\boldsymbol{\rho} = m' r\boldsymbol{\rho} = m' \boldsymbol{\rho}'. \qquad (\text{B.41})$$

B.3 Getting the wave equation

Since this term is the generalization of the invariant term of the wave of the electron, which is also the mass term of the Lagrangian density, this density is, in the case of the electron+neutrino pair, the scalar part $\mathcal{L} = \langle L \rangle_0$ of

$$L = \widetilde{\Psi} \mathbf{D} \Psi \gamma_{012} + m\boldsymbol{\rho}, \qquad (\text{B.42})$$

where $\mathbf{D}$ is the covariant derivative Eq. (6.22). We shall use for the pair electron+neutrino:

$$\Psi_L = P_+(\Psi) = \begin{pmatrix} \phi_L & \phi_n \\ \widehat{\phi}_n & \widehat{\phi}_L \end{pmatrix}, \tag{B.43}$$

$$\Psi_R = P_-(\Psi) = \begin{pmatrix} \phi_R & 0 \\ 0 & \widehat{\phi}_R \end{pmatrix},$$

$$\Psi = \Psi_R + \Psi_L, \tag{B.44}$$

$$P_0(\Psi) = (\Psi_L + 2\Psi_R)\gamma_{21}, \tag{B.45}$$

and we have

$$P_0(\Psi) = (\Psi_R + \Psi_L)\gamma_{21} + \Psi_R\gamma_{21} = (\Psi_L + 2\Psi_R)\gamma_{21}, \tag{B.46}$$

$$\Psi_L + 2\Psi_R = \begin{pmatrix} \phi_L + 2\phi_R & \phi_n \\ \widehat{\phi}_n & \widehat{\phi}_L + 2\widehat{\phi}_R \end{pmatrix}. \tag{B.47}$$

We have also

$$\mathbf{D}\Psi = \boldsymbol{\partial}\Psi + \frac{g_1}{2}\mathbf{B}P_0(\Psi) + \frac{g_2}{2}\mathbf{W}^j P_j(\Psi) \tag{B.48}$$

$$= \boldsymbol{\partial}\Psi + \frac{g_1}{2}\mathbf{B}(\Psi_L + 2\Psi_R)\gamma_{21} + \frac{g_2}{2}[\mathbf{W}^1\Psi_L\gamma_3\mathbf{i} + \mathbf{W}^2\Psi_L\gamma_3 + \mathbf{W}^3\Psi_L(-\mathbf{i})].$$

This gives

$$\mathbf{D}\Psi\gamma_{012} = \boldsymbol{\partial}\Psi\gamma_{012} + \frac{g_1}{2}\mathbf{B}(\Psi_L + 2\Psi_R)\gamma_0 \tag{B.49}$$

$$- \frac{g_2}{2}[\mathbf{W}^1\Psi_L + \mathbf{W}^2\Psi_L\mathbf{i} + \mathbf{W}^3\Psi_L\gamma_3].$$

Next we have

$$\boldsymbol{\partial}\Psi\gamma_{012} = \begin{pmatrix} 0 & \nabla \\ \widehat{\nabla} & 0 \end{pmatrix} \begin{pmatrix} \phi_e & \phi_n \\ \widehat{\phi}_n & \widehat{\phi}_e \end{pmatrix} \begin{pmatrix} 0 & -i\sigma_3 \\ -i\sigma_3 & 0 \end{pmatrix}, \tag{B.50}$$

$$= \begin{pmatrix} -i\nabla(\widehat{\phi}_R + \widehat{\phi}_L)\sigma_3 & -i\nabla\widehat{\phi}_n\sigma_3 \\ -i\widehat{\nabla}\phi_n\sigma_3 & -i\widehat{\nabla}(\phi_R + \phi_L)\sigma_3 \end{pmatrix},$$

$$\boldsymbol{\partial}\Psi\gamma_{012} = -i\begin{pmatrix} \nabla(-\widehat{\phi}_R + \widehat{\phi}_L) & \nabla\widehat{\phi}_n \\ -\widehat{\nabla}\phi_n & \widehat{\nabla}(\phi_R - \phi_L) \end{pmatrix}. \tag{B.51}$$

And we get for B

$$\mathbf{B}(\Psi_L + 2\Psi_R)\gamma_0 = \begin{pmatrix} 0 & B \\ \widehat{B} & 0 \end{pmatrix} \begin{pmatrix} \phi_L + 2\phi_R & \phi_n \\ \widehat{\phi}_n & \widehat{\phi}_L + 2\widehat{\phi}_R \end{pmatrix} \begin{pmatrix} 0 & I \\ I & 0 \end{pmatrix}$$

$$= \begin{pmatrix} 0 & B \\ \widehat{B} & 0 \end{pmatrix} \begin{pmatrix} \phi_n & \phi_e + \phi_R \\ \widehat{\phi}_e + \widehat{\phi}_R & \widehat{\phi}_n \end{pmatrix}, \tag{B.52}$$

$$\mathbf{B}P_0(\Psi)\gamma_{012} = \begin{pmatrix} B(\widehat{\phi}_e + \widehat{\phi}_R) & B\widehat{\phi}_n \\ \widehat{B}\phi_n & \widehat{B}(\phi_e + \phi_R) \end{pmatrix}. \tag{B.53}$$

Similarly we have

$$\mathbf{W}^1\Psi_L = \begin{pmatrix} 0 & W^1 \\ \widehat{W}^1 & 0 \end{pmatrix}\begin{pmatrix} \phi_L & \phi_n \\ \widehat{\phi}_n & \widehat{\phi}_L \end{pmatrix} = \begin{pmatrix} W^1\widehat{\phi}_n & W^1\widehat{\phi}_L \\ \widehat{W}^1\phi_L & \widehat{W}^1\phi_n \end{pmatrix}, \tag{B.54}$$

$$\mathbf{W}^2\Psi_L\mathbf{i} = \begin{pmatrix} 0 & W^2 \\ \widehat{W}^2 & 0 \end{pmatrix}\begin{pmatrix} \phi_L & \phi_n \\ \widehat{\phi}_n & \widehat{\phi}_L \end{pmatrix}\begin{pmatrix} i & 0 \\ 0 & -i \end{pmatrix} = \begin{pmatrix} iW^2\widehat{\phi}_n & -iW^2\widehat{\phi}_L \\ i\widehat{W}^2\phi_L & -i\widehat{W}^2\phi_n \end{pmatrix}, \tag{B.55}$$

$$\mathbf{W}^3\Psi_L\gamma_3 = \begin{pmatrix} 0 & W^3 \\ \widehat{W}^3 & 0 \end{pmatrix}\begin{pmatrix} \phi_L & \phi_n \\ \widehat{\phi}_n & \widehat{\phi}_L \end{pmatrix}\begin{pmatrix} 0 & \sigma_3 \\ -\sigma_3 & 0 \end{pmatrix}$$

$$= \begin{pmatrix} -W^3\widehat{\phi}_L\sigma_3 & W^3\widehat{\phi}_n\sigma_3 \\ -\widehat{W}^3\phi_n\sigma_3 & \widehat{W}^3\phi_L\sigma_3 \end{pmatrix} = \begin{pmatrix} -W^3\widehat{\phi}_L & W^3\widehat{\phi}_n \\ \widehat{W}^3\phi_n & -\widehat{W}^3\phi_L \end{pmatrix}. \tag{B.56}$$

We then get

$$\mathbf{W}^1\Psi_L + \mathbf{W}^2\Psi_L\mathbf{i} + \mathbf{W}^3\Psi_L\gamma_3$$

$$= \begin{pmatrix} (W^1+iW^2)\widehat{\phi}_n - W^3\widehat{\phi}_L & (W^1-iW^2)\widehat{\phi}_L + W^3\widehat{\phi}_n \\ (\widehat{W}^1+i\widehat{W}^2)\phi_L + \widehat{W}^3\phi_n & (\widehat{W}^1-i\widehat{W}^2)\phi_n - \widehat{W}^3\phi_L \end{pmatrix}. \tag{B.57}$$

Next we get

$$\widetilde{\Psi}\mathbf{D}\Psi\gamma_{012} = \widetilde{\Psi}\partial\Psi\gamma_{012} + \frac{g_1}{2}\widetilde{\Psi}\mathbf{B}(\Psi_L + 2\Psi_R)\gamma_0$$

$$- \frac{g_2}{2}\widetilde{\Psi}[\mathbf{W}^1\Psi_L + \mathbf{W}^2\Psi_L\mathbf{i} + \mathbf{W}^3\Psi_L\gamma_3], \tag{B.58}$$

$$\widetilde{\Psi}\partial\Psi\gamma_{012} = -i\begin{pmatrix} \overline{\phi}_e & \phi_n^\dagger \\ \overline{\phi}_n & \phi_e^\dagger \end{pmatrix}\begin{pmatrix} \nabla(-\widehat{\phi}_R + \widehat{\phi}_L) & \nabla\widehat{\phi}_n \\ -\widehat{\nabla}\phi_n & \widehat{\nabla}(\phi_R - \phi_L) \end{pmatrix} \tag{B.59}$$

$$= -i\begin{pmatrix} \overline{\phi}_e\nabla(-\widehat{\phi}_R + \widehat{\phi}_L) - \phi_n^\dagger\widehat{\nabla}\phi_n & \overline{\phi}_e\nabla\widehat{\phi}_n + \phi_n^\dagger\widehat{\nabla}(\phi_R - \phi_L) \\ \overline{\phi}_n\nabla(-\widehat{\phi}_R + \widehat{\phi}_L) - \phi_e^\dagger\widehat{\nabla}\phi_n & \overline{\phi}_n\nabla\widehat{\phi}_n + \phi_e^\dagger\widehat{\nabla}(\phi_R - \phi_L) \end{pmatrix}.$$

With the matrix representation Eq. (1.75) the real part of any multivector of space-time is the real part of the scalar part of the matrix. We then get

$$\Re(\widetilde{\Psi}\partial\Psi\gamma_{012}) = \Re[i\overline{\phi}_e\nabla(\widehat{\phi}_R - \widehat{\phi}_L) + i\phi_n^\dagger\widehat{\nabla}\phi_n]. \tag{B.60}$$

Next we get

$$\widetilde{\Psi}\mathbf{B}P_0(\Psi)\gamma_{012} = \begin{pmatrix} \overline{\phi}_e & \phi_n^\dagger \\ \overline{\phi}_n & \phi_e^\dagger \end{pmatrix}\begin{pmatrix} B(\widehat{\phi}_R + \widehat{\phi}_e) & B\widehat{\phi}_n \\ \widehat{B}\phi_n & \widehat{B}(\phi_R + \phi_e) \end{pmatrix} \tag{B.61}$$

$$= \begin{pmatrix} \overline{\phi}_eB(\widehat{\phi}_R + \widehat{\phi}_e) + \phi_n^\dagger\widehat{B}\phi_n & \overline{\phi}_eB\widehat{\phi}_n + \phi_n^\dagger\widehat{B}(\phi_R + \phi_e) \\ \overline{\phi}_nB(\widehat{\phi}_R + \widehat{\phi}_e) + \phi_e^\dagger\widehat{B}\phi_n & \overline{\phi}_nB\widehat{\phi}_n + \phi_e^\dagger\widehat{B}(\phi_R + \phi_e) \end{pmatrix},$$

$$\Re[\widetilde{\Psi}\mathbf{B}P_0(\Psi)\gamma_{012}] = \Re[\overline{\phi}_eB(\widehat{\phi}_R + \widehat{\phi}_e) + \phi_n^\dagger\widehat{B}\phi_n]. \tag{B.62}$$

And we get

$$\widetilde{\Psi}[\mathbf{W}^1\Psi_L + \mathbf{W}^2\Psi_L\mathbf{i} + \mathbf{W}^3\Psi_L\gamma_3] \tag{B.63}$$

$$= \begin{pmatrix} \overline{\phi}_e & \phi_n^\dagger \\ \overline{\phi}_n & \phi_e^\dagger \end{pmatrix} \begin{pmatrix} (W^1 + iW^2)\widehat{\phi}_n - W^3\widehat{\phi}_L & (W^1 - iW^2)\widehat{\phi}_L + W^3\widehat{\phi}_n \\ (\widehat{W}^1 + i\widehat{W}^2)\phi_L + \widehat{W}^3\phi_n & (\widehat{W}^1 - i\widehat{W}^2)\phi_n - \widehat{W}^3\phi_L \end{pmatrix} = \begin{pmatrix} U & V \\ \widehat{V} & \widehat{U} \end{pmatrix},$$

$$U = \overline{\phi}_e[(W^1 + iW^2)\widehat{\phi}_n - W^3\widehat{\phi}_L] + \phi_n^\dagger[(\widehat{W}^1 + i\widehat{W}^2)\phi_L + \widehat{W}^3\phi_n]. \tag{B.64}$$

This gives

$$\Re\left[\widetilde{\Psi}[\mathbf{W}^1\Psi_L + \mathbf{W}^2\Psi_L\mathbf{i} + \mathbf{W}^3\Psi_L\gamma_3]\right] = \Re(\overline{\phi}_e W^1\widehat{\phi}_n + \phi_n^\dagger\widehat{W}^1\phi_L) \tag{B.65}$$
$$+ \Re(i\overline{\phi}_e W^2\widehat{\phi}_n + i\phi_n^\dagger\widehat{W}^2\phi_L)$$
$$+ \Re(-\overline{\phi}_e W^3\widehat{\phi}_L + \phi_n^\dagger\widehat{W}^3\phi_n).$$

Next we have

$$2\Re[i\overline{\phi}_e\nabla(\widehat{\phi}_R - \widehat{\phi}_L) + i\phi_n^\dagger\widehat{\nabla}\phi_n]$$
$$= -i\eta_1^*\partial_0\eta_1 + i\eta_1\partial_0\eta_1^* - i\eta_2^*\partial_0\eta_2 + i\eta_2\partial_0\eta_2^* + i\xi_2\partial_0\xi_2^* - i\xi_2^*\partial_0\xi_2$$
$$+ i\xi_1\partial_0\xi_1^* - i\xi_1^*\partial_0\xi_1 + i\zeta_2\partial_0\zeta_2^* - i\zeta_2^*\partial_0\zeta_2 + i\zeta_1\partial_0\zeta_1^* - i\zeta_1^*\partial_0\zeta_1$$
$$+ i\eta_1^*\partial_1\eta_2 - i\eta_1\partial_1\eta_2^* + i\eta_2^*\partial_1\eta_1 - i\eta_2\partial_1\eta_1^* + i\xi_2\partial_1\xi_1^* - i\xi_2^*\partial_1\xi_1$$
$$+ i\xi_1\partial_1\xi_2^* - i\xi_1^*\partial_1\xi_2 - i\zeta_2\partial_1\zeta_1^* + i\zeta_2^*\partial_1\zeta_1 - i\zeta_1\partial_1\zeta_2^* + i\zeta_1^*\partial_1\zeta_2$$
$$+ \eta_1^*\partial_2\eta_2 + \eta_1\partial_2\eta_2^* - \eta_2^*\partial_2\eta_1 - \eta_2\partial_2\eta_1^* + \xi_2\partial_2\xi_1^* + \xi_2^*\partial_2\xi_1$$
$$- \xi_1\partial_2\xi_2^* - \xi_1^*\partial_2\xi_2 - \zeta_2\partial_2\zeta_1^* - \zeta_2^*\partial_2\zeta_1 + \zeta_1\partial_2\zeta_2^* + \zeta_1^*\partial_2\zeta_2$$
$$+ i\eta_1^*\partial_3\eta_1 - i\eta_1\partial_3\eta_1^* - i\eta_2^*\partial_3\eta_2 + i\eta_2\partial_3\eta_2^* - i\xi_2\partial_3\xi_2^* + i\xi_2^*\partial_3\xi_2$$
$$+ i\xi_1\partial_3\xi_1^* - i\xi_1^*\partial_3\xi_1 + i\zeta_2\partial_3\zeta_2^* - i\zeta_2^*\partial_3\zeta_2 - i\zeta_1\partial_3\zeta_1^* + i\zeta_1^*\partial_3\zeta_1. \tag{B.66}$$

From Eq. (B.62) we have

$$\Re\ [\overline{\phi}_e B(\widehat{\phi}_R + \widehat{\phi}_e) + \phi_n^\dagger\widehat{B}\phi_n]$$
$$= B_0(\eta_1\eta_1^* + \eta_2\eta_2^* + 2\xi_1\xi_1^* + 2\xi_2\xi_2^* + \zeta_1\zeta_1^* + \zeta_2\zeta_2^*)$$
$$+ B_1(-\eta_1\eta_2^* - \eta_2\eta_1^* + 2\xi_1\xi_2^* + 2\xi_2\xi_1^* - \zeta_1\zeta_2^* - \zeta_2\zeta_1^*)$$
$$+ B_2 i(-\eta_1\eta_2^* + \eta_2\eta_1^* + 2\xi_1\xi_2^* - 2\xi_2\xi_1^* - \zeta_1\zeta_2^* + \zeta_2\zeta_1^*)$$
$$+ B_3(-\eta_1\eta_1^* + \eta_2\eta_2^* + 2\xi_1\xi_1^* - 2\xi_2\xi_2^* - \zeta_1\zeta_1^* + \zeta_2\zeta_2^*) \tag{B.67}$$

and from Eq. (B.64) we have

$$\Re(\overline{\phi}_e W^1\widehat{\phi}_n + \phi_n^\dagger\widehat{W}^1\phi_L) = W_0^1(\eta_1\zeta_1^* + \eta_2\zeta_2^* + \zeta_1\eta_1^* + \zeta_2\eta_2^*)$$
$$+ W_1^1(-\eta_1\zeta_2^* - \eta_2\zeta_1^* - \zeta_1\eta_2^* - \zeta_2\eta_1^*)$$
$$+ W_2^1 i(-\eta_1\zeta_2^* + \eta_2\zeta_1^* - \zeta_1\eta_2^* + \zeta_2\eta_1^*)$$
$$+ W_3^1(-\eta_1\zeta_1^* + \eta_2\zeta_2^* - \zeta_1\eta_1^* + \zeta_2\eta_2^*), \tag{B.68}$$

$$\Re(i\overline{\phi}_e W^2 \widehat{\phi}_n + i\phi_n^\dagger \widehat{W}^2 \phi_L) = W_0^2 i(-\eta_1\zeta_1^* - \eta_2\zeta_2^* + \zeta_1\eta_1^* + \zeta_2\eta_2^*)$$
$$+ W_1^2 i(\eta_1\zeta_2^* + \eta_2\zeta_1^* - \zeta_1\eta_2^* - \zeta_2\eta_1^*)$$
$$+ W_2^2 (-\eta_1\zeta_2^* + \eta_2\zeta_1^* + \zeta_1\eta_2^* - \zeta_2\eta_1^*)$$
$$+ W_3^2 i(-\eta_1\zeta_1^* + \eta_2\zeta_2^* - \zeta_1\eta_1^* + \zeta_2\eta_2^*), \tag{B.69}$$

$$\Re(-\overline{\phi}_e W^3 \widehat{\phi}_L + \phi_n^\dagger \widehat{W}^3 \phi_n) = W_0^3 (-\eta_1\eta_1^* - \eta_2\eta_2^* + \zeta_1\zeta_1^* + \zeta_2\zeta_2^*)$$
$$+ W_1^3 (+\eta_1\eta_2^* + \eta_2\eta_1^* - \zeta_1\zeta_2^* - \zeta_2\zeta_1^*)$$
$$+ W_2^3 i(+\eta_1\eta_2^* - \eta_2\eta_1^* - \zeta_1\zeta_2^* + \zeta_2\zeta_1^*)$$
$$+ W_3^3 (\eta_1\eta_1^* - \eta_2\eta_2^* - \zeta_1\zeta_1^* + \zeta_2\zeta_2^*). \tag{B.70}$$

Therefore the Lagrangian density is

$$\mathcal{L}_l = \mathcal{L}_0 + g_1\mathcal{L}_1 + g_2\mathcal{L}_2 + m\boldsymbol{\rho}, \tag{B.71}$$

$$\mathcal{L}_0 = \Re[-i(\eta_e^\dagger \sigma^\mu \partial_\mu \eta_e + \xi_e^\dagger \widehat{\sigma}^\mu \partial_\mu \xi_e + \eta_n^\dagger \sigma^\mu \partial_\mu \eta_n)], \tag{B.72}$$

$$\mathcal{L}_1 = B_\mu (\frac{1}{2}\eta_e^\dagger \sigma^\mu \eta_e + \xi_e^\dagger \widehat{\sigma}^\mu \xi_e + \frac{1}{2}\eta_n^\dagger \sigma^\mu \eta_n), \tag{B.73}$$

$$\mathcal{L}_2 = -\Re[(W_\mu^1 + iW_\mu^2)\eta_e^\dagger \sigma^\mu \eta_n] + \frac{W_\mu^3}{2}(\eta_e^\dagger \sigma^\mu \eta_e - \eta_n^\dagger \sigma^\mu \eta_n). \tag{B.74}$$

The Lagrange equation $\dfrac{\partial \mathcal{L}}{\partial \xi_1^*} = \partial_\mu \left(\dfrac{\partial \mathcal{L}}{\partial(\partial_\mu \xi_1^*)}\right)$ gives

$$0 = -i[(\partial_0 + \partial_3)\xi_1 + (\partial_1 - i\partial_2)\xi_2], \tag{B.75}$$
$$+ g_1[(B_0 + B_3)\xi_1 + (B_1 - iB_2)\xi_2] + \frac{m}{\rho}(\mathbf{a}\eta_1 + \mathbf{c}\zeta_1).$$

The Lagrange equation $\dfrac{\partial \mathcal{L}}{\partial \xi_2^*} = \partial_\mu \left(\dfrac{\partial \mathcal{L}}{\partial(\partial_\mu \xi_2^*)}\right)$ gives

$$0 = -i[(\partial_1 + i\partial_2)\xi_1 + (\partial_0 - \partial_3)\xi_2], \tag{B.76}$$
$$+ g_1[(B_1 + iB_2)\xi_1 + (B_0 - B_3)\xi_2] + \frac{m}{\rho}(\mathbf{a}\eta_2 + \mathbf{c}\zeta_2).$$

Together these two equations read

$$0 = -i \begin{pmatrix} \partial_0 + \partial_3 & \partial_1 - i\partial_2 \\ \partial_1 + i\partial_2 & \partial_0 - \partial_3 \end{pmatrix} \begin{pmatrix} \xi_1 & 0 \\ \xi_2 & 0 \end{pmatrix} \tag{B.77}$$
$$+ g_1 \begin{pmatrix} B_0 + B_3 & B_1 - iB_2 \\ B_1 + iB_2 & B_0 - B_3 \end{pmatrix} \begin{pmatrix} \xi_1 & 0 \\ \xi_2 & 0 \end{pmatrix} + \frac{m}{\rho}\left[\mathbf{a}\begin{pmatrix} \eta_1 & 0 \\ \eta_2 & 0 \end{pmatrix} + \mathbf{c}\begin{pmatrix} \zeta_1 & 0 \\ \zeta_2 & 0 \end{pmatrix}\right].$$

Multiplying by $\sqrt{2}$ we get

$$-i\widehat{\nabla}\phi_R + g_1\widehat{B}\phi_R + \frac{m}{\rho}(\mathbf{a}\widehat{\phi}_L + \mathbf{c}\widehat{\phi}_n) = 0. \tag{B.78}$$

Since $\phi_R \sigma_3 = \phi_R$ and $\widehat{\phi}_L \sigma_3 = \widehat{\phi}_L$, this reads

$$\widehat{\nabla}\phi_R \sigma_{21} + g_1 \widehat{B}\phi_R + \frac{m}{\rho}(\mathbf{a}\widehat{\phi}_L + \mathbf{c}\widehat{\phi}_n) = 0. \tag{B.79}$$

Then using the conjugation $M \mapsto \widehat{M}$ we get

$$\nabla\widehat{\phi}_R \sigma_{21} + g_1 B\widehat{\phi}_R + \frac{m}{\rho}(\mathbf{a}^*\phi_L + \mathbf{c}^*\phi_n) = 0. \tag{B.80}$$

The Lagrange equation $\dfrac{\partial \mathcal{L}}{\partial \eta_1^*} = \partial_\mu \left(\dfrac{\partial \mathcal{L}}{\partial(\partial_\mu \eta_1^*)} \right)$ gives

$$0 = -i[(\partial_0 - \partial_3)\eta_1 + (-\partial_1 + i\partial_2)\eta_2] + \frac{g_1}{2}[(B_0 - B_3)\eta_1 + (-B_1 + iB_2)\eta_2]$$
$$- \frac{g_2}{2}\begin{pmatrix} (W_0^1 - W_3^1)\zeta_1 + (-W_1^1 + iW_2^1)\zeta_2 \\ +i[(W_0^2 - W_3^2)\zeta_1 + (-W_1^2 + iW_2^2)\zeta_2] \\ -(W_0^3 - W_3^3)\eta_1 - (-W_1^3 + iW_2^3)\eta_2 \end{pmatrix} + \frac{m}{\rho}(\mathbf{a}^*\xi_1 + \mathbf{b}\zeta_2^*). \tag{B.81}$$

The Lagrange equation $\dfrac{\partial \mathcal{L}}{\partial \eta_2^*} = \partial_\mu \left(\dfrac{\partial \mathcal{L}}{\partial(\partial_\mu \eta_2^*)} \right)$ gives

$$0 = -i[(-\partial_1 - i\partial_2)\eta_1 + (\partial_0 + \partial_3)\eta_2] + \frac{g_1}{2}[(-B_1 - iB_2)\eta_1 + (B_0 + B_3)\eta_2]$$
$$- \frac{g_2}{2}\begin{pmatrix} (-W_1^1 - iW_2^1)\zeta_1 + (W_0^1 + W_3^1)\zeta_2 \\ +i[(-W_1^2 - iW_2^2)\zeta_1 + (W_0^2 + W_3^2)\zeta_2] \\ -(-W_1^3 - iW_2^3)\eta_1 - (W_0^3 + W_3^3)\eta_2 \end{pmatrix} + \frac{m}{\rho}(\mathbf{a}^*\xi_2 - \mathbf{b}\zeta_1^*). \tag{B.82}$$

Together these equations give

$$0 = -i\begin{pmatrix} \partial_0 - \partial_3 & -\partial_1 + i\partial_2 \\ -\partial_1 - i\partial_2 & \partial_0 + \partial_3 \end{pmatrix}\begin{pmatrix} \eta_1 & 0 \\ \eta_2 & 0 \end{pmatrix}$$
$$+ \frac{g_1}{2}\begin{pmatrix} B_0 - B_3 & -B_1 + iB_2 \\ -B_1 - iB_2 & B_0 + B_3 \end{pmatrix}\begin{pmatrix} \eta_1 & 0 \\ \eta_2 & 0 \end{pmatrix} \tag{B.83}$$
$$- \frac{g_2}{2}\left[\begin{pmatrix} W_0^1 - W_3^1 & -W_1^1 + iW_2^1 \\ -W_1^1 - iW_2^1 & W_0^1 + W_3^1 \end{pmatrix}\begin{pmatrix} \zeta_1 & 0 \\ \zeta_2 & 0 \end{pmatrix} \right.$$
$$+ i\begin{pmatrix} W_0^2 - W_3^2 & -W_1^2 + iW_2^2 \\ -W_1^2 - iW_2^2 & W_0^2 + W_3^2 \end{pmatrix}\begin{pmatrix} \zeta_1 & 0 \\ \zeta_2 & 0 \end{pmatrix}$$
$$\left. - \begin{pmatrix} W_0^3 - W_3^3 & -W_1^3 + iW_2^3 \\ -W_1^3 - iW_2^3 & W_0^3 + W_3^3 \end{pmatrix}\begin{pmatrix} \eta_1 & 0 \\ \eta_2 & 0 \end{pmatrix} \right]$$
$$+ \frac{m}{\rho}\left[\mathbf{a}^*\begin{pmatrix} \xi_1 & 0 \\ \xi_2 & 0 \end{pmatrix} + \mathbf{b}\begin{pmatrix} \zeta_2^* & 0 \\ -\zeta_1^* & 0 \end{pmatrix} \right].$$

Multiplying by $\sqrt{2}$ this reads

$$0 = \nabla\widehat{\phi}_L \sigma_{21} + \frac{g_1}{2} B \widehat{\phi}_L + \frac{g_2}{2} [-(W^1 + iW^2)\widehat{\phi}_n + W^3 \widehat{\phi}_L] \qquad (\text{B.84})$$
$$+ \frac{m}{\rho}(\mathbf{a}^* \phi_R - \mathbf{b}\phi_n \sigma_1).$$

Adding Eq. (B.80) to Eq. (B.84) we get the wave equation

$$0 = \nabla\widehat{\phi}_e \sigma_{21} + \frac{g_1}{2} B(\widehat{\phi}_L + 2\widehat{\phi}_R) + \frac{g_2}{2} [-(W^1 + iW^2)\widehat{\phi}_n + W^3 \widehat{\phi}_L] \qquad (\text{B.85})$$
$$+ \frac{m}{\rho}(\mathbf{a}^* \phi_e - \mathbf{b}\phi_n \sigma_1 + \mathbf{c}^* \phi_n).$$

Without its mass term, this equation is the wave equation of the electron Eq. (6.57).

The Lagrange equation $\dfrac{\partial \mathcal{L}}{\partial \zeta_1^*} = \partial_\mu \left(\dfrac{\partial \mathcal{L}}{\partial(\partial_\mu \zeta_1^*)} \right)$ gives

$$0 = -i[(\partial_0 - \partial_3)\zeta_1 + (-\partial_1 + i\partial_2)\zeta_2] + \frac{g_1}{2}[(B_0 - B_3)\zeta_1 + (-B_1 + iB_2)\zeta_2]$$
$$+ \frac{g_2}{2}\begin{pmatrix} -[(W_0^1 - W_3^1)\eta_1 + (-W_1^1 + iW_2^1)\eta_2] \\ +i[(W_0^2 - W_3^2)\eta_1 + (-W_1^2 + iW_2^2)\eta_2] \\ -(W_0^3 - W_3^3)\zeta_1 - (-W_1^3 + iW_2^3)\zeta_2 \end{pmatrix} + \frac{m}{\rho}(\mathbf{c}^*\xi_1 - \mathbf{b}\eta_2^*).$$
$$(\text{B.86})$$

The Lagrange equation $\dfrac{\partial \mathcal{L}}{\partial \zeta_2^*} = \partial_\mu \left(\dfrac{\partial \mathcal{L}}{\partial(\partial_\mu \zeta_2^*)} \right)$ gives

$$0 = -i[(-\partial_1 - i\partial_2)\zeta_1 + (\partial_0 + \partial_3)\zeta_2] + \frac{g_1}{2}[(-B_1 - iB_2)\zeta_1 + (B_0 + B_3)\zeta_2]$$
$$+ \frac{g_2}{2}\begin{pmatrix} -[(-W_1^1 - iW_2^1)\eta_1 + (W_0^1 + W_3^1)\eta_2] \\ +i[(-W_1^2 - iW_2^2)\eta_1 + (W_0^2 + W_3^2)\eta_2] \\ -(-W_1^3 - iW_2^3)\zeta_1 - (W_0^3 + W_3^3)\zeta_2 \end{pmatrix} + \frac{m}{\rho}(\mathbf{c}^*\xi_2 + \mathbf{b}\eta_1^*).$$
$$(\text{B.87})$$

Together these equations read

$$0 = -i\begin{pmatrix} \partial_0 - \partial_3 & -\partial_1 + i\partial_2 \\ -\partial_1 - i\partial_2 & \partial_0 + \partial_3 \end{pmatrix}\begin{pmatrix} \zeta_1 & 0 \\ \zeta_2 & 0 \end{pmatrix} + \frac{g_1}{2}\begin{pmatrix} B_0 - B_3 & -B_1 + iB_2 \\ -B_1 - iB_2 & B_0 + B_3 \end{pmatrix}\begin{pmatrix} \zeta_1 & 0 \\ \zeta_2 & 0 \end{pmatrix}$$
$$+ \frac{g_2}{2}\left[-\begin{pmatrix} W_0^1 - W_3^1 & -W_1^1 + iW_2^1 \\ -W_1^1 - iW_2^1 & W_0^1 + W_3^1 \end{pmatrix}\begin{pmatrix} \eta_1 & 0 \\ \eta_2 & 0 \end{pmatrix} \right. \qquad (\text{B.88})$$
$$+ i\begin{pmatrix} W_0^2 - W_3^2 & -W_1^2 + iW_2^2 \\ -W_1^2 - iW_2^2 & W_0^2 + W_3^2 \end{pmatrix}\begin{pmatrix} \eta_1 & 0 \\ \eta_2 & 0 \end{pmatrix}$$
$$\left. - \begin{pmatrix} W_0^3 - W_3^3 & -W_1^3 + iW_2^3 \\ -W_1^3 - iW_2^3 & W_0^3 + W_3^3 \end{pmatrix}\begin{pmatrix} \zeta_1 & 0 \\ \zeta_2 & 0 \end{pmatrix} \right] + \frac{m}{\rho}\left[\mathbf{b}\begin{pmatrix} -\eta_2^* & 0 \\ \eta_1^* & 0 \end{pmatrix} + \mathbf{c}^*\begin{pmatrix} \xi_1 & 0 \\ \xi_2 & 0 \end{pmatrix} \right].$$

Multiplying by $\sqrt{2}$ this reads

$$0 = \nabla \widehat{\phi}_n \sigma_{21} + \frac{g_1}{2} B \widehat{\phi}_n + \frac{g_2}{2} [(-W^1 + iW^2)\widehat{\phi}_L - W^3 \widehat{\phi}_n]$$
$$+ \frac{m}{\rho}(\mathbf{c}^* \phi_R + \mathbf{b}\phi_L \sigma_1). \tag{B.89}$$

Without its mass term, this equation is the wave equation Eq. (6.58) of the electronic neutrino. The system made of the two wave equations Eq. (B.85) and Eq. (B.89) is equivalent to the equation:

$$\mathbf{D}\Psi\gamma_{012} + m\boldsymbol{\rho}\chi_l = 0, \tag{B.90}$$

where

$$\chi_l = \frac{1}{\rho^2}\begin{pmatrix} \mathbf{a}^*\phi_e - \mathbf{b}\phi_n\sigma_1 + \mathbf{c}^*\phi_n & \mathbf{b}\phi_L\sigma_1 + \mathbf{c}^*\phi_R \\ -\mathbf{b}^*\widehat{\phi}_L\sigma_1 + \mathbf{c}\widehat{\phi}_R & \mathbf{a}\widehat{\phi}_e + \mathbf{b}^*\widehat{\phi}_n\sigma_1 + \mathbf{c}\widehat{\phi}_n \end{pmatrix}, \tag{B.91}$$

or to the invariant equation

$$\widetilde{\Psi}\mathbf{D}\Psi\gamma_{012} + m\boldsymbol{\rho}\widetilde{\Psi}\chi_l = 0. \tag{B.92}$$

B.4   Invariances

From Eq. (B.91) we get

$$\rho^2\widetilde{\Psi}\chi_l = \begin{pmatrix} \overline{\phi}_e & \phi_n^\dagger \\ \overline{\phi}_n & \phi_e^\dagger \end{pmatrix}\begin{pmatrix} \mathbf{a}^*\phi_e - \mathbf{b}\phi_n\sigma_1 + \mathbf{c}^*\phi_n & \mathbf{b}\phi_L\sigma_1 + \mathbf{c}^*\phi_R \\ -\mathbf{b}^*\widehat{\phi}_L\sigma_1 + \mathbf{c}\widehat{\phi}_R & \mathbf{a}\widehat{\phi}_e + \mathbf{b}^*\widehat{\phi}_n\sigma_1 + \mathbf{c}\widehat{\phi}_n \end{pmatrix}$$
$$= \begin{pmatrix} U & V \\ \widehat{V} & \widehat{U} \end{pmatrix}, \tag{B.93}$$

$$U = \mathbf{a}^*\overline{\phi}_e\phi_e - \mathbf{b}\overline{\phi}_e\phi_n\sigma_1 + \mathbf{c}^*\overline{\phi}_e\phi_n - \mathbf{b}^*\phi_n^\dagger\widehat{\phi}_L\sigma_1 + \mathbf{c}\phi_n^\dagger\widehat{\phi}_R. \tag{B.94}$$

We have

$$\mathbf{a}^*\overline{\phi}_e\phi_e = \mathbf{a}\mathbf{a}^* \tag{B.95}$$

$$-\mathbf{b}\overline{\phi}_e\phi_n\sigma_1 - \mathbf{b}^*\phi_n^\dagger\widehat{\phi}_L\sigma_1 = \begin{pmatrix} \mathbf{b}\mathbf{b}^* & 0 \\ -\mathbf{b}\mathbf{c} & \mathbf{b}\mathbf{b}^* \end{pmatrix} = \mathbf{b}\mathbf{b}^* + \frac{\mathbf{b}\mathbf{c}}{2}(-\sigma_1 + i\sigma_2) \tag{B.96}$$

$$\mathbf{c}^*\overline{\phi}_e\phi_n + \mathbf{c}\phi_n^\dagger\widehat{\phi}_R = \begin{pmatrix} 0 & -\mathbf{b}^*\mathbf{c}^* \\ 0 & 2\mathbf{c}\mathbf{c}^* \end{pmatrix} = \mathbf{c}\mathbf{c}^*(1 - \sigma_3) + \frac{\mathbf{b}^*\mathbf{c}^*}{2}(-\sigma_1 - i\sigma_2)$$
$$\tag{B.97}$$

$$U = \rho^2 - \Re(\mathbf{b}\mathbf{c})\sigma_1 - \Im(\mathbf{b}\mathbf{c})\sigma_2 - \mathbf{c}\mathbf{c}^*\sigma_3 \tag{B.98}$$

$$\Re(m\boldsymbol{\rho}\widetilde{\Psi}\chi_l) = m\boldsymbol{\rho}. \tag{B.99}$$

Therefore the Lagrangian density Eq. (B.71) is also the real part of the invariant wave equation Eq. (B.90). The double link between the wave equation and the Lagrangian density, that we have first encountered for the electron alone, exists also in the electron+neutrino case.

The V term is simpler because we get

$$V = \mathbf{ac}^*. \tag{B.100}$$

B.4.1 *Form invariance*

Let M be any invertible element defining Eq. (1.42). We have

$$\nabla = \overline{M}\nabla'\widehat{M} \; ; \quad \Psi' = N\Psi, \tag{B.101}$$

$$\mathbf{D} = \widetilde{N}\mathbf{D}'N \; ; \quad N = \begin{pmatrix} M & 0 \\ 0 & \widehat{M} \end{pmatrix} \; ; \quad \widetilde{N} = \begin{pmatrix} \overline{M} & 0 \\ 0 & M^\dagger \end{pmatrix}, \tag{B.102}$$

$$0 = \widetilde{\Psi}\mathbf{D}\Psi\gamma_{012} + m\rho\widetilde{\Psi}\chi_l = \widetilde{\Psi}\widetilde{N}\mathbf{D}'N\Psi\gamma_{012} + m\rho\widetilde{\Psi}\chi_l$$
$$= \widetilde{\Psi}'\mathbf{D}'\Psi'\gamma_{012} + m'\rho'\widetilde{\Psi}\chi_l. \tag{B.103}$$

We shall get the form invariance of the wave equation if and only if

$$\widetilde{\Psi}\chi_l = \widetilde{\Psi}'\chi_l' = \widetilde{\Psi}\widetilde{N}\chi_l', \tag{B.104}$$

$$\chi_l' = \widetilde{N}^{-1}\chi_l. \tag{B.105}$$

And we have

$$\frac{1}{\rho'^2}(\mathbf{a}'^*\phi_e' - \mathbf{b}'\phi_n'\sigma_1 + \mathbf{c}'^*\phi_n')$$

$$= \frac{1}{r^2\rho^2}(re^{-i\theta}\mathbf{a}^*M\phi_e - re^{-i\theta}\mathbf{b}M\phi_n\sigma_1 + re^{-i\theta}\mathbf{c}^*M\phi_n)$$

$$= \frac{M}{re^{i\theta}}\frac{1}{\rho^2}(\mathbf{a}^*\phi_e - \mathbf{b}\phi_n\sigma_1 + \mathbf{c}^*\phi_n)$$

$$= \overline{M}^{-1}\frac{1}{\rho^2}(\mathbf{a}^*\phi_e - \mathbf{b}\phi_n\sigma_1 + \mathbf{c}^*\phi_n). \tag{B.106}$$

$$\frac{1}{\rho'^2}(\mathbf{b}'\phi_L'\sigma_1 + \mathbf{c}'^*\phi_n')$$

$$= \frac{1}{r^2\rho^2}(re^{-i\theta}\mathbf{b}M\phi_L\sigma_1 + re^{-i\theta}\mathbf{c}^*M\phi_n)$$

$$= \frac{M}{re^{i\theta}}\frac{1}{\rho^2}(\mathbf{b}\phi_L\sigma_1 + \mathbf{c}^*\phi_n) = \overline{M}^{-1}\frac{1}{\rho^2}(\mathbf{b}\phi_L\sigma_1 + \mathbf{c}^*\phi_n). \tag{B.107}$$

This gives Eq. (B.104) because

$$\chi_l' = \begin{pmatrix} \overline{M}^{-1} & 0 \\ 0 & M^{\dagger -1} \end{pmatrix}\chi_l = \widetilde{N}^{-1}\chi_l. \tag{B.108}$$

And the wave is invariant under Cl_3^*, therefore it is relativistic invariant.

B.4.2 *Gauge invariance — group generated by P_0*

We use the following form of P_0

$$P_0(\Psi) = \frac{\mathbf{i}}{2}\Psi + \frac{3}{2}\Psi\gamma_{21}, \tag{B.109}$$

$$\Psi' = [\exp(\theta P_0)](\Psi) = e^{\frac{\theta}{2}\mathbf{i}}\Psi e^{\frac{3\theta}{2}\gamma_{21}}. \tag{B.110}$$

From Eq. (B.14) we deduce

$$\mathbf{D}'\Psi' = e^{-\frac{\theta}{2}\mathbf{i}}(\mathbf{D}\Psi)e^{\frac{3\theta}{2}\gamma_{21}}, \tag{B.111}$$

$$\mathbf{D}'\Psi'\gamma_{012} = e^{-\frac{\theta}{2}\mathbf{i}}(\mathbf{D}\Psi\gamma_{012})e^{\frac{3\theta}{2}\gamma_{21}}. \tag{B.112}$$

Equation Eq. (B.110) reads

$$\begin{pmatrix} \phi'_e & \phi'_n \\ \widehat{\phi}'_n & \widehat{\phi}'_e \end{pmatrix} = \begin{pmatrix} e^{i\frac{\theta}{2}} & 0 \\ 0 & e^{-i\frac{\theta}{2}} \end{pmatrix} \begin{pmatrix} \phi_e & \phi_n \\ \widehat{\phi}_n & \widehat{\phi}_e \end{pmatrix} \begin{pmatrix} e^{\frac{3\theta}{2}i\sigma_3} & 0 \\ 0 & e^{\frac{3\theta}{2}i\sigma_3} \end{pmatrix}$$

$$= \begin{pmatrix} e^{i\frac{\theta}{2}}\phi_e e^{\frac{3\theta}{2}i\sigma_3} & e^{i\frac{\theta}{2}}\phi_n e^{\frac{3\theta}{2}i\sigma_3} \\ e^{-i\frac{\theta}{2}}\widehat{\phi}_n e^{\frac{3\theta}{2}i\sigma_3} & e^{-i\frac{\theta}{2}}\widehat{\phi}_e e^{\frac{3\theta}{2}i\sigma_3} \end{pmatrix}. \tag{B.113}$$

This reads

$$\begin{pmatrix} \xi'_1 & -\eta'^*_2 \\ \xi'_2 & \eta'^*_1 \end{pmatrix} = \begin{pmatrix} e^{2i\theta}\xi_1 & -e^{-i\theta}\eta^*_2 \\ e^{2i\theta}\xi_2 & e^{-i\theta}\eta^*_1 \end{pmatrix}$$

$$\begin{pmatrix} 0 & -\zeta'^*_2 \\ 0 & \zeta'^*_1 \end{pmatrix} = \begin{pmatrix} 0 & -e^{-i\theta}\zeta^*_2 \\ 0 & e^{-i\theta}\zeta^*_1 \end{pmatrix}, \tag{B.114}$$

and we get

$$\begin{pmatrix} \xi'_1 \\ \xi'_2 \end{pmatrix} = e^{2i\theta}\begin{pmatrix} \xi_1 \\ \xi_2 \end{pmatrix} ; \quad \begin{pmatrix} \eta'_1 \\ \eta'_2 \end{pmatrix} = e^{i\theta}\begin{pmatrix} \eta_1 \\ \eta_2 \end{pmatrix} ; \quad \begin{pmatrix} \zeta'_1 \\ \zeta'_2 \end{pmatrix} = e^{i\theta}\begin{pmatrix} \zeta_1 \\ \zeta_2 \end{pmatrix}. \tag{B.115}$$

We then get for **a, b, c**:

$$\mathbf{a}' = e^{i\theta}\mathbf{a} ; \quad \mathbf{a}'\mathbf{a}'^* = \mathbf{a}\mathbf{a}^*, \tag{B.116}$$

$$\mathbf{b}' = e^{2i\theta}\mathbf{b} ; \quad \mathbf{b}'\mathbf{b}'^* = \mathbf{b}\mathbf{b}^*, \tag{B.117}$$

$$\mathbf{c}' = e^{i\theta}\mathbf{c} ; \quad \mathbf{c}'\mathbf{c}'^* = \mathbf{c}\mathbf{c}^*, \tag{B.118}$$

$$\rho' = \rho. \tag{B.119}$$

We need to study

$$\chi'_l = \frac{1}{\rho'^2} \times \begin{pmatrix} \mathbf{a}'^*\phi'_e - \mathbf{b}'\phi'_n\sigma_1 + \mathbf{c}'^*\phi'_n & \mathbf{b}'\phi'_L\sigma_1 + \mathbf{c}'^*\phi'_R \\ -\mathbf{b}'^*\widehat{\phi}'_L\sigma_1 + \mathbf{c}'\widehat{\phi}'_R & \mathbf{a}'\widehat{\phi}'_e + \mathbf{b}'^*\widehat{\phi}'_n\sigma_1 + \mathbf{c}'\widehat{\phi}'_n \end{pmatrix}. \tag{B.120}$$

We get

$$\mathbf{a'}^* \phi'_e - \mathbf{b'} \phi'_n \sigma_1 + \mathbf{c'}^* \phi'_n$$

$$= e^{-i\theta}\mathbf{a}^* e^{i\frac{\theta}{2}} \phi_e e^{\frac{3\theta}{2}i\sigma_3} - e^{2i\theta}\mathbf{b} e^{i\frac{\theta}{2}} \phi_n e^{\frac{3\theta}{2}i\sigma_3} + e^{-i\theta}\mathbf{c}^* e^{i\frac{\theta}{2}} \phi_n e^{\frac{3\theta}{2}i\sigma_3} \quad \text{(B.121)}$$

$$= e^{-i\frac{\theta}{2}}(\mathbf{a}^* \phi_e - e^{3i\theta}\mathbf{b} \phi_n e^{3i\theta\sigma_3}\sigma_1 + \mathbf{c}^* \phi_n)e^{\frac{3\theta}{2}i\sigma_3}$$

$$= e^{-i\frac{\theta}{2}}(\mathbf{a}^* \phi_e - e^{3i\theta}\mathbf{b} \phi_n e^{-3i\theta}\sigma_1 + \mathbf{c}^* \phi_n)e^{\frac{3\theta}{2}i\sigma_3}$$

$$= e^{-i\frac{\theta}{2}}(\mathbf{a}^* \phi_e - \mathbf{b}\phi_n\sigma_1 + \mathbf{c}^* \phi_n)e^{\frac{3\theta}{2}i\sigma_3}, \quad \text{(B.122)}$$

and similarly

$$\mathbf{b'}\phi'_L\sigma_1 + \mathbf{c'}^* \phi'_R = e^{2i\theta}\mathbf{b} e^{i\frac{\theta}{2}} \phi_L e^{\frac{3\theta}{2}i\sigma_3}\sigma_1 + e^{-i\theta}\mathbf{c}^* e^{i\frac{\theta}{2}} \phi_R e^{\frac{3\theta}{2}i\sigma_3}$$

$$= e^{i\frac{5\theta}{2}}\mathbf{b}\phi_L e^{3i\theta\sigma_3}\sigma_1 e^{\frac{3\theta}{2}i\sigma_3} + e^{-i\frac{\theta}{2}}\mathbf{c}^* \phi_R e^{\frac{3\theta}{2}i\sigma_3}$$

$$= (e^{i\frac{5\theta}{2}}\mathbf{b}\phi_L e^{-3i\theta}\sigma_1 + e^{-i\frac{\theta}{2}}\mathbf{c}^* \phi_R)e^{\frac{3\theta}{2}i\sigma_3}$$

$$= e^{-i\frac{\theta}{2}}(\mathbf{b}\phi_L\sigma_1 + \mathbf{c}^* \phi_R)e^{\frac{3\theta}{2}i\sigma_3}. \quad \text{(B.123)}$$

This gives

$$\chi'_l = \frac{1}{\rho^2}\begin{pmatrix} e^{-i\frac{\theta}{2}}(\mathbf{a}^* \phi_e - \mathbf{b}\phi_n\sigma_1 + \mathbf{c}^* \phi_n)e^{\frac{3\theta}{2}i\sigma_3} & e^{-i\frac{\theta}{2}}(\mathbf{b}\phi_L\sigma_1 + \mathbf{c}^* \phi_R)e^{\frac{3\theta}{2}i\sigma_3} \\ e^{i\frac{\theta}{2}}(-\mathbf{b}^* \widehat{\phi}_L\sigma_1 + \mathbf{c}\widehat{\phi}_R)e^{\frac{3\theta}{2}i\sigma_3} & e^{i\frac{\theta}{2}}(\mathbf{a}\widehat{\phi}_e + \mathbf{b}\widehat{\phi}_n\sigma_1 + \mathbf{c}\widehat{\phi}_n)e^{\frac{3\theta}{2}i\sigma_3} \end{pmatrix}, \quad \text{(B.124)}$$

which reads

$$\chi'_l = e^{-\frac{\theta}{2}\mathbf{i}}\chi_l e^{\frac{3\theta}{2}\gamma_{21}}, \quad \text{(B.125)}$$

and we finally get

$$\mathbf{D'}\Psi'\gamma_{012} + m\rho'\chi'_l = e^{-\frac{\theta}{2}\mathbf{i}}(\mathbf{D}\Psi\gamma_{012} + m\rho\chi_l)e^{\frac{3\theta}{2}\gamma_{21}} = 0. \quad \text{(B.126)}$$

This allows us to say that the wave equation of the electron+neutrino pair is gauge invariant under the group generated by P_0.

B.4.3 *Gauge invariance — group generated by P_3*

This generator acts only on left waves: we get

$$\xi'_1 = \xi_1 \; ; \; \xi'_2 = \xi_2. \quad \text{(B.127)}$$

And with left waves we have

$$\Psi'_L = \Psi_L e^{-\theta\mathbf{i}}, \quad \text{(B.128)}$$

$$\begin{pmatrix} \phi'_L & \phi'_n \\ \widehat{\phi}'_n & \widehat{\phi}'_L \end{pmatrix} = \begin{pmatrix} \phi_L & \phi_n \\ \widehat{\phi}_n & \widehat{\phi}_L \end{pmatrix}\begin{pmatrix} e^{-i\theta} & 0 \\ 0 & e^{i\theta} \end{pmatrix}, \quad \text{(B.129)}$$

$$\phi'_L = e^{-i\theta}\phi_L, \quad \text{(B.130)}$$

$$\phi'_n = e^{i\theta}\phi_n, \quad \text{(B.131)}$$

which reads

$$\eta_1' = e^{i\theta}\eta_1 \ ; \quad \eta_2' = e^{i\theta}\eta_2 \ ; \quad \zeta_1' = e^{-i\theta}\zeta_1 \ ; \quad \zeta_2' = e^{-i\theta}\zeta_2. \tag{B.132}$$

We then get for **a**, **b**, **c**:

$$\mathbf{a}' = e^{-i\theta}\mathbf{a} \ ; \quad \mathbf{a}'\mathbf{a}'^* = \mathbf{aa}^*, \tag{B.133}$$

$$\mathbf{b}' = \mathbf{b} \ ; \quad \mathbf{b}'\mathbf{b}'^* = \mathbf{bb}^*, \tag{B.134}$$

$$\mathbf{c}' = e^{i\theta}\mathbf{c} \ ; \quad \mathbf{c}'\mathbf{c}'^* = \mathbf{cc}^*, \tag{B.135}$$

$$\rho' = \rho. \tag{B.136}$$

The covariant derivative is here reduced to

$$\mathbf{D} = \partial + \frac{g_2}{2}\mathbf{W}^3 P_3. \tag{B.137}$$

We let

$$[\partial\Psi + \frac{g_2}{2}\mathbf{W}^3 P_3(\Psi)]\gamma_{012} + m\rho\chi_l = \begin{pmatrix} A & B \\ \hat{B} & \hat{A} \end{pmatrix}. \tag{B.138}$$

We have

$$A = (\nabla\widehat{\phi}_e + i\frac{g_2}{2}\mathbf{W}^3\widehat{\phi}_L)\sigma_{21} + \frac{m}{\rho}(\mathbf{a}^*\phi_e - \mathbf{b}\phi_n\sigma_1 + \mathbf{c}^*\phi_n), \tag{B.139}$$

$$B = (\nabla\widehat{\phi}_n - i\frac{g_2}{2}\mathbf{W}^3\widehat{\phi}_n)\sigma_{21} + \frac{m}{\rho}(\mathbf{b}\phi_L\sigma_1 + \mathbf{c}^*\phi_R). \tag{B.140}$$

Only the left column of B is not zero, this gives the simple result:

$$B' = e^{-i\theta}B. \tag{B.141}$$

For A which has a right column and a left column we note

$$A = A_L + A_R, \tag{B.142}$$

$$A_L = (\nabla\widehat{\phi}_L + i\frac{g_2}{2}\mathbf{W}^3\widehat{\phi}_L)\sigma_{21} + \frac{m}{\rho}(\mathbf{a}^*\phi_R - \mathbf{b}\phi_n\sigma_1), \tag{B.143}$$

$$A_R = \nabla\widehat{\phi}_R\sigma_{21} + \frac{m}{\rho}(\mathbf{a}^*\phi_L + \mathbf{c}^*\phi_n). \tag{B.144}$$

We then get

$$A_L' = e^{i\theta}A_L \ ; \quad A_R' = A_R, \tag{B.145}$$

$$A' = A \begin{pmatrix} e^{i\theta} & 0 \\ 0 & 1 \end{pmatrix}. \tag{B.146}$$

Since the same term multiplies the differential term and the mass term of the wave equation, we may say that the equation is invariant under the gauge group generated by P_3.

B.4.4 Gauge invariance — group generated by P_1

This generator acts also only on left waves: we have

$$\xi_1' = \xi_1 \; ; \quad \xi_2' = \xi_2. \tag{B.147}$$

And with these left waves we have

$$\Psi_L' = \begin{pmatrix} \phi_L' & \phi_n' \\ \widehat{\phi}_n' & \widehat{\phi}_L' \end{pmatrix} = \Psi_L e^{\theta\gamma_3 \mathbf{i}} = \begin{pmatrix} \phi_L & \phi_n \\ \widehat{\phi}_n & \widehat{\phi}_L \end{pmatrix} \begin{pmatrix} \cos(\theta) & -i\sin(\theta)\sigma_3 \\ -i\sin(\theta)\sigma_3 & \cos(\theta) \end{pmatrix},$$
$$\tag{B.148}$$

$$\phi_L' = \cos(\theta)\phi_L - i\sin(\theta)\phi_n\sigma_3 \; ; \quad \phi_n' = \cos(\theta)\phi_n - i\sin(\theta)\phi_L\sigma_3, \tag{B.149}$$

which reads with $C = \cos(\theta)$ and $S = \sin(\theta)$:

$$\phi_L' = C\phi_L + iS\phi_n \; ; \quad \widehat{\phi}_L' = C\widehat{\phi}_L - iS\widehat{\phi}_n, \tag{B.150}$$

$$\phi_n' = C\phi_n + iS\phi_L \; ; \quad \widehat{\phi}_n' = C\widehat{\phi}_n - iS\widehat{\phi}_L, \tag{B.151}$$

$$\eta_1' = C\eta_1 - iS\zeta_1 \; ; \quad \eta_2' = C\eta_1 - iS\zeta_2, \tag{B.152}$$

$$\zeta_1' = C\zeta_1 - iS\eta_1 \; ; \quad \zeta_2' = C\zeta_2 - iS\eta_2. \tag{B.153}$$

We then get for $\mathbf{a}$, $\mathbf{b}$, $\mathbf{c}$:

$$\mathbf{a}' = C\mathbf{a} + iS\mathbf{c}, \tag{B.154}$$

$$\mathbf{b}' = \mathbf{b}, \tag{B.155}$$

$$\mathbf{c}' = C\mathbf{c} + iS\mathbf{a}, \tag{B.156}$$

$$\rho'^2 = \mathbf{a}'\mathbf{a}'^* + \mathbf{b}'\mathbf{b}'^* + \mathbf{c}'\mathbf{c}'^* = \mathbf{a}\mathbf{a}^* + \mathbf{b}\mathbf{b}^* + \mathbf{c}\mathbf{c}^* = \rho^2. \tag{B.157}$$

The covariant derivative is now reduced to

$$\mathbf{D} = \partial + \frac{g_2}{2}\mathbf{W}^1 P_1. \tag{B.158}$$

We let

$$[\partial\Psi + \frac{g_2}{2}\mathbf{W}^1 P_1(\Psi)]\gamma_{012} + m\boldsymbol{\rho}\chi_l = \begin{pmatrix} A & B \\ \widehat{B} & \widehat{A} \end{pmatrix}. \tag{B.159}$$

We have

$$A = \nabla\widehat{\phi}_e\sigma_{21} - \frac{g_2}{2}W^1\widehat{\phi}_n + \frac{m}{\rho}(\mathbf{a}^*\phi_e - \mathbf{b}\phi_n\sigma_1 + \mathbf{c}^*\phi_n), \tag{B.160}$$

$$B = \nabla\widehat{\phi}_n\sigma_{21} - \frac{g_2}{2}W^1\widehat{\phi}_L + \frac{m}{\rho}(\mathbf{b}\phi_L\sigma_1 + \mathbf{c}^*\phi_R). \tag{B.161}$$

As previously, for A, which has a left and a right column, we note:

$$A = A_L + A_R, \tag{B.162}$$

$$A_L = (\nabla\widehat{\phi}_L + i\frac{g_2}{2}W^1\widehat{\phi}_L)\sigma_{21} + \frac{m}{\rho}(\mathbf{a}^*\phi_R - \mathbf{b}\phi_n\sigma_1), \tag{B.163}$$

$$A_R = \nabla\widehat{\phi}_R\sigma_{21} + \frac{m}{\rho}(\mathbf{a}^*\phi_L + \mathbf{c}^*\phi_n). \tag{B.164}$$

We then get

$$A'_L = CA_L - iSB, \tag{B.165}$$

$$A'_R = A_R, \tag{B.166}$$

$$B' = CB - iSA_L. \tag{B.167}$$

And since the mass term changes in exactly the same way as the differential term, we may say that the wave equation is gauge invariant under the group generated by P_1. The invariance under the group generated by P_2 is demonstrated similarly.

B.5   Complete wave

B.5.1   *Scalar densities*

There are now $6 \times 5/2 = 15$ new densities:

$$s_1 = 2(\xi_{1\bar{u}r}\eta^*_{1ug} + \xi_{2\bar{u}r}\eta^*_{2ug}) = 2(\eta^*_{2ur}\eta^*_{1ug} - \eta^*_{1ur}\eta^*_{2ug}), \tag{B.168}$$

$$s_2 = 2(\xi_{1\bar{u}g}\eta^*_{1ub} + \xi_{2\bar{u}g}\eta^*_{2ub}) = 2(\eta^*_{2ug}\eta^*_{1ub} - \eta^*_{1ug}\eta^*_{2ub}), \tag{B.169}$$

$$s_3 = -2(\xi_{1\bar{u}r}\eta^*_{1ub} + \xi_{2\bar{u}r}\eta^*_{2ub}) = 2(\eta^*_{2ub}\eta^*_{1ur} - \eta^*_{1ub}\eta^*_{2ur}), \tag{B.170}$$

$$s_4 = 2(\xi_{1\bar{d}r}\eta^*_{1dg} + \xi_{2\bar{d}r}\eta^*_{2dg}) = 2(\eta^*_{2dr}\eta^*_{1dg} - \eta^*_{1dr}\eta^*_{2dg}), \tag{B.171}$$

$$s_5 = 2(\xi_{1\bar{d}g}\eta^*_{1db} + \xi_{2\bar{d}g}\eta^*_{2db}) = 2(\eta^*_{2dg}\eta^*_{1db} - \eta^*_{1dg}\eta^*_{2db}), \tag{B.172}$$

$$s_6 = -2(\xi_{1\bar{d}r}\eta^*_{1db} + \xi_{2\bar{d}r}\eta^*_{2db}) = 2(\eta^*_{2db}\eta^*_{1dr} - \eta^*_{1db}\eta^*_{2dr}), \tag{B.173}$$

$$s_7 = 2(\xi_{1\bar{u}r}\eta^*_{1dr} + \xi_{2\bar{u}r}\eta^*_{2dr}) = 2(\eta^*_{2ur}\eta^*_{1dr} - \eta^*_{1ur}\eta^*_{2dr}), \tag{B.174}$$

$$s_8 = 2(\xi_{1\bar{u}g}\eta^*_{1dg} + \xi_{2\bar{u}g}\eta^*_{2dg}) = 2(\eta^*_{2ug}\eta^*_{1dg} - \eta^*_{1ug}\eta^*_{2dg}), \tag{B.175}$$

$$s_9 = 2(\xi_{1\bar{u}b}\eta^*_{1db} + \xi_{2\bar{u}b}\eta^*_{2db}) = 2(\eta^*_{2ub}\eta^*_{1db} - \eta^*_{1ub}\eta^*_{2db}), \tag{B.176}$$

$$s_{10} = 2(\xi_{1\bar{u}r}\eta^*_{1dg} + \xi_{2\bar{u}r}\eta^*_{2dg}) = 2(\eta^*_{2ur}\eta^*_{1dg} - \eta^*_{1ur}\eta^*_{2dg}), \tag{B.177}$$

$$s_{11} = 2(\xi_{1\bar{u}g}\eta^*_{1db} + \xi_{2\bar{u}g}\eta^*_{2db}) = 2(\eta^*_{2ug}\eta^*_{1db} - \eta^*_{1ug}\eta^*_{2db}), \tag{B.178}$$

$$s_{12} = -2(\xi_{1\bar{d}r}\eta^*_{1ub} + \xi_{2\bar{d}r}\eta^*_{2ub}) = 2(\eta^*_{2ub}\eta^*_{1dr} - \eta^*_{1ub}\eta^*_{2dr}), \tag{B.179}$$

$$s_{13} = 2(\xi_{1\bar{u}r}\eta^*_{1db} + \xi_{2\bar{u}r}\eta^*_{2db}) = 2(\eta^*_{2ur}\eta^*_{1db} - \eta^*_{1ur}\eta^*_{2db}), \tag{B.180}$$

$$s_{14} = -2(\xi_{1\bar{d}r}\eta^*_{1ug} + \xi_{2\bar{d}r}\eta^*_{2ug}) = 2(\eta^*_{2ug}\eta^*_{1dr} - \eta^*_{1ug}\eta^*_{2dr}), \tag{B.181}$$

$$s_{15} = -2(\xi_{1\bar{d}g}\eta^*_{1ub} + \xi_{2\bar{d}g}\eta^*_{2ub}) = 2(\eta^*_{2ub}\eta^*_{1dg} - \eta^*_{1ub}\eta^*_{2dg}). \tag{B.182}$$

B.5.2 *Mass term*

We used in [21] and in Eq. (B.91):

$$\chi_l = \frac{1}{\rho_1^2} \begin{pmatrix} a_1^*\phi_e + a_2^*\phi_n\sigma_1 + a_3^*\phi_n & -a_2^*\phi_{eL}\sigma_1 + a_3^*\phi_{eR} \\ a_2\widehat{\phi}_{eL}\sigma_1 + a_3\widehat{\phi}_{eR} & a_1\widehat{\phi}_e - a_2\widehat{\phi}_n\sigma_1 + a_3\widehat{\phi}_n \end{pmatrix}, \quad \text{(B.183)}$$

with $\phi_{eR} = \phi_e(1+\sigma_3)/2$ and $\phi_{eL} = \phi_e(1-\sigma_3)/2$, and we shall need also:

$$\rho_2^2\chi_r = \begin{pmatrix} \begin{pmatrix} s_4^*\phi_{dg} - s_6^*\phi_{db} - s_7^*\phi_{ur} \\ -s_{12}^*\phi_{ub} - s_{14}^*\phi_{ug} \end{pmatrix}\sigma_1 & \begin{pmatrix} s_1^*\phi_{ug} - s_3^*\phi_{ub} + s_7^*\phi_{dr} \\ +s_{10}^*\phi_{dg} + s_{13}^*\phi_{db} \end{pmatrix}\sigma_1 \\ \begin{pmatrix} -s_1\widehat{\phi}_{ug} + s_3\widehat{\phi}_{ub} - s_7\widehat{\phi}_{dr} \\ -s_{10}\widehat{\phi}_{dg} - s_{13}\widehat{\phi}_{db} \end{pmatrix}\sigma_1 & \begin{pmatrix} -s_4\widehat{\phi}_{dg} + s_6\widehat{\phi}_{db} + s_7\widehat{\phi}_{ur} \\ +s_{12}\widehat{\phi}_{ub} + s_{14}\widehat{\phi}_{ug} \end{pmatrix}\sigma_1 \end{pmatrix},$$

$$\text{(B.184)}$$

$$\rho_2^2\chi_g = \begin{pmatrix} \begin{pmatrix} s_5^*\phi_{db} - s_4^*\phi_{dr} - s_8^*\phi_{ug} \\ -s_{10}^*\phi_{ur} - s_{15}^*\phi_{ub} \end{pmatrix}\sigma_1 & \begin{pmatrix} s_2^*\phi_{ub} - s_1^*\phi_{ur} + s_8^*\phi_{dg} \\ +s_{11}^*\phi_{db} + s_{14}^*\phi_{dr} \end{pmatrix}\sigma_1 \\ \begin{pmatrix} -s_2\widehat{\phi}_{ub} + s_1\widehat{\phi}_{ur} - s_8\widehat{\phi}_{dg} \\ -s_{11}\widehat{\phi}_{db} - s_{14}\widehat{\phi}_{dr} \end{pmatrix}\sigma_1 & \begin{pmatrix} -s_5\widehat{\phi}_{db} + s_4\widehat{\phi}_{dr} + s_8\widehat{\phi}_{ug} \\ +s_{10}\widehat{\phi}_{ur} + s_{15}\widehat{\phi}_{ub} \end{pmatrix}\sigma_1 \end{pmatrix},$$

$$\text{(B.185)}$$

$$\rho_2^2\chi_b = \begin{pmatrix} \begin{pmatrix} s_6^*\phi_{dr} - s_5^*\phi_{dg} - s_9^*\phi_{ub} \\ -s_{11}^*\phi_{ug} - s_{13}^*\phi_{ur} \end{pmatrix}\sigma_1 & \begin{pmatrix} s_3^*\phi_{ur} - s_2^*\phi_{ug} + s_9^*\phi_{db} \\ +s_{12}^*\phi_{dr} + s_{15}^*\phi_{dg} \end{pmatrix}\sigma_1 \\ \begin{pmatrix} -s_3\widehat{\phi}_{ur} + s_2\widehat{\phi}_{ug} - s_9\widehat{\phi}_{db} \\ -s_{12}\widehat{\phi}_{dr} - s_{15}\widehat{\phi}_{dg} \end{pmatrix}\sigma_1 & \begin{pmatrix} -s_6\widehat{\phi}_{dr} + s_5\widehat{\phi}_{dg} + s_9\widehat{\phi}_{ub} \\ +s_{11}\widehat{\phi}_{ug} + s_{13}\widehat{\phi}_{ur} \end{pmatrix}\sigma_1 \end{pmatrix}.$$

$$\text{(B.186)}$$

B.5.3 *Group generated by $\underline{P}_0$*

We have here

$$\underline{P}_0(\Psi^c) = \Psi^c(-\frac{1}{3}L_{21}), \quad \text{(B.187)}$$

$$\Psi'^c = [\exp(\theta\underline{P}_0)](\Psi^c) = \Psi^c\exp(-\frac{\theta}{3}L_{21}), \quad \text{(B.188)}$$

$$B'_\mu = B_\mu - \frac{2}{g_1}B_\mu. \quad \text{(B.189)}$$

To get the gauge invariance of the wave equation we must have

$$\chi'^c = \chi^c\exp(-\frac{\theta}{3}L_{21}); \quad \chi'_c = \chi_c\exp(-\frac{\theta}{3}\gamma_{21}), \quad c = r, g, b. \quad \text{(B.190)}$$

This is satisfied because

$$\phi'_{dc} = \phi_{dc} e^{-i\frac{\theta}{3}\sigma_3}; \quad \phi'_{uc} = \phi_{uc} e^{-i\frac{\theta}{3}\sigma_3}, \tag{B.191}$$

$$\eta'^*_{1dc} = e^{i\frac{\theta}{3}}\eta^*_{1dc}; \quad \eta'^*_{1uc} = e^{i\frac{\theta}{3}}\eta^*_{1uc}, \tag{B.192}$$

$$\eta'^*_{2dc} = e^{i\frac{\theta}{3}}\eta^*_{2dc}; \quad \eta'^*_{2uc} = e^{i\frac{\theta}{3}}\eta^*_{2uc}, \tag{B.193}$$

$$s'_j = e^{2i\frac{\theta}{3}}s_j, \quad j = 1, 2, \dots, 15. \tag{B.194}$$

All χ_c terms contain in the upper line $s_j^* \phi_{dc}\sigma_1$ and $s_j^* \phi_{uc}\sigma_1$. We have

$$\phi'_{dc} = \phi_{dc} e^{-i\frac{\theta}{3}\sigma_3} = e^{i\frac{\theta}{3}}\phi_{dc}, \tag{B.195}$$

$$s'^*_j \phi'_{dc}\sigma_1 = e^{-i\frac{\theta}{3}}\phi_{dc}\sigma_1 = \phi_{dc} e^{\frac{\theta}{3}\sigma_{12}}\sigma_1 = \phi_{dc}\sigma_1 e^{-\frac{\theta}{3}\sigma_{12}}, \tag{B.196}$$

$$\chi'_c = \chi_c \exp(-\frac{\theta}{3}\gamma_{21}), \tag{B.197}$$

$$\chi'^c = \chi^c \exp(-\frac{\theta}{3}L_{21}). \tag{B.198}$$

This finally gives

$$(\underline{D}'\Psi'^c)L_{012} + m_2\rho'_2\chi'^c = [(\underline{D}\Psi^c)L_{012} + m_2\rho_2\chi^c]\exp(-\frac{\theta}{3}L_{21}) = 0. \tag{B.199}$$

Therefore the wave equation is invariant under the gauge group generated by $\underline{P}_0$.

B.5.4   *Group generated by $\underline{P}_1$*

We get in this case

$$\underline{P}_1(\Psi^c) = \Psi^c L_{35}, \tag{B.200}$$

$$\Psi'^c = [\exp(\theta\underline{P}_1)](\Psi^c) = \Psi^c \exp(\theta L_{35}), \tag{B.201}$$

$$W'^1_\mu = W^1_\mu - \frac{2}{g_2}\partial_\mu\theta. \tag{B.202}$$

Since $\underline{P}_1(\Psi^c) = \Psi^c L_{35}$ we get

$$\Psi'^c = [\exp(\theta\underline{P}_1)](\Psi^c) = \Psi^c \exp(\theta L_{35}), \tag{B.203}$$

$$\Psi'_c = \Psi_c e^{\theta\gamma_3 \mathbf{i}}, \quad c = r, g, b. \tag{B.204}$$

We let

$$C = \cos(\theta) \ ; \quad S = \sin(\theta). \tag{B.205}$$

Then Eq. (B.204) is equivalent to the system

$$\widehat{\phi}'_{dc} = C\widehat{\phi}_{dc} - iS\widehat{\phi}_{uc}\sigma_3, \tag{B.206}$$

$$\widehat{\phi}'_{uc} = C\widehat{\phi}_{uc} - iS\widehat{\phi}_{dc}\sigma_3, \tag{B.207}$$

or to the system

$$\eta'_{1dc} = C\eta_{1dc} - iS\eta_{1uc}; \quad \eta'^{*}_{1dc} = C\eta^{*}_{1dc} + iS\eta^{*}_{1uc}, \qquad (B.208)$$

$$\eta'_{2dc} = C\eta_{2dc} - iS\eta_{2uc}; \quad \eta'^{*}_{2dc} = C\eta^{*}_{2dc} + iS\eta^{*}_{2uc}, \qquad (B.209)$$

$$\eta'_{1uc} = C\eta_{1uc} - iS\eta_{1dc}; \quad \eta'^{*}_{1uc} = C\eta^{*}_{1uc} + iS\eta^{*}_{1dc}, \qquad (B.210)$$

$$\eta'_{2uc} = C\eta_{2uc} - iS\eta_{2dc}; \quad \eta'^{*}_{2uc} = C\eta^{*}_{2uc} + iS\eta^{*}_{2dc}. \qquad (B.211)$$

We then get

$$s'_1 = C^2 s_1 - S^2 s_4 + iCS(s_{10} - s_{14}), \qquad (B.212)$$

$$s'_4 = C^2 s_4 - S^2 s_1 + iCS(s_{10} - s_{14}), \qquad (B.213)$$

$$s'_{10} = C^2 s_{10} + S^2 s_{14} + iCS(s_1 + s_4), \qquad (B.214)$$

$$s'_{14} = C^2 s_{14} + S^2 s_{10} - iCS(s_1 + s_4). \qquad (B.215)$$

This implies

$$s'_1 s'^{*}_1 + s'_4 s'^{*}_4 + s'_{10} s'^{*}_{10} + s'_{14} s'^{*}_{14} = s_1 s^{*}_1 + s_4 s^{*}_4 + s_{10} s^{*}_{10} + s_{14} s^{*}_{14}. \quad (B.216)$$

Similarly, permuting colors we get

$$s'_2 = C^2 s_2 - S^2 s_5 + iCS(s_{11} - s_{15}), \qquad (B.217)$$

$$s'_5 = C^2 s_5 - S^2 s_2 + iCS(s_{11} - s_{15}), \qquad (B.218)$$

$$s'_{11} = C^2 s_{11} + S^2 s_{15} + iCS(s_2 + s_5), \qquad (B.219)$$

$$s'_{15} = C^2 s_{15} + S^2 s_{11} - iCS(s_2 + s_5). \qquad (B.220)$$

This implies

$$s'_2 s'^{*}_2 + s'_5 s'^{*}_5 + s'_{11} s'^{*}_{11} + s'_{15} s'^{*}_{15} = s_2 s^{*}_2 + s_5 s^{*}_5 + s_{11} s^{*}_{11} + s_{15} s^{*}_{15}. \quad (B.221)$$

and also

$$s'_3 = C^2 s_3 - S^2 s_6 + iCS(s_{12} - s_{13}), \qquad (B.222)$$

$$s'_6 = C^2 s_6 - S^2 s_3 + iCS(s_{12} - s_{13}), \qquad (B.223)$$

$$s'_{12} = C^2 s_{12} + S^2 s_{13} + iCS(s_3 + s_6), \qquad (B.224)$$

$$s'_{13} = C^2 s_{13} + S^2 s_{12} - iCS(s_3 + s_6). \qquad (B.225)$$

This implies

$$s'_3 s'^{*}_3 + s'_6 s'^{*}_6 + s'_{12} s'^{*}_{12} + s'_{13} s'^{*}_{13} = s_3 s^{*}_3 + s_6 s^{*}_6 + s_{12} s^{*}_{12} + s_{13} s^{*}_{13}. \quad (B.226)$$

Moreover we have

$$s'_7 = s_7; \quad s'_8 = s_8; \quad s'_9 = s_9. \qquad (B.227)$$

We then get

$$\rho' = \rho. \qquad (B.228)$$

Next we have

$$\chi_r = \begin{pmatrix} A & B \\ \widehat{B} & \widehat{A} \end{pmatrix} ; \; \chi'_r = \begin{pmatrix} A' & B' \\ \widehat{B'} & \widehat{A'} \end{pmatrix}, \tag{B.229}$$

$$\widehat{A} = (-s_4\widehat{\phi}_{dg} + s_6\widehat{\phi}_{db} + s_7\widehat{\phi}_{ur} + s_{12}\widehat{\phi}_{ub} + s_{14}\widehat{\phi}_{ug})\sigma_1, \tag{B.230}$$

$$\widehat{B} = (-s_1\widehat{\phi}_{ug} + s_3\widehat{\phi}_{ub} - s_7\widehat{\phi}_{dr} - s_{10}\widehat{\phi}_{dg} - s_{13}\widehat{\phi}_{db})\sigma_1. \tag{B.231}$$

And we get

$$\widehat{A'} = C\widehat{A} - iS\widehat{B}\sigma_3, \tag{B.232}$$

$$\widehat{B'} = C\widehat{B} - iS\widehat{A}\sigma_3, \tag{B.233}$$

$$\chi'_r = \chi_r \begin{pmatrix} C & -iS\sigma_3 \\ -iS\sigma_3 & C \end{pmatrix} = \chi_r e^{\theta\gamma_3 \mathbf{i}}. \tag{B.234}$$

Since we get the same relation for colors g and b we finally have

$$\chi'^c = \chi^c \exp(\theta L_{35}),$$
$$(\underline{D'}\Psi'^c)L_{012} + m_2\rho'_2\chi'^c = (\underline{D}\Psi^c)\exp(\theta L_{35})L_{012} + m_2\rho_2\chi'^c$$
$$= [(\underline{D}\Psi^c)L_{012} + m_2\rho_2\chi^c]\exp(\theta L_{35}) = 0. \tag{B.235}$$

The wave equation with mass term is then gauge invariant under the group generated by $\underline{P}_1$.

B.5.5 *Group generated by $\underline{P}_2$*

We have here

$$\underline{P}_2(\Psi^c) = \Psi^c L_{0125}, \tag{B.236}$$

$$\Psi'^c = [\exp(\theta\underline{P}_2)](\Psi^c) = \Psi^c \exp(\theta L_{0125}), \tag{B.237}$$

$$W'^2_\mu = W^2_\mu - \frac{2}{g_2}\partial_\mu\theta. \tag{B.238}$$

Since $\underline{P}_2(\Psi^c) = \Psi^c L_{0125}$ we have

$$\Psi'^c = [\exp(\theta\underline{P}_2)](\Psi^c) = \Psi^c \exp(\theta L_{0125}), \tag{B.239}$$

$$\Psi'_c = \Psi_c e^{\theta\gamma_3}, \; c = r, g, b. \tag{B.240}$$

We let

$$C = \cos(\theta) ; \; S = \sin(\theta). \tag{B.241}$$

Then Eq. (B.240) is equivalent to the system

$$\widehat{\phi}'_{dc} = C\widehat{\phi}_{dc} + S\widehat{\phi}_{uc}, \tag{B.242}$$

$$\widehat{\phi}'_{uc} = C\widehat{\phi}_{uc} - S\widehat{\phi}_{dc}, \tag{B.243}$$

or to the system

$$\eta'_{1dc} = C\eta_{1dc} + S\eta_{1uc}; \quad \eta'^{*}_{1dc} = C\eta^{*}_{1dc} + S\eta^{*}_{1uc}, \tag{B.244}$$

$$\eta'_{2dc} = C\eta_{2dc} + S\eta_{2uc}; \quad \eta'^{*}_{2dc} = C\eta^{*}_{2dc} + S\eta^{*}_{2uc}, \tag{B.245}$$

$$\eta'_{1uc} = C\eta_{1uc} - S\eta_{1dc}; \quad \eta'^{*}_{1uc} = C\eta^{*}_{1uc} - S\eta^{*}_{1dc}, \tag{B.246}$$

$$\eta'_{2uc} = C\eta_{2uc} - S\eta_{2dc}; \quad \eta'^{*}_{2uc} = C\eta^{*}_{2uc} - S\eta^{*}_{2dc}. \tag{B.247}$$

We then have

$$s'_1 = C^2 s_1 + S^2 s_4 - CS s_{10} + CS s_{14}, \tag{B.248}$$

$$s'_4 = C^2 s_4 + S^2 s_1 + CS s_{10} - CS s_{14}, \tag{B.249}$$

$$s'_{10} = C^2 s_{10} + S^2 s_{14} + CS s_1 - CS s_4, \tag{B.250}$$

$$s'_{14} = C^2 s_{14} + S^2 s_{10} - CS s_1 + CS s_4. \tag{B.251}$$

This implies

$$s'_1 s'^{*}_1 + s'_4 s'^{*}_4 + s'_{10} s'^{*}_{10} + s'_{14} s'^{*}_{14} = s_1 s^{*}_1 + s_4 s^{*}_4 + s_{10} s^{*}_{10} + s_{14} s^{*}_{14}. \tag{B.252}$$

Similarly permuting colors we get

$$s'_2 = C^2 s_2 + S^2 s_5 - CS s_{11} + CS s_{15}, \tag{B.253}$$

$$s'_5 = C^2 s_5 + S^2 s_2 + CS s_{11} - CS s_{15}, \tag{B.254}$$

$$s'_{11} = C^2 s_{11} + S^2 s_{15} + CS s_2 - CS s_5, \tag{B.255}$$

$$s'_{15} = C^2 s_{15} + S^2 s_{11} - CS s_2 + CS s_5. \tag{B.256}$$

This implies

$$s'_2 s'^{*}_2 + s'_5 s'^{*}_5 + s'_{11} s'^{*}_{11} + s'_{15} s'^{*}_{15} = s_2 s^{*}_2 + s_5 s^{*}_5 + s_{11} s^{*}_{11} + s_{15} s^{*}_{15}. \tag{B.257}$$

and also

$$s'_3 = C^2 s_3 + S^2 s_6 - CS s_{12} + CS s_{13}, \tag{B.258}$$

$$s'_6 = C^2 s_6 + S^2 s_3 + CS s_{12} - CS s_{13}, \tag{B.259}$$

$$s'_{12} = C^2 s_{12} + S^2 s_{13} + CS s_3 - CS s_6, \tag{B.260}$$

$$s'_{13} = C^2 s_{13} + S^2 s_{12} - CS s_3 + CS s_6. \tag{B.261}$$

This implies

$$s'_3 s'^{*}_3 + s'_6 s'^{*}_6 + s'_{12} s'^{*}_{12} + s'_{13} s'^{*}_{13} = s_3 s^{*}_3 + s_6 s^{*}_6 + s_{12} s^{*}_{12} + s_{13} s^{*}_{13}. \tag{B.262}$$

Moreover we get

$$s'_7 = s_7; \quad s'_8 = s_8; \quad s'_9 = s_9. \tag{B.263}$$

We then get

$$\rho' = \rho. \tag{B.264}$$

Next we have with Eq. (B.241)

$$\widehat{A}' = C\widehat{A} - S\widehat{B}\sigma_3, \tag{B.265}$$

$$\widehat{B}' = C\widehat{B} + S\widehat{A}\sigma_3, \tag{B.266}$$

$$\chi_r' = \chi_r \begin{pmatrix} C & -S\sigma_3 \\ S\sigma_3 & C \end{pmatrix} = \chi_r e^{-\theta\gamma_3}. \tag{B.267}$$

Since we get the same relation for g and b colors we finally get

$$\chi'^c = \chi^c \exp(-\theta L_{0125}),$$
$$(\underline{D}'\Psi'^c)L_{012} + m_2\rho_2'\chi'^c = (\underline{D}\Psi^c)\exp(\theta L_{0125})L_{012} + m_2\rho_2'\chi'^c$$
$$= [(\underline{D}\Psi^c)L_{012} + m_2\rho_2\chi^c]\exp(-\theta L_{0125}) = 0. \tag{B.268}$$

The complete wave equation with mass term is then gauge invariant under the group generated by $\underline{P}_2$.

B.5.6   *Group generated by $\underline{P}_3$*

We have here

$$\underline{P}_3(\Psi^c) = \Psi^c L_{3012}, \tag{B.269}$$

$$\Psi'^c = [\exp(\theta\underline{P}_3)](\Psi^c) = \Psi^c \exp(\theta L_{3012}), \tag{B.270}$$

$$W_\mu'^3 = W_\mu^3 - \frac{2}{g_2}\partial_\mu\theta. \tag{B.271}$$

Since $\underline{P}_3(\Psi^c) = \Psi^c L_{3012}$ we get

$$\Psi'^c = [\exp(\theta\underline{P}_3)](\Psi^c) = \Psi^c \exp(\theta L_{3012}), \tag{B.272}$$

$$\Psi_c' = \Psi_c e^{\theta\gamma_{3012}}, \quad c = r, g, b. \tag{B.273}$$

This is equivalent to the system

$$\widehat{\phi}_{dc}' = e^{i\theta}\widehat{\phi}_{dc}, \tag{B.274}$$

$$\widehat{\phi}_{uc}' = e^{-i\theta}\widehat{\phi}_{uc}, \tag{B.275}$$

or to the system

$$\eta_{1dc}' = e^{i\theta}\eta_{1dc}; \quad \eta_{1dc}'^* = e^{-i\theta}\eta_{1dc}^*, \tag{B.276}$$

$$\eta_{2dc}' = e^{i\theta}\eta_{2dc}; \quad \eta_{2dc}'^* = e^{-i\theta}\eta_{2dc}^*, \tag{B.277}$$

$$\eta_{1uc}' = e^{-i\theta}\eta_{1uc}; \quad \eta_{1uc}'^* = e^{i\theta}\eta_{1uc}^*, \tag{B.278}$$

$$\eta_{2uc}' = e^{-i\theta}\eta_{2uc}; \quad \eta_{2uc}'^* = e^{i\theta}\eta_{2uc}^*. \tag{B.279}$$

We then get

$$s_1' = e^{2i\theta} s_1; \quad s_2' = e^{2i\theta} s_2; \quad s_3' = e^{2i\theta} s_3, \tag{B.280}$$

$$s_4' = e^{-2i\theta} s_4; \quad s_5' = e^{-2i\theta} s_5; \quad s_6' = e^{-2i\theta} s_6, \tag{B.281}$$

$$s_7' = s_7; \quad s_8' = s_8; \quad s_9' = s_9, \tag{B.282}$$

$$s_{10}' = s_{10}; \quad s_{11}' = s_{11}; \quad s_{12}' = s_{12}, \tag{B.283}$$

$$s_{13}' = s_{13}; \quad s_{14}' = s_{14}; \quad s_{15}' = s_{15}. \tag{B.284}$$

This implies

$$\rho' = \rho. \tag{B.285}$$

Next we have with Eq. (B.229)

$$\widehat{A}' = e^{-i\theta} \widehat{A} \; ; \quad A' = e^{i\theta} A, \tag{B.286}$$

$$\widehat{B}' = e^{i\theta} \widehat{B} \; ; \quad B' = e^{-i\theta} B, \tag{B.287}$$

$$\chi_r' = \chi_r \begin{pmatrix} e^{i\theta} & 0 \\ 0 & e^{-i\theta} \end{pmatrix} = \chi_r e^{\theta i}. \tag{B.288}$$

Since we have the same relation for g and b colors we finally get

$$\chi'^c = \chi^c \exp(-\theta L_{3012}),$$
$$(\underline{D}'\Psi'^c)L_{012} + m_2\rho_2'\chi'^c = (\underline{D}\Psi^c)\exp(\theta L_{3012})L_{012} + m_2\rho_2'\chi'^c$$
$$= [(\underline{D}\Psi^c)L_{012} + m_2\rho_2\chi^c]\exp(-\theta L_{3012}) = 0. \tag{B.289}$$

The complete wave equation with mass term is then gauge invariant under the group generated by $\underline{P}_3$.

B.5.7 Group generated by Γ_1

We now use the transformation

$$\Psi_r' = C\Psi_r + Si\Psi_g; \quad C = \cos(\theta); \quad S = \sin(\theta), \tag{B.290}$$

$$\Psi_g' = C\Psi_g + Si\Psi_r, \tag{B.291}$$

$$\Psi_b' = \Psi_b. \tag{B.292}$$

We may then forget here Ψ_b which does not change. The gauge invariance signifies that the system

$$\partial\Psi_r = -\frac{g_3}{2}\mathbf{G}^1 i\Psi_g + m_2\rho_2\chi_r\gamma_{012},$$

$$\partial\Psi_g = -\frac{g_3}{2}\mathbf{G}^1 i\Psi_r + m_2\rho_2\chi_g, \gamma_{012} \tag{B.293}$$

must be equivalent to the system

$$\partial\Psi'_r = -\frac{g_3}{2}\mathbf{G}'^1 i\Psi'_g + m_2\rho'_2\chi'_r\gamma_{012},$$

$$\partial\Psi'_g = -\frac{g_3}{2}\mathbf{G}'^1 i\Psi'_r + m_2\rho'_2\chi'_g\gamma_{012}. \tag{B.294}$$

By using Eq. (B.290) and Eq. (B.291) the Eq. (B.293) system is equivalent to the Eq. (B.294) system if and only if

$$\mathbf{G}'^1 = \mathbf{G}^1 - \frac{2}{g_3}\partial\theta. \tag{B.295}$$

We name f_1 the gauge transformation:

$$f_1 : \Psi^c \mapsto \underline{i}\Gamma_1(\Psi^c) = \begin{pmatrix} 0 & i\Psi_g \\ i\Psi_r & 0 \end{pmatrix}, \tag{B.296}$$

which implies with $C = \cos(\theta)$ and $S = \sin(\theta)$

$$[\exp(\theta f_1)](\Psi^c) = \begin{pmatrix} 0 & C\Psi_r + Si\Psi_g \\ C\Psi_g + Si\Psi_r & \Psi_b \end{pmatrix} = \begin{pmatrix} 0 & \Psi'_r \\ \Psi'_g & \Psi'_b \end{pmatrix}, \tag{B.297}$$

$$\Psi'_r = C\Psi_r + Si\Psi_g, \tag{B.298}$$

$$\Psi'_g = C\Psi_g + Si\Psi_r, \tag{B.299}$$

$$\Psi'_b = \Psi_b. \tag{B.300}$$

The equality Eq. (B.290) is equivalent to the system

$$\eta'^*_{1dr} = C\eta^*_{1dr} + iS\eta^*_{1dg}; \quad \eta'^*_{1ur} = C\eta^*_{1ur} + iS\eta^*_{1ug}, \tag{B.301}$$

$$\eta'^*_{2dr} = C\eta^*_{2dr} + iS\eta^*_{2dg}; \quad \eta'^*_{2ur} = C\eta^*_{2ur} + iS\eta^*_{2ug}. \tag{B.302}$$

The equality Eq. (B.291) is equivalent to the system

$$\eta'^*_{1dg} = C\eta^*_{1dg} + iS\eta^*_{1dr}; \quad \eta'^*_{1ug} = C\eta^*_{1ug} + iS\eta^*_{1ur}, \tag{B.303}$$

$$\eta'^*_{2dg} = C\eta^*_{2dg} + iS\eta^*_{2dr}; \quad \eta'^*_{2ug} = C\eta^*_{2ug} + iS\eta^*_{2ur}. \tag{B.304}$$

This gives for the scalars s_j

$$s'_1 = s_1; \quad s'_4 = s_4; \quad s'_9 = s_9, \tag{B.305}$$

$$s'_2 = Cs_2 - iSs_3; \quad s'_3 = Cs_3 - iSs_2, \tag{B.306}$$

$$s'_5 = Cs_5 - iSs_6; \quad s'_6 = Cs_6 - iSs_5, \tag{B.307}$$

$$s'_{11} = Cs_{11} + iSs_{13}; \quad s'_{13} = Cs_{13} + iSs_{11}, \tag{B.308}$$

$$s'_{12} = Cs_{12} + iSs_{15}; \quad s'_{15} = Cs_{15} + iSs_{12}, \tag{B.309}$$

$$s'_7 = C^2s_7 - S^2s_8 + iCSs_{10} + iCSs_{14}, \tag{B.310}$$

$$s'_8 = C^2s_8 - S^2s_7 + iCSs_{14} + iCSs_{10}, \tag{B.311}$$

$$s'_{10} = C^2s_{10} - S^2s_{14} + iCSs_7 + iCSs_8, \tag{B.312}$$

$$s'_{14} = C^2s_{14} - S^2s_{10} + iCSs_8 + iCSs_7. \tag{B.313}$$

We then have

$$s'_2 s'^*_2 + s'_3 s'^*_3 = s_2 s^*_2 + s_3 s^*_3, \qquad \text{(B.314)}$$

$$s'_5 s'^*_5 + s'_6 s'^*_6 = s_5 s^*_5 + s_6 s^*_6, \qquad \text{(B.315)}$$

$$s'_{11} s'^*_{11} + s'_{13} s'^*_{13} = s_{11} s^*_{11} + s_{13} s^*_{13}, \qquad \text{(B.316)}$$

$$s'_{12} s'^*_{12} + s'_{15} s'^*_{15} = s_{12} s^*_{12} + s_{15} s^*_{15}, \qquad \text{(B.317)}$$

$$s'_7 s'^*_7 + s'_8 s'^*_8 + s'_{10} s'^*_{10} + s'_{14} s'^*_{14} = s_7 s^*_7 + s_8 s^*_8 + s_{10} s^*_{10} + s_{14} s^*_{14}, \qquad \text{(B.318)}$$

$$\rho' = \rho. \qquad \text{(B.319)}$$

Next we let

$$\chi_r = \begin{pmatrix} A_r & B_r \\ \widehat{B}_r & \widehat{A}_r \end{pmatrix}; \; \chi'_r = \begin{pmatrix} A'_r & B'_r \\ \widehat{B}'_r & \widehat{A}'_r \end{pmatrix}, \qquad \text{(B.320)}$$

$$\chi_g = \begin{pmatrix} A_g & B_g \\ \widehat{B}_g & \widehat{A}_g \end{pmatrix}; \; \chi'_g = \begin{pmatrix} A'_g & B'_g \\ \widehat{B}'_g & \widehat{A}'_g \end{pmatrix}, \qquad \text{(B.321)}$$

and we get

$$A'_r = CA_r - iSA_g; \; B'_r = CB_r - iSB_g, \qquad \text{(B.322)}$$

$$A'_g = CA_g - iSA_r; \; B'_g = CB_g - iSB_r. \qquad \text{(B.323)}$$

This gives the awaited result:

$$\rho' = \rho, \qquad \text{(B.324)}$$

$$\chi'_r = C\chi_r - Si\chi_g, \qquad \text{(B.325)}$$

$$\chi'_g = C\chi_g - Si\chi_r. \qquad \text{(B.326)}$$

The change of sign between Eq. (B.290) and Eq. (B.325) comes from the anticommutation between $\mathbf{i}$ and ∂.

B.5.8 *Group generated by* Γ_k , $k > 1$

We use with $k = 2$ the gauge transformation

$$\Psi'_r = C\Psi_r + S\Psi_g; \; C = \cos(\theta); \; S = \sin(\theta), \qquad \text{(B.327)}$$

$$\Psi'_g = C\Psi_g - S\Psi_r, \qquad \text{(B.328)}$$

$$\Psi'_b = \Psi_b. \qquad \text{(B.329)}$$

The gauge invariance means that the system

$$\partial\Psi_r = -\frac{g_3}{2}\mathbf{G}^2\Psi_g + m_2\rho_2\chi_r\gamma_{012},$$

$$\partial\Psi_g = \frac{g_3}{2}\mathbf{G}^2\Psi_r + m_2\rho_2\chi_g\gamma_{012}, \qquad \text{(B.330)}$$

must be equivalent to the system

$$\partial \Psi_r' = -\frac{g_3}{2}\mathbf{G}'^2\Psi_g' + m_2\rho_2'\chi_r'\gamma_{012},$$

$$\partial \Psi_g' = \frac{g_3}{2}\mathbf{G}'^2\Psi_r' + m_2\rho_2'\chi_g'\gamma_{012}. \tag{B.331}$$

By using the relations Eq. (B.327) and Eq. (B.328) the Eq. (B.330) system is equivalent to Eq. (B.331) if and only if

$$\mathbf{G}'^2 = \mathbf{G}^2 - \frac{2}{g_3}\partial\theta, \tag{B.332}$$

because we have

$$\rho' = \rho, \tag{B.333}$$

$$\chi_r' = C\chi_r + S\chi_g, \tag{B.334}$$

$$\chi_g' = C\chi_g - S\chi_r. \tag{B.335}$$

The $k = 3$ case will be detailed in Sec. B.5.9 and the $k = 8$ case will be detailed in Sec. B.5.10. The $k = 4$ and $k = 6$ cases are similar to $k = 1$ and the $k = 5$ and $k = 7$ cases are similar to $k = 2$ by permuting color indices.

B.5.9   *Group generated by* Γ_3

We name f_3 the gauge transformation:

$$f_3 : \Psi^c \mapsto \mathbf{i}\Gamma_3(\Psi^c) = \begin{pmatrix} 0 & \mathbf{i}\Psi_r \\ -\mathbf{i}\Psi_g & 0 \end{pmatrix}, \tag{B.336}$$

which implies

$$[\exp(\theta f_3)](\Psi^c) = \begin{pmatrix} 0 & e^{\theta\mathbf{i}}\Psi_r \\ e^{-\theta\mathbf{i}}\Psi_g & \Psi_b \end{pmatrix} = \begin{pmatrix} 0 & \Psi_r' \\ \Psi_g' & \Psi_b' \end{pmatrix}, \tag{B.337}$$

$$\Psi_r' = e^{\theta\mathbf{i}}\Psi_r, \tag{B.338}$$

$$\Psi_g' = e^{-\theta\mathbf{i}}\Psi_g, \tag{B.339}$$

$$\Psi_b' = \Psi_b. \tag{B.340}$$

The equality Eq. (B.338) is equivalent to

$$\begin{pmatrix} \phi_{dr}' & \phi_{ur}' \\ \widehat{\phi}_{ur}' & \widehat{\phi}_{dr}' \end{pmatrix} = \begin{pmatrix} e^{i\theta} & 0 \\ 0 & e^{-i\theta} \end{pmatrix} \begin{pmatrix} \phi_{dr} & \phi_{ur} \\ \widehat{\phi}_{ur} & \widehat{\phi}_{dr} \end{pmatrix}. \tag{B.341}$$

The equality Eq. (B.339) is equivalent to

$$\begin{pmatrix} \phi_{dg}' & \phi_{ug}' \\ \widehat{\phi}_{ug}' & \widehat{\phi}_{dg}' \end{pmatrix} = \begin{pmatrix} e^{-i\theta} & 0 \\ 0 & e^{i\theta} \end{pmatrix} \begin{pmatrix} \phi_{dg} & \phi_{ug} \\ \widehat{\phi}_{ug} & \widehat{\phi}_{dg} \end{pmatrix}. \tag{B.342}$$

We get

$$\eta'^*_{1dr} = e^{-i\theta}\eta^*_{1dr}; \; \eta'^*_{1ur} = e^{-i\theta}\eta^*_{1ur}, \tag{B.343}$$

$$\eta'^*_{2dr} = e^{-i\theta}\eta^*_{2dr}; \; \eta'^*_{2ur} = e^{-i\theta}\eta^*_{2ur}, \tag{B.344}$$

$$\eta'^*_{1dg} = e^{i\theta}\eta^*_{1dg}; \; \eta'^*_{1ug} = e^{i\theta}\eta^*_{1ug}, \tag{B.345}$$

$$\eta'^*_{2dg} = e^{i\theta}\eta^*_{2dg}; \; \eta'^*_{2ug} = e^{i\theta}\eta^*_{2ug}. \tag{B.346}$$

This gives

$$s'_1 = s_1 \; ; \; s'_2 = e^{-i\theta}s_2 \; ; \; s'_3 = e^{i\theta}s_3, \tag{B.347}$$

$$s'_4 = s_4 \; ; \; s'_5 = e^{-i\theta}s_5 \; ; \; s'_6 = e^{i\theta}s_6, \tag{B.348}$$

$$s'_9 = s_9 \; ; \; s'_8 = e^{-2i\theta}s_8 \; ; \; s'_7 = e^{2i\theta}s_7, \tag{B.349}$$

$$s'_{10} = s_{10} \; ; \; s'_{11} = e^{-i\theta}s_{11} \; ; \; s'_{12} = e^{i\theta}s_{12}, \tag{B.350}$$

$$s'_{14} = s_{14} \; ; \; s'_{15} = e^{-i\theta}s_{15} \; ; \; s'_{13} = e^{i\theta}s_{13}, \tag{B.351}$$

so we deduce

$$s'_j s'^*_j = s_j s^*_j, \; j = 1, 2, \ldots, 15, \tag{B.352}$$

$$\rho' = \rho, \tag{B.353}$$

$$\chi'_r = e^{-i\theta}\chi_r, \tag{B.354}$$

$$\chi'_g = e^{i\theta}\chi_g, \tag{B.355}$$

These relations are the awaited ones because

$$\boldsymbol{\partial}\Psi'_r = \boldsymbol{\partial}(e^{i\theta}\Psi_r) = e^{-i\theta}(-i\boldsymbol{\partial}\theta\Psi_r + \boldsymbol{\partial}\Psi_r), \tag{B.356}$$

$$\boldsymbol{\partial}\Psi'_g = \boldsymbol{\partial}(e^{-i\theta}\Psi_g) = e^{i\theta}(i\boldsymbol{\partial}\theta\Psi_g + \boldsymbol{\partial}\Psi_g), \tag{B.357}$$

$$\mathbf{G}'^3 = \mathbf{G}^3 - \frac{2}{g_3}\boldsymbol{\partial}\theta. \tag{B.358}$$

B.5.10 *Group generated by* Γ_8

We name f_8 the gauge transformation

$$f_8 : \Psi^c \mapsto \underline{i}\Gamma_8(\Psi^c) = \begin{pmatrix} 0 & \frac{i}{\sqrt{3}}\Psi_r \\ \frac{i}{\sqrt{3}}\Psi_g & -\frac{2i}{\sqrt{3}}\Psi_b \end{pmatrix}, \tag{B.359}$$

which implies

$$[\exp(\theta f_8)](\Psi^c) = \begin{pmatrix} 0 & e^{\frac{\theta i}{\sqrt{3}}}\Psi_r \\ e^{\frac{\theta i}{\sqrt{3}}}\Psi_g & e^{-\frac{2\theta i}{\sqrt{3}}}\Psi_b \end{pmatrix} = \begin{pmatrix} 0 & \Psi'_r \\ \Psi'_g & \Psi'_b \end{pmatrix}, \tag{B.360}$$

$$\Psi'_r = \exp(\frac{\theta i}{\sqrt{3}})\Psi_r, \tag{B.361}$$

$$\Psi'_g = \exp(\frac{\theta i}{\sqrt{3}})\Psi_g, \tag{B.362}$$

$$\Psi'_b = \exp(-\frac{2\theta i}{\sqrt{3}})\Psi_b. \tag{B.363}$$

This gives

$$\phi'_{dr} = \exp(\frac{i\theta}{\sqrt{3}})\phi_{dr}; \ \phi'_{ur} = \exp(\frac{i\theta}{\sqrt{3}})\phi_{ur}, \qquad (\text{B.364})$$

$$\phi'_{dg} = \exp(\frac{i\theta}{\sqrt{3}})\phi_{dg}; \ \phi'_{ug} = \exp(\frac{i\theta}{\sqrt{3}})\phi_{ug}, \qquad (\text{B.365})$$

$$\phi'_{db} = \exp(-\frac{2i\theta}{\sqrt{3}})\phi_{db}; \ \phi'_{ub} = \exp(-\frac{2i\theta}{\sqrt{3}})\phi_{ub}. \qquad (\text{B.366})$$

We then get

$$\eta'^*_{1dr} = \exp(\frac{i\theta}{\sqrt{3}})\eta^*_{1dr}; \ \eta'^*_{1dg} = \exp(\frac{i\theta}{\sqrt{3}})\eta^*_{1dg}; \ \eta'^*_{1db} = \exp(-\frac{2i\theta}{\sqrt{3}})\eta^*_{1dg},$$
$$(\text{B.367})$$

$$\eta'^*_{2dr} = \exp(\frac{i\theta}{\sqrt{3}})\eta^*_{2dr}; \ \eta'^*_{2dg} = \exp(\frac{i\theta}{\sqrt{3}})\eta^*_{2dg}; \ \eta'^*_{2db} = \exp(-\frac{2i\theta}{\sqrt{3}})\eta^*_{2dg},$$
$$(\text{B.368})$$

$$\eta'^*_{1ur} = \exp(\frac{i\theta}{\sqrt{3}})\eta^*_{1ur}; \ \eta'^*_{1ug} = \exp(\frac{i\theta}{\sqrt{3}})\eta^*_{1ug}; \ \eta'^*_{1ub} = \exp(-\frac{2i\theta}{\sqrt{3}})\eta^*_{1ug},$$
$$(\text{B.369})$$

$$\eta'^*_{2ur} = \exp(\frac{i\theta}{\sqrt{3}})\eta^*_{2ur}; \ \eta'^*_{2ug} = \exp(\frac{i\theta}{\sqrt{3}})\eta^*_{2ug}; \ \eta'^*_{2ub} = \exp(-\frac{2i\theta}{\sqrt{3}})\eta^*_{2ug}$$
$$(\text{B.370})$$

This implies

$$s'_1 = \exp(\frac{2i\theta}{\sqrt{3}})s_1; \ s'_2 = \exp(-\frac{i\theta}{\sqrt{3}})s_2; \ s'_3 = \exp(-\frac{i\theta}{\sqrt{3}})s_3, \qquad (\text{B.371})$$

$$s'_4 = \exp(\frac{2i\theta}{\sqrt{3}})s_4; \ s'_5 = \exp(-\frac{i\theta}{\sqrt{3}})s_5; \ s'_6 = \exp(-\frac{i\theta}{\sqrt{3}})s_6, \qquad (\text{B.372})$$

$$s'_7 = \exp(\frac{2i\theta}{\sqrt{3}})s_7; \ s'_8 = \exp(\frac{2i\theta}{\sqrt{3}})s_8; \ s'_9 = \exp(-\frac{4i\theta}{\sqrt{3}})s_9, \qquad (\text{B.373})$$

$$s'_{10} = \exp(\frac{2i\theta}{\sqrt{3}})s_{10}; \ s'_{11} = \exp(-\frac{i\theta}{\sqrt{3}})s_{11}; \ s'_{12} = \exp(-\frac{i\theta}{\sqrt{3}})s_{12}, \quad (\text{B.374})$$

$$s'_{13} = \exp(-\frac{i\theta}{\sqrt{3}})s_{13}; \ s'_{14} = \exp(\frac{2i\theta}{\sqrt{3}})s_{14}; \ s'_{15} = \exp(-\frac{i\theta}{\sqrt{3}})s_{15}. \quad (\text{B.375})$$

We then get the awaited results:

$$s'_j s'^*_j = s_j s^*_j, \ j = 1, 2, \ldots, 15 \ ; \ \rho' = \rho, \qquad (\text{B.376})$$

$$\chi'_r = \exp(-\frac{i\theta}{\sqrt{3}})\chi_r; \ \chi'_g = \exp(-\frac{i\theta}{\sqrt{3}})\chi_g; \ \chi'_b = \exp(\frac{2i\theta}{\sqrt{3}})\chi_b. \qquad (\text{B.377})$$

We deduce:

$$\mathbf{D}'\Psi' = \gamma^\mu D'_\mu \Psi' = \gamma^\mu e^{a^0 \frac{b}{2}\mathbf{i}}(D_\mu\Psi)e^{a^0(a+\frac{b}{2})\gamma_{21}}$$

$$= e^{-a^0 \frac{b}{2}\mathbf{i}}(\mathbf{D}\Psi)e^{a^0(a+\frac{b}{2})\gamma_{21}}, \tag{B.378}$$

because $\mathbf{i}$ anti-commutes with each γ^μ. Next we have

$$\widetilde{\Psi}' = e^{a^0(a+\frac{b}{2})\widetilde{\gamma}_{21}}\widetilde{\Psi}e^{a^0 \frac{b}{2}\widetilde{\mathbf{i}}}$$

$$= e^{-a^0(a+\frac{b}{2})\gamma_{21}}\widetilde{\Psi}e^{a^0 \frac{b}{2}\mathbf{i}}, \tag{B.379}$$

$$\widetilde{\Psi}'\mathbf{D}'\Psi' = e^{-a^0(a+\frac{b}{2})\gamma_{21}}\widetilde{\Psi}e^{a^0 \frac{b}{2}\mathbf{i}}e^{-a^0 \frac{b}{2}\mathbf{i}}(\mathbf{D}\Psi)e^{a^0(a+\frac{b}{2})\gamma_{21}}$$

$$= e^{-a^0(a+\frac{b}{2})\gamma_{21}}\widetilde{\Psi}(\mathbf{D}\Psi)e^{a^0(a+\frac{b}{2})\gamma_{21}}. \tag{B.380}$$

Appendix C

The hydrogen atom

We study the resolution of the homogeneous nonlinear equation for the hydrogen atom. Our resolution uses a method separating the variables in spherical coordinates. The solutions are very near particular solutions of the Dirac equation which are not the usual ones, and which have a Yvon–Takabayasi angle that is everywhere defined and small.

The hydrogen atom is the jewel of the Dirac theory. The solutions calculated by C. G. Darwin [6], which we may also find in newer reports [55], are proper values of an ad hoc operator, coming from the non-relativistic theory, that is not the total angular momentum operator. These solutions give the expected number of states, the true formula for the energy levels, and have the expected non-relativistic approximations. This was considered very satisfying. Most of Darwin's solutions suffer the disadvantage that they have a Yvon–Takabayasi angle that is not everywhere defined and small. Therefore they cannot be linear approximations of the solutions to our homogeneous nonlinear equation.

We got previously [9] other solutions in the linear case, which have a Yvon–Takabayasi angle everywhere defined and small, and so those may be the linear approximations of the solutions to our nonlinear equation.

C.1   Separating variables

To solve the Dirac equation Eq. (2.21) or the homogeneous nonlinear equation Eq. (3.1), in the case of the hydrogen atom, two methods exist. We shall use here, not the initial method based on the non-relativistic wave equations, but the new method invented more recently by H. Krüger [44], a classic method from the mathematical point of view for an equation with partial derivatives, separating the variables in spherical coordinates:

$$x^1 = r\sin\theta\cos\varphi \ ; \ \ x^2 = r\sin\theta\sin\varphi \ ; \ \ x^3 = r\cos\theta. \qquad (C.1)$$

We use [7] the following notations:

$$i_1 = \sigma_{23} = i\sigma_1 \ ; \ \ i_2 = \sigma_{31} = i\sigma_2 \ ; \ \ i_3 = \sigma_{12} = i\sigma_3, \qquad (C.2)$$

$$S = e^{-\frac{\varphi}{2}i_3}e^{-\frac{\theta}{2}i_2} \ ; \ \ \Omega = r^{-1}(\sin\theta)^{-\frac{1}{2}}S, \qquad (C.3)$$

$$\vec{\partial}' = \sigma_3\partial_r + \frac{1}{r}\sigma_1\partial_\theta + \frac{1}{r\sin\theta}\sigma_2\partial_\varphi. \qquad (C.4)$$

H. Krüger got the remarkable identity:

$$\vec{\partial} = \Omega\vec{\partial}'\Omega^{-1} \qquad (C.5)$$

that, with:

$$\nabla' = \partial_0 - \vec{\partial}' = \partial_0 - (\sigma_3\partial_r + \frac{1}{r}\sigma_1\partial_\theta + \frac{1}{r\sin\theta}\sigma_2\partial_\varphi), \qquad (C.6)$$

also gives

$$\Omega^{-1}\nabla = \nabla'\Omega^{-1}. \qquad (C.7)$$

In the wave equations Eq. (2.21) or Eq. (3.1), to separate the temporal variable $x^0 = ct$ and the angular variable φ from the radial variable r and the angular variable θ, we let:

$$\phi = \Omega X e^{(\lambda\varphi - Ex^0 + \delta)i_3}, \qquad (C.8)$$

where X is a function, with value into the Pauli algebra, of r and θ alone, $\hbar cE$ is the energy of the electron, and δ is an arbitrary phase which plays no role as the equations Eq. (2.21) and Eq. (3.1) are electric gauge invariant. λ is a real constant. We get then

$$\Omega^{-1}\phi = X e^{(\lambda\varphi - Ex^0 + \delta)i_3}, \qquad (C.9)$$

$$\Omega^{-1}\widehat{\phi} = \widehat{X} e^{(\lambda\varphi - Ex^0 + \delta)i_3}. \qquad (C.10)$$

We also have:

$$\rho e^{i\beta} = \det(\phi) = \det(\Omega)\det(X)\det[e^{(\lambda\varphi - Ex^0 + \delta)i_3}]$$

$$\det(\Omega) = r^{-2}(\sin\theta)^{-1} \ ; \ \ \det[e^{(\lambda\varphi - Ex^0 + \delta)i_3}] = 1$$

$$\rho e^{i\beta} = \frac{\det(X)}{r^2\sin\theta}. \qquad (C.11)$$

---

7. S has nothing to do with the tensor S_3 and Ω must not be confused with the relativistic invariants Ω_1 and Ω_2 studied in chapter 2.

So, if we let:

$$\rho_X e^{i\beta_X} = \det(X), \tag{C.12}$$

we get:

$$\rho = \frac{\rho_X}{r^2 \sin\theta} \ ; \quad \beta = \beta_X. \tag{C.13}$$

Thus with Eq. (C.8) for the wave, the Yvon–Takabayasi angle does not depend on time or on the φ angle, only on r and θ. It is why the separation of variables, in the linear case or in the nonlinear case, may begin in the same way. We have

$$\nabla'\Omega^{-1}\widehat{\phi} = (\partial_0 - \sigma_3\partial_r - \frac{1}{r}\sigma_1\partial_\theta - \frac{1}{r\sin\theta}\sigma_2\partial_\varphi)[\widehat{X}e^{(\lambda\varphi - Ex^0 + \delta)i_3}], \tag{C.14}$$

$$\partial_0(\widehat{X}e^{(\lambda\varphi - Ex^0 + \delta)i_3}) = -E\widehat{X}i_3 e^{(\lambda\varphi - Ex^0 + \delta)i_3}, \tag{C.15}$$

$$\partial_r(\widehat{X}e^{(\lambda\varphi - Ex^0 + \delta)i_3}) = (\partial_r\widehat{X})e^{(\lambda\varphi - Ex^0 + \delta)i_3}, \tag{C.16}$$

$$\partial_\theta(\widehat{X}e^{(\lambda\varphi - Ex^0 + \delta)i_3}) = (\partial_\theta\widehat{X})e^{(\lambda\varphi - Ex^0 + \delta)i_3}, \tag{C.17}$$

$$\partial_\varphi(\widehat{X}e^{(\lambda\varphi - Ex^0 + \delta)i_3}) = \lambda\widehat{X}i_3 e^{(\lambda\varphi - Ex^0 + \delta)i_3}. \tag{C.18}$$

We get then:

$$\nabla\widehat{\phi} = \Omega(-E\widehat{X}i_3 - \sigma_3\partial_r\widehat{X} - \frac{1}{r}\sigma_1\partial_\theta\widehat{X} - \frac{\lambda}{r\sin\theta}\sigma_2\widehat{X}i_3)e^{(\lambda\varphi - Ex^0 + \delta)i_3}. \tag{C.19}$$

For the hydrogen atom, we have:

$$qA = qA^0 = -\frac{\alpha}{r} \ ; \quad \alpha = \frac{e^2}{\hbar c}, \tag{C.20}$$

where α is the fine structure constant. We have:

$$qA\widehat{\phi}\sigma_{12} = -\frac{\alpha}{r}\widehat{\phi}i_3 = -\frac{\alpha}{r}\Omega\widehat{X}e^{(\lambda\varphi - Ex^0 + \delta)i_3}i_3$$
$$= \Omega(-\frac{\alpha}{r}\widehat{X}i_3)e^{(\lambda\varphi - Ex^0 + \delta)i_3}. \tag{C.21}$$

So the homogeneous nonlinear equation Eq. (3.1) becomes

$$-E\widehat{X}i_3 - \sigma_3\partial_r\widehat{X} - \frac{1}{r}\sigma_1\partial_\theta\widehat{X} - \frac{\lambda}{r\sin\theta}\sigma_2\widehat{X}i_3 - \frac{\alpha}{r}\widehat{X}i_3 + me^{-i\beta}Xi_3 = 0, \tag{C.22}$$

that is to say:

$$(E + \frac{\alpha}{r})\widehat{X}i_3 + \sigma_3\partial_r\widehat{X} + \frac{1}{r}\sigma_1\partial_\theta\widehat{X} + \frac{\lambda}{r\sin\theta}\sigma_2\widehat{X}i_3 = me^{-i\beta}Xi_3, \tag{C.23}$$

while the Dirac equation gives:

$$(E + \frac{\alpha}{r})\widehat{X}i_3 + \sigma_3\partial_r\widehat{X} + \frac{1}{r}\sigma_1\partial_\theta\widehat{X} + \frac{\lambda}{r\sin\theta}\sigma_2\widehat{X}i_3 = mXi_3. \tag{C.24}$$

We let now:

$$X = \begin{pmatrix} \mathbf{a} & -\mathbf{b}^* \\ \mathbf{c} & \mathbf{d}^* \end{pmatrix}, \tag{C.25}$$

where $\mathbf{a}$, $\mathbf{b}$, $\mathbf{c}$, $\mathbf{d}$ are functions with complex value of the real variables r and θ. We get then:

$$\widehat{X} = \begin{pmatrix} \mathbf{d} & -\mathbf{c}^* \\ \mathbf{b} & \mathbf{a}^* \end{pmatrix}. \tag{C.26}$$

We get then:

$$e^{-i\beta} X i_3 = i e^{-i\beta} X \sigma_3 = i e^{-i\beta} \begin{pmatrix} \mathbf{a} & \mathbf{b}^* \\ \mathbf{c} & -\mathbf{d}^* \end{pmatrix}, \tag{C.27}$$

$$\widehat{X} i_3 = \begin{pmatrix} \mathbf{d} & -\mathbf{c}^* \\ \mathbf{b} & \mathbf{a}^* \end{pmatrix} \begin{pmatrix} i & 0 \\ 0 & -i \end{pmatrix} = \begin{pmatrix} i\mathbf{d} & i\mathbf{c}^* \\ i\mathbf{b} & -i\mathbf{a}^* \end{pmatrix}, \tag{C.28}$$

$$\sigma_3 \partial_r \widehat{X} = \begin{pmatrix} 1 & 0 \\ 0 & -1 \end{pmatrix} \begin{pmatrix} \partial_r \mathbf{d} & -\partial_r \mathbf{c}^* \\ \partial_r \mathbf{b} & \partial_r \mathbf{a}^* \end{pmatrix} = \begin{pmatrix} \partial_r \mathbf{d} & -\partial_r \mathbf{c}^* \\ -\partial_r \mathbf{b} & -\partial_r \mathbf{a}^* \end{pmatrix}, \tag{C.29}$$

$$\sigma_1 \partial_\theta \widehat{X} = \begin{pmatrix} 0 & 1 \\ 1 & 0 \end{pmatrix} \begin{pmatrix} \partial_\theta \mathbf{d} & -\partial_\theta \mathbf{c}^* \\ \partial_\theta \mathbf{b} & \partial_\theta \mathbf{a}^* \end{pmatrix} = \begin{pmatrix} \partial_\theta \mathbf{b} & \partial_\theta \mathbf{a}^* \\ \partial_\theta \mathbf{d} & -\partial_\theta \mathbf{c}^* \end{pmatrix}, \tag{C.30}$$

$$\sigma_2 \widehat{X} i_3 = i_2 \widehat{X} \sigma_3 = \begin{pmatrix} 0 & 1 \\ -1 & 0 \end{pmatrix} \begin{pmatrix} \mathbf{d} & -\mathbf{c}^* \\ \mathbf{b} & \mathbf{a}^* \end{pmatrix} \begin{pmatrix} 1 & 0 \\ 0 & -1 \end{pmatrix} = \begin{pmatrix} \mathbf{b} & -\mathbf{a}^* \\ -\mathbf{d} & -\mathbf{c}^* \end{pmatrix}. \tag{C.31}$$

Consequently the nonlinear equation Eq. (3.1) becomes:

$$(E + \frac{\alpha}{r}) \begin{pmatrix} i\mathbf{d} & i\mathbf{c}^* \\ i\mathbf{b} - i\mathbf{a}^* \end{pmatrix} + \begin{pmatrix} \partial_r \mathbf{d} & -\partial_r \mathbf{c}^* \\ -\partial_r \mathbf{b} & -\partial_r \mathbf{a}^* \end{pmatrix} \tag{C.32}$$

$$+ \frac{1}{r} \begin{pmatrix} \partial_\theta \mathbf{b} & \partial_\theta \mathbf{a}^* \\ \partial_\theta \mathbf{d} & -\partial_\theta \mathbf{c}^* \end{pmatrix} + \frac{\lambda}{r \sin\theta} \begin{pmatrix} \mathbf{b} & -\mathbf{a}^* \\ -\mathbf{d} & -\mathbf{c}^* \end{pmatrix} = ime^{-i\beta} \begin{pmatrix} \mathbf{a} & \mathbf{b}^* \\ \mathbf{c} & -\mathbf{d}^* \end{pmatrix}.$$

Conjugating equations with *, we get the system:

$$i(E + \frac{\alpha}{r})\mathbf{d} + \partial_r \mathbf{d} + \frac{1}{r}(\partial_\theta + \frac{\lambda}{\sin\theta})\mathbf{b} = ime^{-i\beta}\mathbf{a},$$

$$-i(E + \frac{\alpha}{r})\mathbf{c} - \partial_r \mathbf{c} + \frac{1}{r}(\partial_\theta - \frac{\lambda}{\sin\theta})\mathbf{a} = -ime^{i\beta}\mathbf{b}, \tag{C.33}$$

$$i(E + \frac{\alpha}{r})\mathbf{b} - \partial_r \mathbf{b} + \frac{1}{r}(\partial_\theta - \frac{\lambda}{\sin\theta})\mathbf{d} = ime^{-i\beta}\mathbf{c},$$

$$-i(E + \frac{\alpha}{r})\mathbf{a} + \partial_r \mathbf{a} + \frac{1}{r}(\partial_\theta + \frac{\lambda}{\sin\theta})\mathbf{c} = -ime^{i\beta}\mathbf{d}.$$

In addition we have:

$$\rho e^{i\beta} = \det(\phi) = \frac{\det(X)}{r^2 \sin\theta} = \frac{\mathbf{ad}^* + \mathbf{cb}^*}{r^2 \sin\theta},$$ (C.34)

so we get:

$$e^{i\beta} = \frac{\mathbf{ad}^* + \mathbf{cb}^*}{|\mathbf{ad}^* + \mathbf{cb}^*|}.$$ (C.35)

For the four Eq. (C.33) there are only two angular operators, so we let:

$$\mathbf{a} = AU \; ; \quad \mathbf{b} = BV \; ; \quad \mathbf{c} = CV \; ; \quad \mathbf{d} = DU,$$ (C.36)

where A, B, C and D are functions of r whilst U and V are functions of θ. The system Eq. (C.33) becomes:

$$i(E + \frac{\alpha}{r})DU + D'U + \frac{1}{r}(V' + \frac{\lambda}{\sin\theta}V)B = ime^{-i\beta}AU,$$

$$-i(E + \frac{\alpha}{r})CV - C'V + \frac{1}{r}(U' - \frac{\lambda}{\sin\theta}U)A = -ime^{i\beta}BV,$$ (C.37)

$$i(E + \frac{\alpha}{r})BV - B'V + \frac{1}{r}(U' - \frac{\lambda}{\sin\theta}U)D = ime^{-i\beta}CV,$$

$$-i(E + \frac{\alpha}{r})AU + A'U + \frac{1}{r}(V' + \frac{\lambda}{\sin\theta}V)C = -ime^{i\beta}DU.$$

So if a κ constant exists such as:

$$U' - \frac{\lambda}{\sin\theta}U = -\kappa V \; ; \quad V' + \frac{\lambda}{\sin\theta}V = \kappa U,$$ (C.38)

the system Eq. (C.37) becomes:

$$i(E + \frac{\alpha}{r})D + D' + \frac{\kappa}{r}B = ime^{-i\beta}A,$$

$$-i(E + \frac{\alpha}{r})C - C' - \frac{\kappa}{r}A = -ime^{i\beta}B,$$ (C.39)

$$i(E + \frac{\alpha}{r})B - B' - \frac{\kappa}{r}D = ime^{-i\beta}C,$$

$$-i(E + \frac{\alpha}{r})A + A' + \frac{\kappa}{r}C = -ime^{i\beta}D.$$

To get the system equivalent to the Dirac equation, from the same process, it is enough to replace β by 0, this does not change the angular system Eq. (C.38), while in the place of Eq. (C.39) we get the system:

$$i(E + \frac{\alpha}{r})D + D' + \frac{\kappa}{r}B = imA,$$

$$-i(E + \frac{\alpha}{r})C - C' - \frac{\kappa}{r}A = -imB,$$ (C.40)

$$i(E + \frac{\alpha}{r})B - B' - \frac{\kappa}{r}D = imC,$$

$$-i(E + \frac{\alpha}{r})A + A' + \frac{\kappa}{r}C = -imD.$$

C.2   Angular momentum operators

We established in [9] the form that, in space-time algebra, the angular momentum operators take. With the Pauli algebra, we have (a detailed calculation is in [15] A.3):

$$J_1\phi = (d_1 + \frac{1}{2}\sigma_{23})\phi\sigma_{21} \; ; \quad d_1 = x^2\partial_3 - x^3\partial_2 = -\sin\varphi\ \partial_\theta - \frac{\cos\varphi}{\tan\theta}\partial_\varphi,$$
$$(C.41)$$

$$J_2\phi = (d_2 + \frac{1}{2}\sigma_{31})\phi\sigma_{21} \; ; \quad d_2 = x^3\partial_1 - x^1\partial_3 = \cos\varphi\ \partial_\theta - \frac{\sin\varphi}{\tan\theta}\partial_\varphi,$$
$$(C.42)$$

$$J_3\phi = (d_3 + \frac{1}{2}\sigma_{12})\phi\sigma_{21} \; ; \quad d_3 = x^1\partial_2 - x^2\partial_1 = \partial_\varphi.$$
$$(C.43)$$

Of course we also have

$$J^2 = J_1^2 + J_2^2 + J_3^2.$$
$$(C.44)$$

We get then

$$J_3\phi = \lambda\phi \iff \phi = \phi(x^0,\ r,\ \theta)e^{\lambda\varphi i_3}.$$
$$(C.45)$$

So the wave ϕ satisfying Eq. (C.8) is a proper vector of J_3 and λ is the magnetic quantum number. Moreover, always for a ϕ wave satisfying Eq. (C.8), we have:

$$J^2\phi = j(j+1)\phi,$$
$$(C.46)$$

if and only if

$$\partial_{\theta\theta}^2 X + [(j+\frac{1}{2})^2 - \frac{\lambda^2}{\sin^2\theta}]X - \lambda\frac{\cos\theta}{\sin^2\theta}\sigma_{12}X\sigma_{12} = 0.$$
$$(C.47)$$

But Eq. (C.38) implies at the second order

$$0 = U'' + (\kappa^2 - \frac{\lambda^2}{\sin^2\theta})U + \lambda\frac{\cos\theta}{\sin^2\theta}U,$$
$$(C.48)$$

$$0 = V'' + (\kappa^2 - \frac{\lambda^2}{\sin^2\theta})V - \lambda\frac{\cos\theta}{\sin^2\theta}V,$$
$$(C.49)$$

$$0 = \partial_{\theta\theta}^2 X + (\kappa^2 - \frac{\lambda^2}{\sin^2\theta})X - \lambda\frac{\cos\theta}{\sin^2\theta}\sigma_{12}X\sigma_{12}.$$
$$(C.50)$$

Consequently ϕ is a proper vector of J^2, with the proper value $j(j+1)$, if and only if

$$\kappa^2 = (j+\frac{1}{2})^2 \; ; \quad |\kappa| = j+\frac{1}{2} \; ; \quad j = |\kappa| - \frac{1}{2}.$$
$$(C.51)$$

With Eq. (C.3) and Eq. (C.8) we can see that the change of φ into $\varphi + 2\pi$ conserves the value of the wave if and only if λ has a half-odd value. General results on angular momentum operators imply then:

$$j = \frac{1}{2}, \frac{3}{2}, \cdots; \ \kappa = \pm 1, \pm 2, \cdots; \ \lambda = -j, -j+1, \cdots j-1, j. \quad \text{(C.52)}$$

To solve the angular system, if $\lambda > 0$ we let, with $C = C(\theta)$:

$$U = \sin^\lambda \theta [\sin(\frac{\theta}{2})C' - (\kappa + \frac{1}{2} - \lambda) \cos(\frac{\theta}{2})C],$$
$$V = \sin^\lambda \theta [\cos(\frac{\theta}{2})C' + (\kappa + \frac{1}{2} - \lambda) \sin(\frac{\theta}{2})C]. \quad \text{(C.53)}$$

If $\lambda < 0$ we let:

$$U = \sin^{-\lambda} \theta [\cos(\frac{\theta}{2})C' + (\kappa + \frac{1}{2} + \lambda) \sin(\frac{\theta}{2})C],$$
$$V = \sin^{-\lambda} \theta [-\sin(\frac{\theta}{2})C' + (\kappa + \frac{1}{2} + \lambda) \cos(\frac{\theta}{2})C]. \quad \text{(C.54)}$$

The angular system Eq. (C.38) is then equivalent [7] to the differential equation :

$$0 = C'' + \frac{2|\lambda|}{\tan \theta}C' + [(\kappa + \frac{1}{2})^2 - \lambda^2]C. \quad \text{(C.55)}$$

The change of variable:

$$z = \cos\theta \ ; \ f(z) = C[\theta(z)], \quad \text{(C.56)}$$

gives then the differential equation of the Gegenbauer's [8] polynomials:

$$0 = f''(z) - \frac{1 + 2|\lambda|}{1 - z^2} z f'(z) + \frac{(\kappa + \frac{1}{2})^2 - \lambda^2}{1 - z^2} f(z). \quad \text{(C.57)}$$

And we get, as only integrable solution:

$$\frac{C(\theta)}{C(0)} = \sum_{n=0}^{\infty} \frac{(|\lambda| - \kappa - \frac{1}{2})_n (|\lambda| + \kappa + \frac{1}{2})_n}{(\frac{1}{2} + |\lambda|)_n n!} \sin^{2n}(\frac{\theta}{2}), \quad \text{(C.58)}$$

with:

$$(a)_0 = 1, \quad (a)_n = a(a+1)\ldots(a+n-1). \quad \text{(C.59)}$$

8. When we solve the Dirac equation with Darwin's method, that is to say with the ad-hoc operators, we get Legendre's polynomials and spherical harmonics. Here, working with ϕ, that is to say with the Weyl spinors ξ and η, we get the Gegenbauer's polynomials, and it is the degree of the Gegenbauer's polynomial which is the needed quantum number.

The $C(0)$ factor is a factor of U and V; its phase may be absorbed by the δ in Eq. (C.8), and its amplitude may be transferred on the radial functions. We can take therefore $C(0) = 1$. This gives:

$$C(\theta) = \sum_{n=0}^{\infty} \frac{(|\lambda| - \kappa - \frac{1}{2})_n(|\lambda| + \kappa + \frac{1}{2})_n}{(\frac{1}{2} + |\lambda|)_n n!} \sin^{2n}(\frac{\theta}{2}).$$ (C.60)

Since we have the conditions Eq. (C.52) on λ and κ, an integer n always exists such as

$$|\lambda| + n = |\kappa + \frac{1}{2}|,$$ (C.61)

and this forces the series in Eq. (C.60) to be a finite sum, so U and V are integrable. And since U and V have real values, we have:

$$e^{i\beta} = \frac{AD^*U^2 + CB^*V^2}{|AD^*U^2 + CB^*V^2|}.$$ (C.62)

C.3   Resolution of the linear radial system

We make the change of radial variable:

$$x = mr \; ; \quad \epsilon = \frac{E}{m} \; ; \quad a(x) = A(r) = A(\frac{x}{m}),$$
$$b(x) = B(r) \; ; \quad c(x) = C(r) \; ; \quad d(x) = D(r).$$ (C.63)

The radial system Eq. (C.40) becomes:

$$i(\epsilon + \frac{\alpha}{x})d + d' + \frac{\kappa}{x}b = ia,$$
$$-i(\epsilon + \frac{\alpha}{x})c - c' - \frac{\kappa}{x}a = -ib,$$
$$i(\epsilon + \frac{\alpha}{x})b - b' - \frac{\kappa}{x}d = ic,$$
$$-i(\epsilon + \frac{\alpha}{x})a + a' + \frac{\kappa}{x}c = -id.$$ (C.64)

Adding and subtracting, we get:

$$i(\epsilon + \frac{\alpha}{x})(d - c) + (d - c)' - \frac{\kappa}{x}(a - b) = i(a - b),$$
$$-i(\epsilon + \frac{\alpha}{x})(a - b) + (a - b)' - \frac{\kappa}{x}(d - c) = -i(d - c),$$
$$i(\epsilon + \frac{\alpha}{x})(c + d) + (c + d)' + \frac{\kappa}{x}(a + b) = i(a + b),$$
$$i(\epsilon + \frac{\alpha}{x})(a + b) - (a + b)' - \frac{\kappa}{x}(c + d) = i(c + d).$$ (C.65)

We let then:

$$a - b = F_- + iG_- \; ; \quad a + b = F_+ + iG_+,$$
$$d - c = F_- - iG_- \; ; \quad c + d = F_+ - iG_+,$$
(C.66)

and the radial system becomes:

$$i(\epsilon + \frac{\alpha}{x})(F_- - iG_-) + (F_- - iG_-)' - \frac{\kappa}{x}(F_- + iG_-) = i(F_- + iG_-),$$
$$-i(\epsilon + \frac{\alpha}{x})(F_- + iG_-) + (F_- + iG_-)' - \frac{\kappa}{x}(F_- - iG_-) = -i(F_- - iG_-),$$
$$i(\epsilon + \frac{\alpha}{x})(F_+ - iG_+) + (F_+ - iG_+)' + \frac{\kappa}{x}(F_+ + iG_+) = i(F_+ + iG_+),$$
$$i(\epsilon + \frac{\alpha}{x})(F_+ + iG_+) - (F_+ + iG_+)' - \frac{\kappa}{x}(F_+ - iG_+) = i(F_+ - iG_+).$$
(C.67)

Adding and subtracting in equations Eq. (C.67), then dividing by i the equations where i is a factor, we get the two separated systems:

$$(-1 + \epsilon + \frac{\alpha}{x})F_- - G_-' - \frac{\kappa}{x}G_- = 0,$$
$$(1 + \epsilon + \frac{\alpha}{x})G_- + F_-' - \frac{\kappa}{x}F_- = 0.$$
(C.68)

$$(-1 + \epsilon + \frac{\alpha}{x})F_+ - G_+' + \frac{\kappa}{x}G_+ = 0,$$
$$(1 + \epsilon + \frac{\alpha}{x})G_+ + F_+' + \frac{\kappa}{x}F_+ = 0.$$
(C.69)

These two systems are exchanged by replacing $-$ indices by $+$ indices and vice versa, and by changing κ to $-\kappa$, so it is enough to study one of the two systems. We let now:

$$F_- = \sqrt{1 + \epsilon} \, e^{-\Lambda x}(\varphi_1 + \varphi_2) \; ; \quad \Lambda = \sqrt{1 - \epsilon^2},$$
$$G_- = \sqrt{1 - \epsilon} \, e^{-\Lambda x}(\varphi_1 - \varphi_2).$$
(C.70)

Dividing the first of the two equations Eq. (C.68) by $\sqrt{1 - \epsilon} \, e^{-\Lambda x}$ and the second by $\sqrt{1 + \epsilon} \, e^{-\Lambda x}$, we get:

$$-\Lambda(\varphi_1 + \varphi_2) + \frac{\alpha}{x}\sqrt{\frac{1 + \epsilon}{1 - \epsilon}}(\varphi_1 + \varphi_2) + \Lambda(\varphi_1 - \varphi_2)$$
$$- \varphi_1' + \varphi_2' - \frac{\kappa}{x}(\varphi_1 - \varphi_2) = 0,$$
$$\Lambda(\varphi_1 - \varphi_2) + \frac{\alpha}{x}\sqrt{\frac{1 - \epsilon}{1 + \epsilon}}(\varphi_1 - \varphi_2) - \Lambda(\varphi_1 + \varphi_2)$$
$$+ \varphi_1' + \varphi_2' - \frac{\kappa}{x}(\varphi_1 + \varphi_2) = 0.$$
(C.71)

But we have:

$$\sqrt{\frac{1+\epsilon}{1-\epsilon}} = \frac{1+\epsilon}{\Lambda} \; ; \quad \sqrt{\frac{1-\epsilon}{1+\epsilon}} = \frac{1-\epsilon}{\Lambda},$$

(C.72)

and we let:

$$c_1 = \frac{\alpha}{\Lambda} \; ; \quad c_2 = \frac{\alpha\epsilon}{\Lambda}.$$

(C.73)

We get then, adding and subtracting the equations Eq. (C.71):

$$-2\Lambda\varphi_2 + \frac{c_1 - \kappa}{x}\varphi_1 + \frac{c_2}{x}\varphi_2 + \varphi_2' = 0$$

$$\frac{c_1 + \kappa}{x}\varphi_2 + \frac{c_2}{x}\varphi_1 - \varphi_1' = 0.$$

(C.74)

We make then the change of variable:

$$z = 2\Lambda x \; ; \quad f_1(z) = \varphi_1(x) \; ; \quad f_2(z) = \varphi_2(x).$$

(C.75)

This puts the system in Eq. (C.74) on the form:

$$-f_2 + \frac{c_1 - \kappa}{z}f_1 + \frac{c_2}{z}f_2 + f_2' = 0$$

$$\frac{c_1 + \kappa}{z}f_2 + \frac{c_2}{z}f_1 - f_1' = 0.$$

(C.76)

We develop now in series:

$$f_1(z) = \sum_{m=0}^{\infty} a_m z^{s+m} \; ; \quad f_2(z) = \sum_{m=0}^{\infty} b_m z^{s+m}.$$

(C.77)

The system Eq. (C.76) gives, for the coefficients of z^{s-1}:

$$(c_1 - \kappa)a_0 + (c_2 + s)b_0 = 0,$$

$$(c_2 - s)a_0 + (c_1 + \kappa)b_0 = 0.$$

(C.78)

A non null solution exists only if the determinant of this system is null:

$$0 = \begin{vmatrix} c_1 - \kappa & c_2 + s \\ c_2 - s & c_1 + \kappa \end{vmatrix} = c_1^2 - \kappa^2 - c_2^2 + s^2.$$

(C.79)

But we have, with Eq. (C.70) and Eq. (C.73):

$$c_1^2 - c_2^2 = \alpha^2.$$

(C.80)

So we get:

$$0 = \alpha^2 + s^2 - \kappa^2 \; ; \quad s^2 = \kappa^2 - \alpha^2.$$

(C.81)

We must take:

$$s = \sqrt{\kappa^2 - \alpha^2},$$

(C.82)

to make the wave integrable at the origin. In this case the system Eq. (C.78) is reduced to

$$b_0 = \frac{\kappa - c_1}{c_2 + s} a_0 = \frac{s - c_2}{c_1 + \kappa} a_0. \tag{C.83}$$

The system Eq. (C.76) gives, for the coefficients of z^{s+m-1}, the system:

$$-b_{m-1} + (c_1 - \kappa)a_m + (c_2 + s + m)b_m = 0,$$
$$(c_1 + \kappa)b_m + (c_2 - s - m)a_m = 0. \tag{C.84}$$

This last equation gives:

$$a_m = \frac{c_1 + \kappa}{-c_2 + s + m} b_m, \tag{C.85}$$

and the first one becomes:

$$- b_{m-1} + (c_1 - \kappa)\frac{c_1 + \kappa}{-c_2 + s + m} b_m + (c_2 + s + m)b_m = 0,$$
$$[(c_1 - \kappa)(c_1 + \kappa) + (s + m)^2 - c_2^2])b_m = (-c_2 + s + m)b_{m-1}, \tag{C.86}$$

which, with Eq. (C.79), gives:

$$b_m = \frac{-c_2 + s + m}{(2s + m)m} b_{m-1} = \frac{(-c_2 + s + 1)_m}{(2s + 1)_m m!} b_0. \tag{C.87}$$

And so we have:

$$f_2(z) = b_0 z^s \sum_{m=0}^{\infty} \frac{(-c_2 + s + 1)_m}{(2s + 1)_m m!} z^m = b_0 z^s F(1 + s - c_2, \ 2s + 1, \ z) \tag{C.88}$$

where F is the hypergeometric function. We have also:

$$b_m = \frac{-c_2 + s + m}{c_1 + \kappa} a_m \ ; \quad b_{m-1} = \frac{-c_2 + s + m - 1}{c_1 + \kappa} a_{m-1}. \tag{C.89}$$

The first of the two equations Eq. (C.84) becomes:

$$-\frac{-c_2 + s + m - 1}{c_1 + \kappa} a_{m-1} + (c_1 - \kappa)a_m + (c_2 + s + m)\frac{-c_2 + s + m}{c_1 + \kappa} a_m = 0 \tag{C.90}$$

which implies:

$$a_m = \frac{-c_2 + s - 1 + m}{(2s + m)m} a_{m-1} = \frac{(-c_2 + s)_m}{(2s + 1)_m m!} a_0 \tag{C.91}$$

And so we have:

$$f_1(z) = a_0 z^s \sum_{m=0}^{\infty} \frac{(-c_2 + s)_m}{(2s + 1)_m m!} z^m = a_0 z^s F(s - c_2, \ 2s + 1, \ z) \tag{C.92}$$

This hypergeometric function is integrable only [9] if the series is polynomial, (up a coefficient, it is a Laguerre's polynomial) with degree n, that is to say if an integer n exists such that

$$-c_2 + s + n = 0 \tag{C.93}$$

$$s + n = \frac{\epsilon\alpha}{\Lambda}, \tag{C.94}$$

which gives, by taking the square:

$$(s+n)^2(1-\epsilon^2) = \epsilon^2\alpha^2$$
$$(s+n)^2 = [(s+n)^2 + \alpha^2]\epsilon^2$$

$$\epsilon^2 = \frac{1}{1 + \dfrac{\alpha^2}{(s+n)^2}}. \tag{C.95}$$

And we get the Sommerfeld's formula for the energy levels

$$\epsilon = \frac{1}{\sqrt{1 + \dfrac{\alpha^2}{(s+n)^2}}} \;;\;\; s = \sqrt{\kappa^2 - \alpha^2} \;;\;\; |\kappa| = j + \frac{1}{2}. \tag{C.96}$$

With Eq. (C.75), Eq. (C.83), Eq. (C.88) and Eq. (C.92), we get now:

$$\varphi_1(x) = a_0(2\Lambda x)^s F(-n,\; 2s+1,\; 2\Lambda x), \tag{C.97}$$

$$\varphi_2(x) = \frac{-na_0}{c_1 + \kappa}(2\Lambda x)^s F(1-n,\; 2s+1,\; 2\Lambda x). \tag{C.98}$$

We let, if $n > 0$:

$$P_1 = F(1-n,\; 2s+1,\; 2\Lambda x) \;;\;\; P_2 = F(-n,\; 2s+1,\; 2\Lambda x). \tag{C.99}$$

And we get:

$$F_- = \frac{\sqrt{1+\epsilon}}{c_1 + \kappa}a_0 e^{-\Lambda x}(2\Lambda x)^s[(c_1 + \kappa)P_2 - nP_1], \tag{C.100}$$

$$G_- = \frac{\sqrt{1-\epsilon}}{c_1 + \kappa}a_0 e^{-\Lambda x}(2\Lambda x)^s[(c_1 + \kappa)P_2 + nP_1]. \tag{C.101}$$

We let then:

$$a_1 = \frac{\sqrt{1+\epsilon}}{c_1 + \kappa}a_0(2\Lambda)^s. \tag{C.102}$$

---

9. The integrability of the wave functions is not optional, but compulsory, since we have seen in Sec. 3.4 that the normalization of the wave comes from the physical fact that the energy of the electron is the energy of its wave.

We get finally:

$$F_- = a_1 e^{-\Lambda x} x^s [(c_1 + \kappa) P_2 - n P_1], \tag{C.103}$$

$$G_- = \sqrt{\frac{1-\epsilon}{1+\epsilon}} a_1 e^{-\Lambda x} x^s [(c_1 + \kappa) P_2 + n P_1]. \tag{C.104}$$

Since we go from F_-, G_- to F_+, G_+ by replacing κ by $-\kappa$, we have also:

$$F_+ = a_2 e^{-\Lambda x} x^s [(c_1 - \kappa) P_2 - n P_1], \tag{C.105}$$

$$G_+ = \sqrt{\frac{1-\epsilon}{1+\epsilon}} a_2 e^{-\Lambda x} x^s [(c_1 - \kappa) P_2 + n P_1], \tag{C.106}$$

where a_2 is, as a_1, a complex constant. We get the same condition on the energy when we say that functions F_+ and G_+ must be polynomials to get integrability of the wave, because Eq. (C.96) contains only κ^2. Therefore if the formula Eq. (C.96) is satisfied we get polynomials for the four radial functions and the wave is integrable.

C.4 Calculation of the Yvon–Takabayasi angle

We have with Eq. (C.62) and Eq. (C.63):

$$e^{i\beta} = \frac{ad^* U^2 + cb^* V^2}{|ad^* U^2 + cb^* V^2|}. \tag{C.107}$$

With Eq. (C.66) we get:

$$2a = F_+ + F_- + i(G_+ + G_-) \; ; \quad 2b = F_+ - F_- + i(G_+ - G_-),$$
$$2d = F_+ + F_- - i(G_+ + G_-) \; ; \quad 2c = F_+ - F_- - i(G_+ - G_-). \tag{C.108}$$

And we get:

$$\begin{aligned}
4(ad^* U^2 + cb^* V^2) =& (F_+ F_+^* + F_- F_-^* - G_+ G_+^* - G_- G_-^*)(U^2 + V^2) \\
&+ (F_+ F_-^* + F_- F_+^* - G_+ G_-^* - G_- G_+^*)(U^2 - V^2) \\
&+ i(F_+ G_+^* + F_- G_-^* + G_+ F_+^* + G_- F_-^*)(U^2 - V^2) \\
&+ i(F_+ G_-^* + F_- G_+^* + G_+ F_-^* + G_- F_+^*)(U^2 + V^2).
\end{aligned} \tag{C.109}$$

With Eq. (C.103) to Eq. (C.106), we get then, if $n > 0$:

$$F_+ F_+^* + F_- F_-^* - G_+ G_+^* - G_- G_-^* \tag{C.110}$$
$$= \frac{2}{1+\epsilon} e^{-2\Lambda x} x^{2s} \left(\begin{array}{c} (|a_1|^2 + |a_2|^2)[\epsilon(c_1^2 + \kappa^2) P_2^2 + \epsilon n^2 P_1^2 - 2n c_1 P_1 P_2] \\ + (|a_2|^2 - |a_1|^2) 2\kappa P_2 (-\epsilon c_1 P_2 + n P_1) \end{array} \right),$$

$$F_+ F_-^* + F_- F_+^* - G_+ G_-^* - G_- G_+^* \tag{C.111}$$

$$= e^{-2\Lambda x} x^{2s} (a_1 a_2^* + a_2 a_1^*) \left(\begin{array}{c} [(c_1 - \kappa) P_2 - n P_1][(c_1 + \kappa) P_2 - n P_1] \\ -\frac{1-\epsilon}{1+\epsilon}([(c_1 - \kappa) P_2 + n P_1][(c_1 + \kappa) P_2 + n P_1]) \end{array} \right),$$

$$F_+ G_+^* + F_- G_-^* + G_+ F_+^* + G_- F_-^* \tag{C.112}$$

$$= 2\sqrt{\frac{1-\epsilon}{1+\epsilon}} e^{-2\Lambda x} x^{2s} \left(\begin{array}{c} |a_2|^2[(c_1 - \kappa)^2 P_2^2 - n^2 P_1^2] \\ +|a_1|^2[(c_1 + \kappa)^2 P_2^2 - n^2 P_1^2] \end{array} \right),$$

$$F_+ G_-^* + F_- G_+^* + G_+ F_-^* + G_- F_+^* \tag{C.113}$$

$$= e^{-2\Lambda x} x^{2s} (a_1 a_2^* + a_2 a_1^*) \sqrt{\frac{1-\epsilon}{1+\epsilon}} \left(\begin{array}{c} [(c_1 - \kappa) P_2 - n P_1][(c_1 + \kappa) P_2 + n P_1] \\ +[(c_1 + \kappa) P_2 - n P_1][(c_1 - \kappa) P_2 + n P_1] \end{array} \right).$$

There is a great simplification, that we will let now, if:

$$a_1 a_2^* + a_2 a_1^* = 0. \tag{C.114}$$

In addition, we have:

$$c_2 = s + n = \frac{\alpha \epsilon}{\Lambda} \; ; \; c_1 = \frac{\alpha}{\Lambda} = \frac{s+n}{\epsilon} = \sqrt{(s+n)^2 + \alpha^2}. \tag{C.115}$$

We have: $s \geqslant 0$, $n \geqslant 0$, therefore $(s+n)^2 \geqslant s^2$, and

$$c_1 = \sqrt{(s+n)^2 + \alpha^2} \geqslant \sqrt{s^2 + \alpha^2} = \sqrt{\kappa^2} = |\kappa| \geqslant \pm\kappa. \tag{C.116}$$

So we always have:

$$c_1 - \kappa \geqslant 0 \; ; \; c_1 + \kappa \geqslant 0. \tag{C.117}$$

If we choose to let:

$$|a_1|^2 = (c_1 - \kappa)k \; ; \; |a_2|^2 = (c_1 + \kappa)k, \tag{C.118}$$

where k is a real positive constant, we get:

$$F_+ F_+^* + F_- F_-^* - G_+ G_+^* - G_- G_-^*$$

$$= \frac{2k}{1+\epsilon} e^{-2\Lambda x} x^{2s} [2\epsilon c_1 (c_1^2 - \kappa^2) P_2^2 + 2\epsilon c_1 n^2 P_1^2 - 4n(c_1^2 - \kappa^2) P_1 P_2],$$

$$\tag{C.119}$$

and since:

$$c_1^2 - \kappa^2 = n(n+2s) \; ; \; \epsilon c_1 = s + n, \tag{C.120}$$

we get:

$$F_+F_+^* + F_-F_-^* - G_+G_+^* - G_-G_-^*$$

$$= \frac{4nk}{1+\epsilon}e^{-2\Lambda x}x^{2s}\left((n+2s)[(s+n)P_2^2 - 2nP_1P_2] + n(s+n)P_1^2\right)$$

$$= \frac{4nk}{1+\epsilon}e^{-2\Lambda x}x^{2s}\left[(n+2s)(\sqrt{s+n}P_2 - \frac{n}{\sqrt{s+n}}P_1)^2 + \frac{ns^2}{s+n}P_1^2\right].$$

$$(C.121)$$

This term, which is the sum of two squares, is always positive, two successive Laguerre's polynomials having no common zero. Then we get

$$F_+G_+^* + F_-G_-^* + G_+F_+^* + G_-F_-^*$$

$$= \frac{2\sqrt{1-\epsilon^2}}{1+\epsilon}e^{-2\Lambda x}x^{2s}\left(\begin{array}{c}|a_2|^2[(c_1-\kappa)^2P_2^2 - n^2P_1^2]\\+|a_1|^2[(c_1+\kappa)^2P_2^2 - n^2P_1^2]\end{array}\right)$$

$$= \frac{4c_1\Lambda k}{1+\epsilon}e^{-2\Lambda x}x^{2s}[(c_1^2-\kappa^2)P_2^2 - n^2P_1^2]$$

$$= \frac{4\alpha nk}{1+\epsilon}e^{-2\Lambda x}x^{2s}[(n+2s)P_2^2 - nP_1^2].$$

$$(C.122)$$

This allows us to write the Yvon–Takabayasi angle as:

$$\tan\beta = \frac{\alpha[(2s+n)P_2^2 - nP_1^2]}{(n+2s)(\sqrt{s+n}P_2 - \frac{n}{\sqrt{s+n}}P_1)^2 + \frac{ns^2}{s+n}P_1^2} \times \frac{U^2-V^2}{U^2+V^2}. \quad (C.123)$$

The denominator contains only sums of squares, which cannot be both null. Consequently, for all the states with a $n > 0$ quantum number, a solution exists such that the Yvon–Takabayasi β angle is everywhere defined. In addition the presence of the fine structure constant, which is small, implies that the β angle is everywhere small. Moreover we now explain why U^2-V^2 is exactly null, for any value of κ and λ, in the plane x^1Ox^2. We start here from the differential equation Eq. (C.57). If f is a solution, then g defined by $g(z) = f(-z)$ is also a solution. Since there is only one polynomial solution with degree n, up to a real factor, we get necessarily $f(-z) = \pm f(z)$ and f is either an even polynomial or an odd polynomial. Therefore C is an even or an odd polynomial of $\cos\theta$. Now from Eq. (C.53) we get

$$U^2 + V^2 = \sin^{2\lambda}\theta[C'^2 + (\kappa + \frac{1}{2} - \lambda)^2C^2], \quad (C.124)$$

$$U^2 - V^2 = \sin^{2\lambda}\theta[-\cos\theta C'^2 - 2(\kappa + \frac{1}{2} - \lambda)\sin\theta C'C + (\kappa + \frac{1}{2} - \lambda)^2\cos\theta C^2].$$

From Eq. (C.54)) we get

$$U^2 + V^2 = \sin^{-2\lambda}\theta[C'^2 + (\kappa + \frac{1}{2} + \lambda)^2C^2], \quad (C.125)$$

$$U^2 - V^2 = \sin^{-2\lambda}\theta[\cos\theta C'^2 + 2(\kappa + \frac{1}{2} + \lambda)\sin\theta C'C - (\kappa + \frac{1}{2} + \lambda)^2\cos\theta C^2].$$

This gives

$$(U^2 - V^2)(\frac{\pi}{2}) = -\frac{\lambda}{|\lambda|}(\kappa + \frac{1}{2} - |\lambda|)(C'C)(\frac{\pi}{2}). \tag{C.126}$$

If C is a constant $C' = 0$. Otherwise either C is an even polynomial of $\cos\theta$ and then C' is odd and the product $C'C$ contains a $\cos\theta$ factor, or C is an odd polynomial of $\cos\theta$ and the product $C'C$ contains also a $\cos\theta$ factor. This factor is null if $\theta = \frac{\pi}{2}$. This proves that

$$\beta(\frac{\pi}{2}) = 0. \tag{C.127}$$

The solutions of the linear Dirac equation satisfying Eq. (C.114) and Eq. (C.118) may therefore be the linear approximations of solutions for the homogeneous nonlinear equation. Now if we have a solution ϕ_0 of the nonlinear homogeneous equation Eq. (3.1) with a not small value β_0 of the Yvon–Takabayasi angle at a point M_0 with coordinates $(x_0, y_0, 0)$, since the nonlinear homogeneous equation is globally gauge invariant under the chiral gauge Eq. (3.21), we let

$$\phi = e^{-i\frac{\beta_0}{2}}\phi_0. \tag{C.128}$$

And we get

$$(\phi\bar{\phi})(M_0) = e^{-i\frac{\beta_0}{2}}\phi_0 e^{-i\frac{\beta_0}{2}}\bar{\phi}_0 = e^{-i\beta_0}\rho_0 e^{i\beta_0} = \rho_0. \tag{C.129}$$

And ϕ has at this point M_0 a null β angle, so the equation Eq. (3.1) at this point is exactly the Dirac equation, we get the separation of variables at this point, we get the angular system Eq. (C.38) and the radial system Eq. (C.39) which is identical to the radial system Eq. (C.40) of the linear equation. Then the β angle is null in all the $z = 0$ plane and the radial system Eq. (C.39) is identical to Eq. (C.40) in all the $z = 0$ plane. Then the necessity of integrability imposes the existence of radial polynomials and we get the quantification of the energy levels and the Sommerfeld's formula Eq. (C.96)

C.5 Radial polynomials with degree 0

To get absolutely all results of the Dirac equation, we have one last thing to explain: how we get $2n^2$ different states with a principal quantum number $n = |\kappa| + n$, and we must return to the particular case where radial polynomials are constants. We start directly from Eq. (C.64), and we let:

$$a = a_0 e^{-\Lambda x} x^s \ ; \quad b = b_0 e^{-\Lambda x} x^s \ ; \quad c = c_0 e^{-\Lambda x} x^s \ ; \quad d = d_0 e^{-\Lambda x} x^s. \tag{C.130}$$

We get from Eq. (C.39):

$$e^{-\Lambda x}\left(i\epsilon d_0 x^s + i\alpha d_0 x^{s-1} - \Lambda d_0 x^s + s d_0 x^{s-1} + \kappa b_0 x^{s-1}\right) = i a_0 e^{-\Lambda x} x^s,$$
$$e^{-\Lambda x}\left(-i\epsilon c_0 x^s - i\alpha c_0 x^{s-1} + \Lambda c_0 x^s - s c_0 x^{s-1} - \kappa a_0 x^{s-1}\right) = -i b_0 e^{-\Lambda x} x^s,$$
$$e^{-\Lambda x}\left(i\epsilon b_0 x^s + i\alpha b_0 x^{s-1} + \Lambda b_0 x^s - s b_0 x^{s-1} - \kappa d_0 x^{s-1}\right) = i c_0 e^{-\Lambda x} x^s,$$
$$e^{-\Lambda x}\left(-i\epsilon a_0 x^s - i\alpha a_0 x^{s-1} - \Lambda a_0 x^s + s a_0 x^{s-1} + \kappa c_0 x^{s-1}\right) = -i d_0 e^{-\Lambda x} x^s.$$

$$(\text{C}.131)$$

This is equivalent to the set formed by the four following systems:

$$\kappa b_0 + (i\alpha + s)d_0 = 0,$$
$$(i\alpha - s)b_0 - \kappa d_0 = 0,$$

$$(\text{C}.132)$$

$$-\kappa a_0 - (i\alpha + s)c_0 = 0,$$
$$-(i\alpha - s)a_0 + \kappa c_0 = 0,$$

$$(\text{C}.133)$$

$$-i a_0 + (i\epsilon - \Lambda)d_0 = 0,$$
$$-(i\epsilon + \Lambda)a_0 + i d_0 = 0,$$

$$(\text{C}.134)$$

$$i b_0 - (i\epsilon - \Lambda)c_0 = 0,$$
$$(i\epsilon + \Lambda)b_0 - i c_0 = 0.$$

$$(\text{C}.135)$$

The cancellation of the determinant in Eq. (C.132) and Eq. (C.133) gives again Eq. (C.81) and Eq. (C.82). The cancellation of the determinant in Eq. (C.134) and Eq. (C.135) is simply equivalent to $\Lambda^2 = 1 - \epsilon^2$, which comes from the definition of Λ. Each system Eq. (C.132) to Eq. (C.135) is then reduced into one equation:

$$\kappa d_0 = (i\alpha - s)b_0,$$
$$\kappa c_0 = (i\alpha - s)a_0,$$
$$d_0 = (\epsilon - i\Lambda)a_0,$$
$$b_0 = (\epsilon + i\Lambda)c_0.$$

$$(\text{C}.136)$$

We get then:

$$\kappa d_0 = \kappa(\epsilon - i\Lambda)a_0 = (i\alpha - s)b_0 = (i\alpha - s)(\epsilon + i\Lambda)c_0 = \frac{(i\alpha - s)^2(\epsilon + i\Lambda)}{\kappa}a_0.$$

$$(\text{C}.137)$$

We have a non-null solution only if:

$$\kappa(\epsilon - i\Lambda) = \frac{(i\alpha - s)^2(\epsilon + i\Lambda)}{\kappa}$$
$$\kappa^2(\epsilon - i\Lambda)^2 = (s - i\alpha)^2$$
$$\kappa(\epsilon - i\Lambda) = \pm(s - i\alpha).$$

$$(\text{C}.138)$$

Since ϵ, s, Λ and α are positive, we finally get

$$|\kappa| = \frac{s}{\epsilon} = \frac{\alpha}{\Lambda}. \tag{C.139}$$

This last equality gives again the formula of energy levels Eq. (C.96) with $n = 0$. Since κ comes with its absolute value, we can equally have $\kappa < 0$ or $\kappa > 0$. But the calculation of solutions by C. G. Darwin, who works with real constants, instead of complex constants at this stage of his computation, forbids κ to be negative, and it is that thing that allows, for a given principal quantum number $\mathbf{n} = n + |\kappa|$, to get $\mathbf{n}(\mathbf{n}+1) + \mathbf{n}(\mathbf{n}-1) = 2\mathbf{n}^2$ states. What really happens is that to change sign in κ also changes V into $-V$. And if we change the sign of κ and V, then $\mathbf{a}$, $\mathbf{b}$, $\mathbf{c}$, $\mathbf{d}$ are invariant if $n = 0$, and the wave is unchanged. To change the sign of κ yields no more solutions and we can use only solutions with $\kappa > 0$, in the case $n = 0$. And this allows us to get the true number of states.

The formula obtained for the energy levels does not account for the Lamb shift, which gives, if $n > 0$, a very small split between energy levels with same other quantum numbers but with opposite signs of κ. If the formula Eq. (C.96) was not the same for two opposite values of κ we should not be able to get four polynomial radial functions with only one condition which gives the quantification of the energy levels. Here also the standard model has already an answer, with the polarization of the vacuum. But the calculation must be revised, both to avoid divergent integrals and to use our solutions instead of Darwin's solutions coming from the non-relativistic Pauli equation.

Bibliography

[1] Bardout, G., Lochak, G. and Fargue, D. (2007). Sur la présence de monopôles légers au pôle nord, *Ann. Fond. Louis de Broglie* **32**, p. 551.

[2] Baylis, W. E. (1996). *Clifford (Geometric) Algebras*, chap. The Paravector Model of Spacetime (Birkhauser, Boston), pp. 237–296.

[3] Boudet, R. (1995). The Takabayasi moving frame, from a potential to the Z boson, in S. Jeffers and J. Vigier (eds.), *The Present Status of the Quantum Theory of the Light* (Kluwer, Dordrecht).

[4] Boudet, R. (2011). *Quantum Mechanics in the Geometry of Space-Time* (Springer, Heidelberg Dordrecht London New York).

[5] Casanova, G. (1976). *L'algèbre vectorielle* (Presses Universitaires de France, Paris).

[6] C.G.Darwin (1928). *Proc. R. Soc. Lond.* **118**, p. 554.

[7] Daviau, C. (1993). *Equation de Dirac non linéaire*, Ph.D. thesis, Université de Nantes.

[8] Daviau, C. (1997a). Solutions of the Dirac equation and of a nonlinear Dirac equation for the hydrogen atom, *Adv. Appl. Clifford Algebras* **7**, (S), pp. 175–194.

[9] Daviau, C. (1997b). Sur l'équation de Dirac dans l'algèbre de Pauli, *Ann. Fond. L. de Broglie* **22**, 1, pp. 87–103.

[10] Daviau, C. (1998). Sur les tenseurs de la théorie de Dirac en algèbre d'espace, *Ann. Fond. Louis de Broglie* **23**, 1.

[11] Daviau, C. (2001). Vers une mécanique quantique sans nombre complexe. *Ann. Fond. L. de Broglie* **26**, special, pp. 149–171.

[12] Daviau, C. (2005). Interprétation cinématique de l'onde de l'électron, *Ann. Fond. L. de Broglie* **30**, 3-4.

[13] Daviau, C. (2011). *L'espace-temps double* (JePublie, Pouillé-les-coteaux).

[14] Daviau, C. (2012a). Cl_3^* invariance of the Dirac equation and of electromagnetism, *Adv. Appl. Clifford Algebras* **22**, 3, pp. 611–623.

[15] Daviau, C. (2012b). *Double Space-Time and more* (JePublie, Pouillé-les-coteaux).

[16] Daviau, C. (2012c). *Nonlinear Dirac Equation, Magnetic Monopoles and Double Space-Time* (CISP, Cambridge UK).

[17] Daviau, C. (2013). Invariant quantum wave equations and double space-time, *Adv. in Imaging and Electron Physics* **179, chapter 1**, pp. 1–137.

[18] Daviau, C. (2015). Gauge group of the standard model in $Cl_{1,5}$, *AACA* **25**, pp. DOI 10.1007/s00006-015-0566-5.

[19] Daviau, C. and Bertrand, J. (2012). Velocity and proper mass of muonic neutrinos, *Ann. Fond. Louis de Broglie* **37**, pp. 129–134.

[20] Daviau, C. and Bertrand, J. (2013). A lepton Dirac equation with additional mass term and a wave equation for a fourth neutrino, *Ann. Fond. Louis de Broglie* **38**.

[21] Daviau, C. and Bertrand, J. (2014a). Relativistic gauge invariant wave equation of the electron-neutrino. *Journal of Modern Physics* **5**, pp. 1001–1022, http://dx.doi.org/10.4236/jmp.2014.511102.

[22] Daviau, C. and Bertrand, J. (2014b). A wave equation including leptons and quarks for the standard model of quantum physics in Clifford algebra, *JMP* **5**, pp. 2149–2173, http://dx.doi.org/10.4236/jmp.2014.518210.

[23] Daviau, C. and Bertrand, J. (2014 and http://hal.archives-ouvertes.fr/hal-00907848). *New Insights in the Standard Model of Quantum Physics in Clifford Algebra* (Je Publie, Pouillé-les-coteaux).

[24] Daviau, C. and Bertrand, J. (2015a). Geometry of the standard model of quantum physics, *Journal of Applied Mathematics and Physics* **3**, pp. 46–61. http://dx.doi.org/10.4236/jamp.2015.31007.

[25] Daviau, C. and Bertrand, J. (2015b). Left chiral solutions for the hydrogen atom of the wave equation for electron + neutrino, *JMP* **5**, p. to be published.

[26] Daviau, C., Fargue, D., Priem, D. and Racineux, G. (2013). Tracks of magnetic monopoles, *Ann. Fond. Louis de Broglie* **38**.

[27] de Beauregard, O. C. (1989). Sur un tenseur encore ininterprété en théorie de Dirac, *Ann. Fond. Louis de Broglie* **14-3**, pp. 335–342.

[28] de Beauregard, O. C. (1997). Induced electromagnetic inertia and physicality of the 4-vector potential, *Physics Essays* **10-4**, pp. 646–650.

[29] de Broglie, L. (1924). Recherches sur la théorie des quantas, *Ann. Fond. Louis de Broglie* **17**, 1.

[30] de Broglie, L. (1934). *L'électron magnétique* (Hermann, Paris).

[31] de Broglie, L. (1940). *La mécanique du photon, Une nouvelle théorie de la lumière : tome 1 La lumière dans le vide* (Hermann, Paris).

[32] de Broglie, L. (1942). *tome 2 Les interactions entre les photons et la matière* (Hermann, Paris).

[33] Deheuvels, R. (1993). *Tenseurs et spineurs* (PUF, Paris).

[34] Dirac, P. (1928). The quantum theory of the electron, *Proc. R. Soc. Lond.* **117**, pp. 610–624.

[35] Doran, C. and Lasenby, A. (2003). *Geometric Algebra* (Cambridge University Press, Cambridge, U.K.).

[36] Einstein, A. (1905). Äœber einen die erzeugung und verwandlung des lichtes betreffenden heuristischen gesichtspunkt, *Annalen der Physik* **17**, pp. 132–148.

[37] Einstein, A. (1925). Einheitliche feldtheorie von gravitation und elektrizität,

Sitzungsberichte der Preussischen Akademie der Wissenschaften (Berlin), Physikalisch-mathematische Klasse , pp. 414–419.

[38] Elbaz, E. (1989). *De l'électromagnétique à l'électro-faible* (Ellipses, Paris).

[39] Hestenes, D. (1966, 1987, 1992). *Space-Time Algebra* (Gordon and Breach, New-York).

[40] Hestenes, D. (1982). Space-time structure of weak and electromagnetic interactions, *Found. of Phys.* **12**, pp. 153–168.

[41] Hestenes, D. (1986). A unified language for Mathematics and Physics and Clifford Algebra and the interpretation of quantum mechanics, in Chisholm and A. Common (eds.), *Clifford Algebras and their applications in Mathematics and Physics* (Reidel, Dordrecht).

[42] H.PoincarÃ© (1896). Remarques sur une expÃ©rience de m. Birkeland, *C.R.A.S.* **123**, pp. 530–533.

[43] Ivoilov, N. (2006). Low energy generation of the strange radiation, *Ann. Fond. Louis de Broglie* **31**, 1, pp. 115–123.

[44] Krüger, H. (1991). New solutions of the Dirac equation for central fields, in D. Hestenes and A. Weingartshofer (eds.), *The Electron* (Kluwer, Dordrecht).

[45] Lasenby, A., Doran, C. and Gull, S. (1993). A multivector derivative approach to lagrangian field theory, *Found. of Phys.* **23**, pp. 1295–1327.

[46] Lochak, G. (1983). Sur un monopôle de masse nulle décrit par l'équation de Dirac et sur une équation générale non linéaire qui contient des monopôles de spin $\frac{1}{2}$, *Ann. Fond. Louis de Broglie* **8**, 4.

[47] Lochak, G. (1984). Sur un monopôle de masse nulle décrit par l'équation de Dirac et sur une équation générale non linéaire qui contient des monopôles de spin $\frac{1}{2}$(partie 2), *Ann. Fond. Louis de Broglie* **9**, 1.

[48] Lochak, G. (1985). Wave equation for a magnetic monopole, *Int. J. of Th. Phys.* **24**, pp. 1019–1050.

[49] Lochak, G. (2001). Liste des publications, *Ann. Fond. Louis de Broglie* **26**, pp. 31–42.

[50] Lochak, G. (2004). Photons électriques et photons magnétiques dans la théorie du photon de Louis de Broglie (un renouvellement possible de la théorie du champ unitaire d'Einstein), *Ann. Fond. Louis de Broglie* **29**, pp. 297–316.

[51] Lochak, G. (2008). "Photons électriques" et "photons magnétiques" dans la théorie du photon de de Broglie, *Ann. Fond. Louis de Broglie* **33**, pp. 107–127.

[52] Lochak, G. (2010). A theory of light with four different photons : electric and magnetic with spin 1 and spin 0. *Ann. Fond. Louis de Broglie* **35**, pp. 1–18.

[53] Naïmark, M. (1962). *Les représentations linéaires du groupe de Lorentz* (Dunod, Paris).

[54] Priem, D., Daviau, C. and Racineux, G. (2009). Transmutations et traces de monopôles obtenues lors de décharges électriques, *Ann. Fond. Louis de Broglie* **34**, p. 103.

[55] Rose, M. E. (1960). *Relativistic electron theory* (John Wiley and sons, New-

York, London).

[56] Schrödinger, E. (1926). *Annalen der Physik (4)* **81**.

[57] Socroun, T. (2015). Clifford to unify general relativity and electromagnetism, *AACA* **25**, pp. DOI 10.1007/s00006–015–0558–5.

[58] Takabayasi, T. (1957). Relativistic hydrodynamics of the Dirac matter, *Theor. Phys. Suppl.* **4**.

[59] Tonnelat, M. A. (1965). *Les théories unitaires de l'électromagnétisme et de la gravitation* (Gauthier-Villars, Paris).

[60] V.F.Mikhailov (1987). Observation of the magnetic charge effect in the experiments with ferromagnetic aerosols, *Ann. Fond. Louis de Broglie* **12**, 4, pp. 491–524.

[61] Weinberg, S. (1967). A model of leptons, *Phys. Rev. Lett.* **19**, pp. 1264–1266.

Index

(non linear homogeneous), vi, 2–5,
    39, 41, 44–47, 52, 58, 66, 91,
    100, 101, 144, 145, 148, 149,
    152, 204
Dirac matrices, 9, 20, 21, 25, 27,
    31–33, 46, 64, 73, 77, 83, 126, 145,
    147, 163
divergence, 15
double pattern, 121, 122

electro-weak interactions, vi, 4, 21,
    26, 29, 75, 85, 86, 90, 93, 94, 98,
    101, 105, 128, 137, 147, 148, 150,
    152, 156, 157
electromagnetic field, v, 50, 51, 54,
    57, 59, 62–64, 67–69, 71, 73, 127,
    141, 145, 150
energy level, 2, 48, 67, 214, 218, 220
even sub-algebra, v, vi, 14, 20

fermion, 64, 73, 93, 99, 141, 144–149,
    152–155, 158, 159

generation, vi, vii, 25, 29, 67, 87, 93,
    99, 101, 109, 128, 147, 148, 152,
    153, 155, 158
geometric transformation, 1, 5, 6, 16,
    18, 27, 32, 33, 45, 49, 51, 52, 56, 65,
    76, 87, 88, 91, 101–104, 130, 133,
    135, 138, 141, 142, 148, 150, 152
gradient, 15
gravitation, vi, 3, 7, 101, 133, 135,
    139–142, 148, 151, 157–159

homomorphism, 17, 19, 135, 141, 155
homothety, 18, 53, 135
hypergeometric function, 213, 214

inertia, vi, 4, 7, 133, 136–139, 141,
    142, 148, 157, 159
integrable, 209, 210, 213–215
invariance, v, 33, 44, 49, 85, 126, 137,
    139, 141, 164, 182
    form invariance, vi, 1, 3, 30, 33–35,
        44, 45, 51, 52, 54, 56, 64,
        86, 91, 98, 103–105, 109,

126, 133, 141, 142, 145, 146,
    149, 152, 153, 155, 157, 183
gauge invariance, vi, 2, 26, 28–30,
    36, 41, 47, 71, 85, 87, 93, 98,
    110, 127, 147, 154, 157, 175,
    184, 185, 187, 189, 195, 197
half-variance, 5, 155
Lorentz invariance, 3, 16, 26, 27,
    30, 43, 56, 60, 85, 135, 145,
    158
inverse, 5, 35, 48, 89, 90, 104, 105,
    144, 146, 153
isomorphism, 10, 21, 153, 155

Lagrangian density, vi, vii, 2, 4–7, 25,
    36, 41, 42, 47, 65, 66, 70, 71, 91,
    107, 108, 126, 137, 139, 141, 147,
    149, 150, 156, 158, 161, 164, 175,
    179, 183
left wave, vi, 5, 28, 76, 77, 81, 82, 85,
    90, 96, 97, 104, 110, 120, 121, 123,
    124, 146, 152, 155, 156, 173, 185,
    187
lepton, 5, 65, 67, 75, 82, 88, 93, 94,
    99, 101, 106–108, 128, 147, 152,
    157, 158, 173
Lie algebra, 3, 19, 29, 32, 144
Lie group
    form invariance, 3, 17, 144
    gauge group, 86, 93, 97, 142, 146,
        147, 150, 152, 154, 157, 158,
        186, 190
    $U(1) \times SU(2)$, vi, 5, 29, 42,
        75, 90, 105, 106, 128, 147,
        148, 152, 153
    $U(1) \times SU(2) \times SU(3)$, vi, 6,
        99, 152, 154, 156
light, v, vi, 9, 49, 53, 59, 64, 91, 112,
    141, 146, 148
Lorentz force, 2, 54, 142
Lorentz group, v, 19, 27, 28, 44, 52,
    64, 133

magnetic monopole, 2, 6, 26, 40, 42,
    55, 59, 72, 85, 111, 113, 118–121,
    125, 127, 128, 131, 151, 158, 173